FERNS

and Fern Allies of Canada

WILLIAM J. CODY · DONALD M. BRITTON

A PATHFINDER BOOK REPRINT EDITION
Printed in the United States of America
ISBN: 978-1951682460

Contains information licensed under the Open Government Licence Canada.

Below: Original publication data from *Ferns and Fern Allies of Canada,*
first published in 1989.

Canadian Cataloguing in Publication Data

Cody, William J., 1922–

    Ferns and fern allies of Canada

(Publication ; 1829)

Issued also in French under title: *Les fougères et
les plantes alliées du Canada.*
Includes index.
Bibliography: p.
Cat. No. A53-1829/1989E
ISBN 0-660-13102-1

1. Ferns–Canada. 2. Ferns–Canada–Nomenclature.
I. Britton, Donald M. II. Canada. Agriculture
Canada. Research Branch. III. Title. IV. Series:
Publication (Canada. Agriculture Canada). English ;
1829.

QK525.C6 1989   587'.31'0971   C89-099204-5

Ferns

and fern allies of Canada

William J Cody
Biosystematics Research Centre, Ottawa, Ontario

Donald M. Britton
University of Guelph, Guelph, Ontario

Research Branch
Agriculture Canada

Publication 1829/E
1989

Royal fern (*Osmunda regalis*), common in swamps, low-lying woods and cedar bogs in eastern Canada.

CONTENTS

ACKNOWLEDGMENTS

The helpful comments of S.G. Aiken and P.M. Catling, who reviewed earlier versions of the manuscript, are gratefully acknowledged. We are particularly appreciative of the comments of T.M.C. Taylor, who took time from his retirement to read the manuscript. Technical support in the preparation of the distribution maps was provided by L.D. Black and W.A. Wojtas. The fine line drawings were done by V. Fulford; her valuable contribution is gratefully acknowledged. We are also indebted to the curators of the various herbaria who kindly lent specimens for our study.

INTRODUCTION

The first treatment of Canadian ferns was undertaken by Macoun and Burgess, who published *Canadian Filicineae* in 1884. As these authors state, "Probably no form of growth throughout the vegetable kingdom attracts more general attention than ferns, which, while appealing strongly to the scientific tastes, have an equally powerful claim upon the artistic. Their distribution over the whole surface of the globe, with the exception of the sterile portions of the polar regions, places at least some forms within the reach of everyone, while, grow in what locality they may, there is none to which they do not lend an added charm ...."

In 1889 George Lawson published *The School Fern-Flora of Canada* as an appendix to Asa Gray's *Botany for Young People and Common Schools: How Plants Grow, A Simple Introduction to Structural Botany, with a Popular Flora*. Robert Campbell in 1898 and 1899 produced a series of fascicles on Canadian ferns in successive issues of *Canadian Horticultural Magazine*. John Macoun (1890) also listed the ferns and fern allies in his *Catalogue of Canadian Plants*.

Since these early works, the only treatments of all the ferns known to occur in Canada have been brief accounts by Boivin (1968) in *Énumération des plantes du Canada* and Scoggan (1978) in *The Flora of Canada*. Local treatments, some including the fern allies, have appeared for Nova Scotia (Roland 1941); Quebec (Marie-Victorin 1923, 1925); Ottawa District (Cody 1978, 1980); and British Columbia (T.M.C. Taylor 1963, 1970); these treatments have been most useful for those regions. Other regional treatments have appeared in various floras such as Calder and Taylor (1968) *Flora of the Queen Charlotte Islands, Part 1*; Erskine (1961) *The Plants of Prince Edward Island*; Fernald (1950) *Gray's Manual of Botany, eighth edition*; Gleason and Cronquist (1963) *Manual of the Vascular Plants of Northeastern United States and Adjacent Canada*; Hitchcock et al. (1969) *Vascular Plants of the Pacific Northwest*; Moss (1959) *Flora of Alberta*; Scoggan (1957) *Flora of Manitoba*; and various local lists, such as Soper (1963) for Manitoulin Island. The present volume is thus the first in over 90 years to combine descriptive information on the various taxa with comments on their relationships, habitats, and distributions. It is hoped that the book will prove to be a useful tool, not only to individuals taking a first look at these interesting plants but also to the dedicated amateur and the professional botanist.

We have found that writing this book has been a singular challenge. We are very aware that books such as this are written by professionals who claim they are writing for amateurs! Our objectives have been to bring together for the first time all the ferns and fern allies (pteridophytes) of Canada; to make readily available references to the literature for those who wish more information on certain taxa; and to supply keys, descriptions, illustrations, and distributions for all the species. We have tried to give the reader an appreciation of or insight into the various approaches of recognizing and naming species. In addition, we have tried to challenge the reader to find gaps in distribution, to extend ranges, and to find rare hybrids. To this end we have included the following:

- Information for correct identification, including easily used field characters.
- Comments on whether a species is easily recognized and has a discrete, limited range or whether much is still to be discovered about the species.
- Comments concerning the aesthetic aspects, popular appeal, and folklore of the plants, although space restrictions have been a limiting factor in this regard.
- References to all the major essential literature.
- An assessment of the taxonomy that appears to be clear and straightforward at this time and that which does not, keeping in mind that we are aware that a great deal of work that is still unpublished is currently being done on many of the species in this book. Taxonomy is not static, and one should therefore expect further change and interpretations of the species that are different from those we have presented.
- An assessment of current trends in the taxonomy of the pteridophytes, i.e., an extrapolation from the past to the present and an indication of the direction in which we are moving.

Latin name

The Latin name, e.g., *Onoclea sensibilis* L., and a few relevant synonyms are used, followed by an accepted common name or names (e.g., sensitive fern). Although synonymy is of great interest to the professional, it usually does not excite the amateur; we have therefore attempted to keep the synonymy as short as possible. Most amateurs are bewildered, or even annoyed, at what appears to be constant name changing, e.g., *Dryopteris carthusiana* for *D. spinulosa* or *Lycopodium digitatum* for *L. flabelliforme*. To increase the amateur's understanding and appreciation of some of the problems, we have included a brief history of some names, as follows:

- *Onoclea sensibilis* L. 1753. The genus name *Onoclea* (initial capital letter) is followed by the species name *sensibilis* (all in lower case). The abbreviation for the author, L., stands for Linnaeus, the father of binomial names (genus plus species), who described the plant and named it in 1753. This name has remained constant, which is an indication of the distinctiveness of the plant.
- *Gymnocarpium dryopteris* (L.) Newman. Linnaeus described the common oak fern in 1753 as *Polypodium dryopteris*, which is accordingly recognized as the basionym. There has been great disagreement through the years as to which genus this species belongs. It has been placed in *Thelypteris*, *Phegopteris*, *Dryopteris*,

Currania, and *Carpogymnia,* among others. It is now recognized as a *Gymnocarpium* and is called *G. dryopteris* (L.) Newman, following work published by Newman in 1851. Because Linnaeus is recognized as the person who originally described the species, his initial appears in parentheses before Newman's name (not all authors' names are abbreviated).

- *Aspidotis densa* (Brack.) Lellinger. This plant has been placed in *Cryptogramma, Cheilanthes,* and *Pellaea* by various authorities. Its affinities with those genera have been questioned and it was placed in *Aspidotis* by Lellinger in 1968.
- *Dryopteris expansa* (C. Presl) Fraser-Jenkins & Jermy. Extensive studies on European populations of *Dryopteris dilatata* showed that there were two species within this taxon, a diploid and a derived allotetraploid. The ancestral diploid species was segregated as *D. assimilis* S. Walker. It was later found that an earlier name, *Nephrodium expansum* C. Presl, 1825, based on a specimen collected by Haenke from Nootka Sound, existed and had priority. Accordingly, *D. assimilis* became invalid and the current name *D. expansa* was created in 1977.
- *Gymnocarpium jessoense* (Koidz.) Koidz. ssp. *parvulum* Sarvela. After studying all the gymnocarpiums of the world, Sarvela decided that material from Alaska, the Yukon, and the Prairie Provinces, which had for the last century been known as *Gymnocarpium robertianum* (Hoffm.) Newman, was related to the Asian *G. jessoense.* He selected a type collected in the Nahanni and described our plant as a new subspecies. At this time, one can find treatments that reduce *G. robertianum* to a mere variety of *G. dryopteris,* e.g., *G. dryopteris* (L.) Newman var. *pumilum* (DC.) Boivin. In addition, many authors recognize both *G. dryopteris* and *G. robertianum* as separate species. In our treatment, based on Sarvela, we recognize a third species, *G. jessoense,* as well as *G. dryopteris* and *G. robertianum.*

Subspecies and varieties

Canada's royal fern, *Osmunda regalis,* is closely related to the European plant. Gray in 1856 described the new world material as varietally distinct, e.g., *O. regalis* (L.) var. *spectabilis* (Willd.) Gray. The story is similar for *Equisetum hyemale* L. The Canadian plant is known as ssp. *affine* (Engelm.) Stone. The situation is less clear for the ostrich fern, *Matteuccia.* Some researchers believe that Canada's species is distinct and should be called *M. pensylvanica,* whereas others treat our plant as a variety of the European *M. struthiopteris* (L.) Todaro.

Some modern treatments would suggest that the above examples describe problems of subspecies, not varieties. A subspecies has no barrier to interbreeding with the species, but it has a distinct morphology and a quite separate distribution, e.g., *Adiantum pedatum* L. ssp. *aleuticum* (Rupr.) Calder & Taylor. An amateur might ask why old varieties have not been logically and uniformly changed to subspecies. Primarily, the reason for this has been to avoid adding complexity to taxonomic literature. Also, the intent of the original author is not always apparent, i.e., we do not always know whether it can be safely assumed that a "variety" was in reality a subspecies. In the current use of categories, variety is a lower rank than subspecies and applies to a group of similar individuals within a subspecies, e.g., *Adiantum pedatum* L. var. *subpumilum* W.H. Wagner.

Taxonomists differ widely in their interpretation of variation in populations. For example, Boivin (1968) recognized *Dryopteris austriaca* without subspecies or varieties. In our opinion, this taxon includes in Canada's flora *D. expansa, D. intermedia, D. campyloptera,* and *D. carthusiana.* After extensive studies on *Botrychium,* W.H. Wagner recognizes *B. oneidense* and *B. minganense* as species, while dismissing *B. obliquum* as a mere form of *B. dissectum* and the var. *europaeum* of *B. virginianum* as not meriting taxonomic rank. Further experimental work should help to clarify some of the widely divergent views that are currently held.

Common names

These vary widely in various parts of the world. Some are well entrenched and difficult to change, even if they are not particularly suitable. For example, ebony spleenwort is not ebony-colored, and Wherry has called it the brown-stemmed spleenwort. The silvery spleenwort is not a spleenwort (*Asplenium*) but an *Athyrium.* We have not attempted to introduce major changes in long-term usage.

Description

The lengths of the frond are given; the divisions (if any) of the blade are described; any scales, hairs, or glands are mentioned; and the important fruiting bodies and indusia (if present) are described. The descriptions are reasonably full, but at the same time are not of the length found in such works as the *Flora of the Pacific Northwest.* At the end of the technical description, certain useful field characters are highlighted.

Cytology

The chromosome number and author of the report are given. An attempt is made to stress reports based on Canadian material. Such Canadian reports are indicated by an asterisk. We have attempted to cite recent literature in the hope that later reports will cite previous records. For those interested in this information, checking the uniformity of opinion, e.g., Löve et al. (1977), is extremely useful. Most chromosome numbers for ferns have been determined from meiotic studies, and so we have usually given the n or gametic number. Where $2n$ numbers are given, we consider that they were determined from somatic material, usually root tips. The designation "n" = $2n$ indicates that there was premeiotic doubling of chromosome number (see Manton 1950), and the species is apogamous.

Habitat

Habitat varies widely for even a single species. We have attempted to give what we consider to be typical habitats. Because some ferns are restricted to basic rocks and others to acidic rocks, we have attempted to indicate these preferences.

Range

Range and occurrence in Canada are indicated primarily by dot maps, but in some cases are highlighted by special comments when the range is unusual. The complete distribution is indicated for North America, and if the taxon is found on other continents, that fact is briefly noted. The Canadian distribution is given from east to west, followed by the United States distribution from east to west. For those interested in visualizing distributions on a shaded map, Mickel (1979) should be consulted.

Remarks

In the Remarks section we attempt to indicate whether the species is clear-cut or not and whether there are problems in its interpretation. Any special features of interest are mentioned.

Segregate species

The most casual observer of the pteridophyte flora of Canada must be impressed and perhaps bewildered by the seeming proliferation of a large number of "new" species, e.g., *Dryopteris expansa*, *Lycopodium digitatum*, *Gymnocarpium jessoense*, and *Botrychium rugulosum*, and might ask how this has occurred and what the final outcome will be.

Segregate species, to a large extent, are the natural outcome of applying the concept of biological species after surveying the variation of populations. For example, if one compares the current treatment of *Dryopteris* with that in Fernald (1950), *Dryopteris spinulosa* is now composed of *D. intermedia, D. expansa, D. campyloptera,* and *D. carthusiana.* One species is now four. The rationale behind this splitting is clear. *Dryopteris intermedia* is a sexual diploid ($n = 41$) with certain limits of variability that can be described, e.g., lacy, subevergreen fronds with glandular indusia, and so on. In terms of evolution, it is considered to be ancestral to the derived tetraploids *D. campyloptera* and *D. carthusiana.* Similarly, *D. expansa* is another sexual ancestral diploid that superficially is a segregate species in Canada's flora, but if the plants in Europe and Japan are considered, *D. expansa* is actually a composite ancestral diploid species because it is extremely wide ranging (circumpolar). We can now include Japan, British Columbia, eastern Canada, and northern Europe in the range of this species. The tetraploids *D. carthusiana* (which has long been known as *D. spinulosa*) and *D. campyloptera* have quite different origins and distributions and are separate interbreeding populations, and therefore are also species. The same situation has been repeated in many other genera, e.g., *Polystichum, Cystopteris, Polypodium,* and *Isoetes,* among others. It is an inevitable trend, resulting from modern research techniques, which is only rarely balanced by the merging of two species that were once considered separate, e.g., *Asplenium ruta-muraria* and *A. cryptolepis, Woodsia alpina* and *W. bellii.* Critics of segregate species have pointed out that more emphasis is placed on differences between species than on similarities. This difference in philosophy has led to the recognition of *Pellaea glabella* var. *simplex.* This variety is certainly close in both morphology and evolution to *P. glabella* var. *glabella,* and it is therefore more appropriately called var. *simplex* rather than *P. suksdorfiana.* This change in name recognizes the morphological differences between var. *simplex* and *P. glabella* var. *glabella.* Another example is *Matteuccia struthiopteris* of Europe versus *M. pensylvanica* of North America. If differences are stressed, they may be considered separate species (although critical experimental breeding data are absent), whereas if similarities are stressed they belong to the same species with two varieties or subspecies.

For many of these plants we do not have the critical evidence from experimental crosses or from naturally occurring hybrids to make an objective decision. In *Botrychium,* for example, the fall *Botrychium* species (such as *B. dissectum* and *B. multifidum*) all have $n = 45$. To date, artificial crosses have failed because of difficulties in growing spores, and even culture of mature plants is difficult. Consequently, we are left with a number of species that are known to grow together. Some researchers have argued that the variation exhibited is an example of the great variability of each species and have recognized only *Botrychium dissectum* and *B. multifidum* as species, whereas others, e.g., W.H. Wagner and Wherry, have stressed that if the plants are growing in close proximity and hence identical environments, then the variation seen is genetic, and they recognize as many as five species. Similar problems arise with various species of *Lycopodium,* e.g., *L. complanatum, L. digitatum,* and *L. tristachyum.*

One could argue in favor of only one collective species for all these—all have the same chromosome number and all apparently can intercross without leading to meiotic irregularities; or conversely one can emphasize the large differences in morphology between extremes and recognize three species.

It would seem to us that the final outcome will inevitably be the recognition of additional species and the further partitioning of variation. We will be able to say that the system is a more "natural" classification, or one that emphasizes evolutionary units. Basic diploids and derived tetraploids will be more clearly delineated.

The methods being used, which include chemical analysis, comparative morphology, and scanning electron microscopy (SEM) of spores, will continue to emphasize differences rather than similarities, and so inevitably, the larger collective species, such as *Athyrium filix-femina* and *Cystopteris fragilis*, will yield more segregates after further study.

Cytology and biosystematics

Manton (1950) brought the methods and philosophy of cytogenetic analysis, which she had used successfully on the Cruciferae, to a study of pteridophytes. By means of artificial hybridization techniques, analysis of naturally occurring hybrids, and analysis of chromosome numbers, she showed that the evolution and phylogeny of the pteridophytes could be greatly clarified. It is impossible to summarize briefly all her results here, but it is important to stress the impact of her work, and of those who followed her, on our classification of the ferns and fern allies. She found, for instance, that all the species of true *Dryopteris* have a basic chromosome number of 41, which is an unusual number not easily manipulated arithmetically because it is not divisible by an integer. Within *Dryopteris*, Manton showed that there were derived tetraploids, e.g., *D. carthusiana* and *D. filix-mas*, and that furthermore, these species initially arose from interspecific hybridization, followed by a doubling of the chromosome number, thus restoring fertility, i.e., by allopolyploidy. This may be shown schematically, as in Diagram 1.

In morphology, *D. carthusiana* is considered to be a blend of the characteristics of its two diploid ancestors, and in breeding behavior, it is a new, derived allotetraploid species. To take this scheme one step further, we think that *D. cristata* is LLBB ($2n = 164$) and that it crossed at one time with *D. goldiana* $2n = 82$, which is GG. The resulting sterile hybrid would be (L)(B)(G), which upon doubling of chromosome number would give rise to the fertile allohexaploid as shown in Diagram 2.

D. intermedia 2n = 82   X   Species B (extinct or not yet found) 2n = 82

II (I = a genome or complete set of 41 chromosomes from D. intermedia)

Hybrid (I)(B) 2n = 82, but since there is no homology between sets I and B, 82 single chromosomes occur at meiosis. When the number of chromosomes is doubled, then the constitution is IIBB, 2n = 164. D. carthusiana with 41 pairs + 41 pairs = 82 pairs of chromosomes at meiosis. One speaks of this species as an allotetraploid because it has two ancestral diploid species in its constitution.

Diagram 1 Derivation of the allotetraploid species *Dryopteris carthusiana*

The allohexaploid is what we call *Dryopteris clintoniana*, and a close scrutiny of the plant indicates the influence of both *D. cristata* and *D. goldiana* in its origin. The message is clear—it is not only the chromosome number, but also the character of the sets of chromosomes (genomes) that determine the final make-up of the species. Patterns such as those above are now well known in *Asplenium, Cystopteris, Polypodium*, and *Polystichum*, for example. It would appear to be a common pattern of evolution for many species.

A different sort of pattern is seen in *Pellaea*. In *P. glabella* var. *nana* the basic chromosome number is 29 and the taxon is sexual, but in *P. glabella* var. *glabella* the somatic chromosome number is 116 and the species is apogamous, i.e., the spores also have 116 chromosomes. The long beech fern is similar in this regard; although the basic chromosome number in *Phegopteris* is 30, *P. connectilis* has a somatic chromosome number of 90 and the viable spores also have 90 chromosomes.

Not all speciation has involved polyploidy. Many of the *Botrychium* species have $n = 45$ and yet some have evolved quite different morphologies, e.g., *B. lunaria* versus *B. multifidum*. At times, the degree of allopolyploidy is in doubt, or analysis may show that we are dealing with an autopolyploid. For example, the diploid hart's tongue (*Phyllitis*) in Europe versus tetraploid hart's tongue in Canada are not markedly different in morphology. Similarly, diploid and tetraploid *Asplenium trichomanes* are superficially very similar in appearance. For amateurs to make such distinctions usually demands more effort in identification than they may wish to expend. They may be forced to measure spore sizes and stomatal sizes or even to determine chromosome number.

Chromosome numbers have also been used as an aid to determine systems of classification at higher levels, such as family. More recent systems of classification have tended to group genera with a basic chromosome number of 29 or 30 into an adiantoid group, e.g., *Adiantum, Cheilanthes, Cryptogramma*, and *Pellaea* (Lovis 1977). These chromosome numbers are quite unlike those for *Dryopteris* (41), *Polystichum* (41), *Cystopteris* (42), and *Athyrium* (40), which in turn are markedly different from *Polypodium*, which has the unusual basic number (X) of 37.

Cytology has been less useful for phylogeny with some groups. *Equisetum*, for instance, appears to be completely uniform with $n = 108$. A different problem arises in *Lycopodium*. Here there are a number of different basic numbers, which suggests that the genus is unnatural or polyphyletic, and some of the species are extremely difficult to analyze, e.g., *L. lucidulum* and *L. selago*.

Modern monographic treatment of genera now includes a whole battery of experimental techniques (Britton 1974) such as cytology, chromatography, electron microscopy, comparative anatomy, and computer analysis. There is even some promise from DNA hybridization studies and the analysis of isoenzymes. It is no wonder,

LL   +   BB   +   GG $2n = 246$

41 pairs   +   41 pairs   +   41 pairs = 123 pairs at meiosis

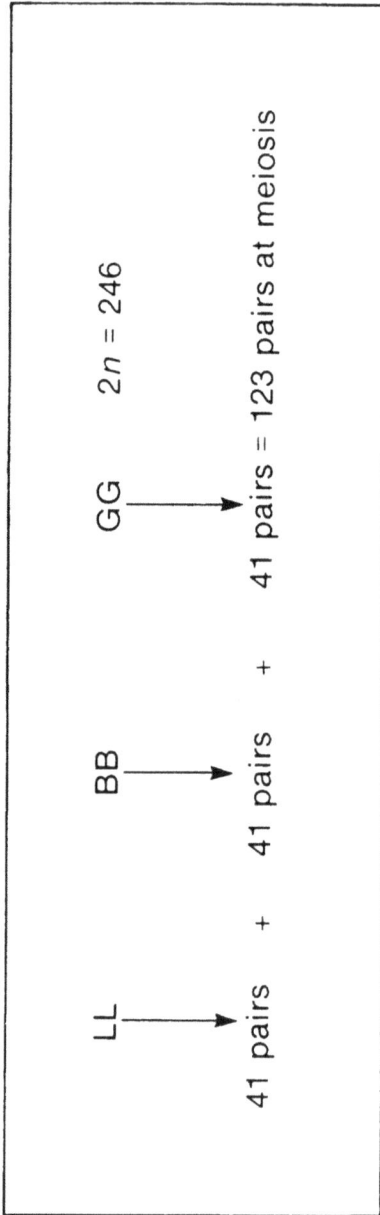

Diagram 2 Derivation of the allohexaploid *Dryopteris clintoniana*

with so many methods of analysis, that the taxonomy of the pteridophytes is slowly changing as the species themselves have evolved through the years. The reader who is interested in these aspects of evolution should consult the review articles in Jermy et al. (1973), Lovis (1977), and Walker (1979).

Hybrids

Interspecific (and intergeneric) hybrids are well known in some genera. Much work has been done on *Asplenium* hybrids (Lovis 1977; W.H. Wagner 1954); *Dryopteris* (W.H. Wagner 1970); *Equisetum* (Hauke 1963); *Lycopodium* (Wilce 1965); *Polypodium* (Shivas 1961); *Polystichum* (D.H. Wagner 1979); and *Phegopteris* (Mulligan and Cody 1979). Knobloch (1976) has published a list of pteridophyte hybrids.

In some other genera, hybrids are either extremely rare or have never been detected, e.g., *Blechnum, Botrychium, Cryptogramma, Osmunda, Pellaea,* and *Thelypteris*.

Interspecific fern hybrids are usually characterized by intermediate morphology, although so-called "one-way hybrids" are not unknown. Even these, e.g., *Dryopteris goldiana* × *D. intermedia* (Evans and Wagner 1964), may superficially resemble one parent (*D. intermedia*), but may prove on analysis to have a blend of characteristics from both parents. In ferns, one expects a high incidence of sterility in these hybrids. For example, *D. goldiana* (GG) × *marginalis* (MM) results in a hybrid with (G)(M). Because there is no homology between the chromosomes at meiosis, one obtains 41 + 41 or 82 single chromosomes at meiosis, which separate very unevenly and the resulting spores abort. On this basis, each resulting hybrid is an end point that is not involved in further crosses. However, it is apparent that some hybrids also produce quite numerous large, spherical spores, a few of which are capable of germinating (DeBenedictis 1969). It is thought that these spores contain all the chromosomes of the parent hybrid plant and are able to germinate and produce genotypes identical to the original hybrid. This would appear to account for the abundance of a few hybrid combinations, e.g., *Dryopteris* × *triploidea, D.* × *boottii, D. filix-mas* × *marginalis, Gymnocarpium* × *intermedium,* and *G. dryopteris* ssp. × *brittonianum*.

Finding and identifying hybrid pteridophytes is a challenge. First, it is necessary to become thoroughly familiar with the parent species in order to assess the full range of variability exhibited by them. Then one should be prepared to examine carefully the spores for abortion and, in some cases, to undertake cytology. It is no longer acceptable to collect a fern with bizarre morphology and blithely pass it off as some rare and exotic hybrid combination. A British "hybrid" said to be *Polypodium vulgare* × *Pteridium aquilinum* reported in 1907 is not taken seriously today. There should be little difficulty in finding such hybrids as *Dryopteris* × *triploidea* and *D.* × *boottii* because they seem to be present wherever the parents are abundant and are growing intermixed. On the other hand, the *Osmunda* hybrid *O.* × *ruggii* Tryon would be a rare find indeed, even though the prospective parents often grow in close proximity over literally thousands of square kilometres.

Throughout the publication, hybrids placed within parentheses are not known to occur in Canada.

Sequence of families, genera, and species

In the present work, the families and genera follow the taxonomic sequence of *Genera Filicum* (Copeland 1947). This follows the modern trend of separating the families Pteridaceae, Aspidiaceae, Blechnaceae and Asplenianeae, which were at one time lumped under the Polypodiaceae. Within genera, the species are placed in such an order that similar species can be compared readily.

Distribution maps

It was not possible for the authors to visit all the herbaria in Canada as well as the major ones in the United States and elsewhere. Our objective was to depict the broad picture of distribution of the species in Canada. At times we have included locations in Alaska and Greenland when this has helped to show the distribution more clearly. Thus, provincial or regional maps with more dots might give a slightly different picture for a smaller area, e.g., the maps for rare plants of the various provinces (Maher et al. 1979; White and Johnson 1980; Douglas et al. 1981). Our maps were prepared initially from specimens in the herbarium of the Biosystematics Research Centre, Ottawa, Ont. (DAO) (Holmgren et al. 1981). These were supplemented by records from 26 Ontario herbaria including National Museum of Natural Sciences, Ottawa, Ont. (CAN); University of Toronto, Toronto, Ont. (TRT); University of Guelph, Guelph, Ont. (OAC); Queen's University, Kingston, Ont. (QK); and University of Western Ontario, London, Ont. (UWO). Additional Quebec and British Columbia records were obtained from the published works of Rousseau (1974), Taylor (1970), Erskine (1961), and Roland and Smith (1969), who relied heavily on herbaria of Laval University, Quebec, Que. (QFA); University of Montreal, Montreal, Que. (MT); University of British Columbia, Vancouver, B.C. (UBC); British Columbia Provincial Museum, Victoria, B.C. (V); Acadia University, Wolfville, N.S. (ACAD); and Nova Scotia Agricultural College, Truro, N.S. (NSAC). Selected specimens have been borrowed from various herbaria, e.g., University of Manitoba, Winnipeg, Man. (WIN), and selected literature citations have been included as well, where they are significant.

ADDENDUM TO THE INTRODUCTION, 1989

In our introduction, we stressed the current trend toward segregate species. This trend is strongly apparent in recent studies on *Botrychium* by W.H. Wagner of the University of Michigan. W.H. Wagner and F.S. Wagner (1983a) have outlined their methods for deciding whether the variation observed in populations is genetic or environmental, and they give their rationale for their description of a large number of new species of *Botrychium* in North America. Five new species have been described recently (W.H. Wagner and F.S. Wagner 1981, 1983b), at least three others are not yet published, and still others may be recognized after further study (W.H. Wagner and F.S. Wagner 1983c). These authors have suggested that what we have called *B. boreale* ssp. *obtusilobum* should be called *B. pinnatum* St. John, and that the name *B. hesperium* (Maxon & Clausen) Wagner & Lellinger should be applied to some populations in Cypress Hills and in Waterton Lakes National Park.

A specimen of *B. paradoxum* W.H. Wagner (W.H. Wagner and F.S. Wagner 1981) was cited from Waterton Lakes National Park. Dr. Wagner reported finding additional specimens in the park in 1982. Elsewhere it has been reported from Montana and British Columbia. This plant is extraordinary in that it lacks a sterile blade. Its relationship to other taxa in the genus is uncertain and needs further study.

The key to the four groups of *Botrychium* subgenus *Botrychium* given by W.H. Wagner and F.S. Wagner (1983a) is perhaps useful, because it bridges the gap between the more classical treatment of Clausen (1938) and the modern, segregation treatment followed by W.H. Wagner and F.S. Wagner. It should be noted that the emphasis is on the species category, at the expense of recognizing subspecies or varieties of any kind. The four groups keyed out encompass six widespread species: *B. simplex*, *B. lunaria*, *B. minganense*, *B. boreale*, *B. lanceolatum*, and *B. matricariaefolium*. F.S. Wagner (1983) states that in western North America there are 13 species and five interspecific hybrids of subgenus *Botrychium*. Of these 13 species, nine are considered to be endemic in the West, two are in the *B. lunaria* group, two in the *B. simplex* group, four in the *B. lanceolatum* group, and *B. paradoxum* is in its own group. W.H. Wagner and F.S. Wagner (1983a) give a key for five species in the *B. lanceolatum* group. Of these five species, only *B. echo* is considered by them to be absent from Canada. It is only known in Colorado, Utah, and Arizona.

W.H. Wagner and F.S. Wagner (1981) state that some specimens annotated by W.H. Wagner to *B. dusenii* are now referable to *B. crenulatum* W.H. Wagner.

However, until all the new entities are described by them, and all the Canadian material has been restudied and the results of their field studies are published, we are unable to state clearly how many species are present in Canada. As an example, W.H. Wagner and F.S. Wagner (1983c) suggest that there are as many as eight species of *Botrychium* in the Cypress Hills of Saskatchewan, and 10 species of the genus in Waterton Lakes National Park. In both cases, these are much larger numbers of species than have been recognized for these areas in the past.

It is interesting to note that R.C. Moran (1983) has produced evidence that the plant we have called *Cystopteris fragilis* var. *mackayii* should be treated as a species, *C. tenuis* (Michx.) Desv., and indeed this was suggested earlier by Lellinger (1981). It is unfortunate that Moran did not see more Canadian material to give a better impression of the northern distribution of this taxon on his map.

Again, V.L. Harms (1983) has reported on the occurrence of *Athyrium filix-femina* in Saskatchewan. He is of the opinion that both var. *sitchense* (*cyclosorum*) and var. *michauxii* occur in that province. *Athyrium filix-femina* is an extremely variable and difficult species to understand. It was studied by Liew Fah Seong at the New York Botanical Garden in the early 1970s, but to our knowledge the study has not been published.

It is of interest, too, to note that the National Museum of Natural Sciences, National Museums of Canada, has published two more volumes in the series on rare plants (Bouchard et al. 1983; Hinds 1983). For Quebec, two fern allies and 26 ferns were considered to be rare. They were not mapped. For New Brunswick, four fern allies and nine ferns were considered to be rare. Distribution maps for the New Brunswick taxa were included in that volume. Also, Argus and White (1983) have mapped 14 species of ferns and fern allies for Ontario. It is worth noting, however, that a large number of species are now withdrawn from the original list of rare vascular plants of Ontario.

KEY TO THE GENERA

A. Stems jointed; nodes covered by sheaths composed of basally united scarious leaves, otherwise leafless; sporangia borne on inner surface of peltate scales of terminal spike-like cones .... *Equisetum* p. 73
A. Stems not conspicuously jointed, bearing green leaves or leaf-like fronds.
   B. Leaves (fronds) small, entire or serrate, very numerous and imbricated, or quill-like and crowning a short corm-like stem; sporangia sessile or subsessile in leaf axils.
      C. Stems elongate, covered with persistent small more or less flattened leaves; plants creeping from rhizomes or decumbent stems.
         D. Leaves without a ligule; strobiles terete, homosporous .... *Lycopodium* p. 19
         D. Leaves ligulate; strobiles 4-sided; sporangia of two kinds, microsporangia containing many minute microspores (male) and macrosporangia containing fewer and larger macrospores (female) .... *Selaginella* p. 49
      C. Stems short, thick, and corm-like, crowned by a rosette of quill-like leaves; spores of two kinds .... *Isoetes* p. 61
   B. Leaves (fronds) usually pinnate or deeply lobed, not closely or only slightly imbricated; sporangia naked, or in sori on the backs or margins of sometimes specially adapted fronds or their divisions, or in sporocarps.
      E. Fronds tiny, bilobed, 2-ranked, floating free on the surface of quiet water; sporangia in sporocarps borne on the underside of the axis .... *Azolla* p. 310
      E. Fronds larger, simple or divided, not bilobed, growing from a persistent rhizome.
         F. Plants aquatic, although sometimes becoming stranded; fronds long-petioled, 4-foliate, floating on the surface; sporangia in hard sporocarps borne on or close to the rhizome .... *Marsilea* p. 307
         F. Plants typically terrestrial; fronds not 4-foliate; sporangia not in sporocarps.
            G. Fronds or portions of them conspicuously dimorphic.
               H. Sterile fronds linear-filiform, blade-less, crowded on a short crown; fertile fronds filiform, tipped by a few tiny crowded finger-like pinnae bearing the sporangia .... *Schizaea* p. 130

H. Sterile fronds with a distinct blade, not crowded; fertile fronds not filiform and tipped by a few tiny pinnae, but much larger.

  I. Sterile part of frond simple; fertile part of frond a long-stalked simple spike with two rows of coherent sporangia .... ***Ophioglossum*** p. 95

  I. Sterile part of frond pinnately divided one or more times.

    J. Sporangia naked.

      K. Fronds fleshy, single, or sometimes two growing from a scarcely developed rhizome and consisting of a sterile lower part and, when present, a more or less upright fertile panicle or spike; sporangia 2-ranked .. .... ***Botrychium*** p. 97

      K. Fronds forming a more or less dense crown at the end of a stout rhizome; sporangia not 2-ranked .... ***Osmunda*** p. 124

    J. Sporangia partly or wholly covered by the rolled-up pinnules, forming globular berry-like divisions of the stiff fertile frond.

      L. Fronds in vase-like clumps; simple pinnate fertile fronds surrounded by tall regularly pinnate sterile ones; sterile fronds oblong-lanceolate; rachis not winged .... ***Matteuccia*** p. 166

      L. Fronds solitary or scattered along the rhizome; sterile fronds deltoid, coarsely pinnatifid; rachis winged

apically; fertile fronds
bipinnate . . . .
. . . . . . ***Onoclea*** p. 168
G. Fertile fronds or fertile portions of fronds
similar to the sterile; fertile fronds
sometimes longer but not hardened and
berry-like (fertile and infertile fronds of
Cryptogramma, Blechnum, and *Aspidotis*
are dimorphic, but of similar texture).
  M. Fronds simple, commonly auricled at
the base.
    N. Fronds long-caudate, sometimes
rooting at the tip . . . .
. . . . ***Camptosorus*** p. 290
    N. Fronds oblong, not attenuate or
rooting at the tip . . . .
. . . . ***Phyllitis*** p. 292
  M. Fronds variously divided.
    O. Fronds small and delicate; blades
one cell thick . . . .
. . . . ***Mecodium*** p. 132
    O. Fronds larger and coarser; blades
more than one cell thick.
      P. Fronds covered beneath by a
conspicuous white to golden
yellow powder . . . .
. . . . . ***Pityrogramma*** p. 155
      P. Fronds not as above . . . .
. . . . Group I

Group I

A. Sori marginal; indusium formed entirely or in part by the
revolute margin of the frond.
  B. Sori distinct, short, mostly not confluent.
    C. Stipe and fronds glabrous . . . . ***Adiantum*** p. 157
    C. Stipe and fronds glandular-hairy . . . .
. . . . ***Dennstaedtia*** p. 133
  B. Sori usually confluent as a marginal band.
    D. Fronds coarse, scattered along stout elongate and
forking rhizomes . . . . ***Pteridium*** p. 135
    D. Fronds finer, tufted from a very short rhizome.
      E. Segments of frond bead-like . . . .
. . . . ***Cheilanthes*** p. 138
      E. Segments of frond not bead-like.
        F. Pinnules and segments of frond jointed at
the base . . . . ***Pellaea*** p. 143

F. Pinnules and segments of frond not jointed at the base.
  G. Stipes herbaceous, green except at the base .... ***Cryptogramma*** p. 150
  G. Stipes wiry, dark, and shiny ....
    .... ***Aspidotis*** p. 140
A. Sori dorsal on the frond or, if marginal, the indusium not formed by the revolute margin.
  H. Sori elongate.
    I. Indusia continuous, attached near the margins of the pinnae .... ***Blechnum*** p. 273
    I. Indusia not continuous.
      J. Sori in chain-like rows, parallel to the midrib ...
      .... ***Woodwardia*** p. 275
      J. Sori parallel to the oblique lateral veins.
        K. Fronds to 1 m long, herbaceous; veins reaching the margin .... ***Athyrium*** p. 263
        K. Fronds smaller, to 40 cm long, evergreen or herbaceous; veins not reaching the margin .... ***Asplenium*** p. 281
  H. Sori round or nearly so.
    L. Indusia present.
      M. Indusia segmented .... ***Woodsia*** p. 170
      M. Indusia not segmented.
        N. Indusium hood-shaped, attached by its base on the side toward the midrib ....
        .... ***Cystopteris*** p. 253
        N. Indusium round or reniform.
          O. Fronds scattered along a thin cord-like rhizome (or tufted from a stout rhizome in *T. limbosperma*) ....
          .... ***Thelypteris*** p. 239
          O. Fronds tufted at the end of a stout rhizome
            P. Indusium reniform or with a deep sinus .... ***Dryopteris*** p. 205
            P. Indusium round, without a deep sinus .... ***Polystichum*** p. 182
    L. Indusia absent.
      Q. Fronds coriaceous, evergreen, simply pinnatifid .... ***Polypodium*** p. 295
      Q. Fronds deciduous, at least pinnate-pinnatifid.
        R. Rhizome stout; fronds forming a crown ....
        .... ***Athyrium*** p. 263
        R. Rhizome cord-like.
          S. Fronds more or less ternate ....
          .... ***Gymnocarpium*** p. 231
          S. Fronds pinnate-pinnatifid ....
          .... ***Phegopteris*** p. 247

FAMILIES AND GENERA

1. LYCOPODIACEAE club-moss family

1. *Lycopodium* L. club-moss

Plants low, evergreen, coarsely moss-like, with simple to much branched stems covered with simple, 1-nerved, 4- to many-ranked, lanceolate or linear leaves. Sporangia in the axils of leaf-like sporophylls similar to the vegetative leaves or segregated in a terminal strobilus or cone. Spores numerous, yellow.

The genus *Lycopodium* has over 400 species throughout the world and occurs on all the continents except Antarctica. There are a large number of tropical species, many of which are epiphytes, although we usually think of Canadian species as being especially adapted to cool, moist, and northern habitats of often bleak and barren places.

In the past 40 years the genus has been examined in some detail (Beitel 1979*b*), and several workers have been impressed by the diversity of sporophyte morphology, gametophyte morphology (Bruce 1976), spore morphology (Wilce 1972), and chromosome numbers (Löve et al. 1977). This diversity has prompted European workers to adopt four generic names (*Lycopodium*, *Diphasiastrum*, *Lycopodiella*, and *Huperzia*) for their species (Jermy et al. 1978). Some researchers have gone so far as to place *Huperzia* in a separate family, Huperziaceae (Löve et al. 1977), whereas others have attempted groupings at a subgeneric or section level, noting that it is a great advantage to have a genus name that is familiar to most workers and typifies plants that can be recognized at a glance (Wilce 1972).

We have retained the name *Lycopodium* here for the 13 species that we recognize as occurring in Canada, but at the same time we have grouped the species into categories corresponding to the genera recognized by others (Holub 1964, 1975).

A. Sporangia in the axils of leaf-like sporophylls.
  B. Leaves sharply erose-serrulate at the apex, flat at the base; leaves in alternating belts, some long, others short . . . .
   . . . . 12. *L. lucidulum*
  B. Leaves entire or nearly so, acuminate, plump, and hollow at the base; all leaves essentially the same length . . . .
   . . . . 13. *L. selago*
A. Sporangia in the axils of modified terminal leafy-bracted strobiles.
  C. Sterile branches horizontal or arching; strobili with green leaf-like bracts . . . . 11. *L. inundatum*
  C. Sterile branches erect or ascending; strobili with firm yellowish scale-like bracts.

D. Strobili sessile at the ends of leafy stems.
    E. Aerial stems erect and tree-like.
        F. Leaves of lower part of stem strongly divergent .... 3. *L. dendroideum*
        F. Leaves of lower part of stem strongly appressed to slightly divergent .... 4. *L. obscurum*
    E. Aerial stems tufted, bushy, or fan-like; branchlets more or less flattened.
        G. Plants to 13–25 cm; leaves 8-ranked, 6–10 mm, not fused to stem .... 2. *L. annotinum*
        G. Plants short, less than 13 cm; leaves 4- to 5-ranked, scale-like, partly fused to stem.
            H. Leaves mostly 4-ranked; leaves of the upper and lower sides unlike the marginal .... 8. *L. alpinum*
            H. Leaves mostly 5-ranked; all the leaves alike .... 9. *L. sitchense*
D. Strobili peduncled.
    I. Leaves linear-subulate with long soft hair-like tips .... 1. *L. clavatum*
    I. Leaves scale-like.
        J. Sterile branchlets not compressed or slightly compressed; leaves uniform and usually in 4 rows ..... 10. *L. sabinifolium*
        J. Sterile branchlets flattened.
            K. Constrictions between seasons' growths conspicuous.
                L. Upright growth habit straggly; branchlets flat .... 5. *L. complanatum*
                L. Upright branches fastigiate; branchlets roundish .... 7. *L. tristachyum*
            K. Constrictions between seasons' growths usually only slightly pronounced; branchlets of the branches arching and fan-like .... 6. *L. digitatum*

Lycopodium *s.s. group*

The leaves in this group are not fused to the stem along their length, and the sporangia are borne in distinct cones or strobili. A typical plant is *L. clavatum*, the common club-moss, staghorn club-moss, or wolf's claw. This species has a rhizome on or near the ground surface and unequally forked upright branches, suggesting antlers of a stag or the paw of a wolf. Other species in this group are *L. annotinum* (stiff club-moss), *L. obscurum* (ground-pine), and *L. dendroideum* (round branched ground-pine). All have chromosome numbers based on $n = 34$ (Löve et al. 1977). The spore morphology is shown accurately, and beautifully, by scanning electron microscopy, and all the spores of the species in this group have a regular reticulum of polygons, i.e., they are in a honeycomb pattern (Wilce 1972). It is, however, a spore pattern similar to that seen in the *L. complanatum* (*Diphasiastrum*) group.

1. **Lycopodium clavatum** L. var. **clavatum**
    L. clavatum L. var. *integerrimum* Spring.
    common club-moss
Fig. 1, habit. Map 1.

Stems elongated, horizontal on the surface of the ground, forking, rooting at intervals; leaves uniform, but lower leaves turned upward. Erect branches at first simple, becoming dichotomous; fertile branches with a leafy-bracted peduncle bearing 2 to several sessile or short-stalked strobiles. Leaves linear-subulate, incurved-spreading, usually tipped with a soft white hair-like bristle. Bracts of strobili yellow, fimbriate-erose, at least the lower with white filiform tips.

Cytology: $n = 34$ (Löve and Löve 1976).

Habitat: Dry woods and clearings.

Range: Circumpolar; in North America from Newfoundland to British Columbia and Alaska, south to North Carolina, Michigan, Minnesota, Idaho, and Washington.

Remarks: Mature fruiting plants present no problems in identification. Young or sterile plants, however, may be confused with *L. annotinum*. The extended soft, hair-like bristles on the leaves are useful for discrimination.

1.1 **Lycopodium clavatum** L. var. **monostachyon** Hook. & Grev.
    L. clavatum L. var. *megastachyon* Fern. & Bissell
    L. clavatum L. var. *brevispicatum* Peck
Fig. 2 (*a*) habit; (*b*) portion of strobilus. Map 2.

Similar to var. *clavatum*, but with the leaves ordinarily ascending or appressed and the cone single on a shorter peduncle.

Cytology: $n = 34$ (Löve and Löve 1966a).

Fig. 1 *Lycopodium clavatum* var. *clavatum*; habit, 3/4 ×.

Lycopodiaceae

Fig. 2 *Lycopodium clavatum* var. *monostachyon*; (a) habit, 1×; (b) portion of strobilus, 8×.

Habitat: Exposed situations, hilltops, alpine and subalpine regions generally north of var. *clavatum* in North America (north of northern Minnesota, Michigan, and northern New England).

Range: Greenland, Labrador, and Newfoundland to Alaska.

Remarks: This taxon, with 1 strobilus per peduncle and more appressed leaves than for typical *L. clavatum*, has been considered a separate species, *L. lagopus* (Laest.) Zinserl. ex Kuzen, as listed in Czerepanov (1981). Others have variously treated the taxon as a subspecies, variety, or form.

2. *Lycopodium annotinum* L.
   bristly club-moss, stiff club-moss
   Fig. 3 (a) habit; (b) portion of strobilus. Map 3.

Stems elongated, prostrate, mostly unbranched, rooting at intervals; leaves uniform but the lower leaves turned upward. Erect stems simple to forked several times, increasing annually to 20 cm or more in height. Leaves 8-ranked, more or less stiff and hard, linear-subulate to linear-oblanceolate, with a sharp spinule. Strobili sessile at the ends of leafy stems.

Several intergrading varieties have been recognized: var. *annotinum*, with leaves of erect stems 6–11 mm long, linear-lanceolate or oblanceolate, coarsely toothed, spreading; var. *acrifolium* Fern., with leaves of erect stems 5.5–7.0 mm long, linear-subulate, spreading or ascending; var. *alpestre* Hartm., with leaves of erect stems 2.5–6.0 mm long, linear-lanceolate to lance-attenuate, thick and hard, dorsally convex, entire, and strongly ascending to appressed; and var. *pungens* (La Pylaie) Desv., with leaves of erect stems 2.5–6.0 mm long, lanceolate to lance-oblong, flat, obscurely serrate and strongly ascending to tightly appressed.

Cytology: $n = 34$ (Löve and Löve 1966a) also for var. *pungens*.

Habitat: Moist woods and clearings, subalpine forests, and exposed rocky and peaty habitats.

Range: Circumpolar; in North America from Greenland and Labrador to Alaska, south to Virginia, Minnesota, and Oregon.

Remarks: The tightly appressed and strongly ascending entire leaves of var. *pungens* seem distinctive from the typical variety, but intergradations occur. There is the usual disparity of views as to whether the varieties mentioned should be species, subspecies, or forms. For example, Czerepanov (1981) includes var. *alpestre* and var.

Fig. 3 *Lycopodium annotinum*; (a) habit, 1 ×; (b) portion of strobilus, 6 ×.

pungens in the species *L. dubium* Zoega, whereas Löve includes var. *pungens* and the species *L. dubium* in ssp. *alpestre* Löve and Löve (Löve et al. 1977).

Lycopodium annotinum is at times confused with *L. lucidulum*, but the latter has no strobili and branches are equal in length.

3. **Lycopodium dendroideum** Michx.
   L. obscurum L. var. *dendroideum* (Michx.) D.C. Eat.
   round-branched ground-pine, tree club-moss
   Fig. 4 (a) habit; (b) position of strobilus. Map 4.

Subterranean stems creeping, branching, and rhizome-like, with broad scale-like leaves. Aerial stems upright, 10–30 cm high, simple below, forking above, constricted between the seasons' growth. Lower leaves strongly divergent; leaves of lateral branchlets in 2 dorsal, 2 ventral, and 2 lateral ranks; leaves strongly decurrent, the free part linear-attenuate. Strobili sessile and terminal on the main axis, or dominant branches and produced in the second, third, or fourth growing season.

Lycopodium dendroideum may be quickly identified by grasping the base of an aerial stem. This will feel distinctly prickly because of the stiff divergent leaves.

Cytology: $n = 34$ (Löve and Löve 1976*).

Habitat: Woods and clearings.

Range: Labrador and Newfoundland to British Columbia and Alaska, south to West Virginia, Michigan, Wisconsin, and Washington; Asia.

Remarks: Hickey (1977) considered this wide-ranging taxon to be a good species and segregated it from *L. obscurum*. The latter used to be considered as consisting of all flat-branched forms (Wherry 1961). Now, however, *L. obscurum* var. *obscurum* is the flat-branched variant, and var. *isophyllum* (equal-leaved ground-pine), with all 6 ranks of leaves of equal size, is a variant of the flat-branched species.

4. **Lycopodium obscurum** L. var. **obscurum**
   ground-pine, tree club-moss
   Fig. 5 (a) habit; (b) portion of branch. Map 5.

---

* Throughout the publication, an asterisk indicates that the count is based on Canadian material.

Fig. 4 *Lycopodium dendroideum*; (a) habit, 2/3 × ; (b) portion of strobilus, 9 × .

Lycopodiaceae

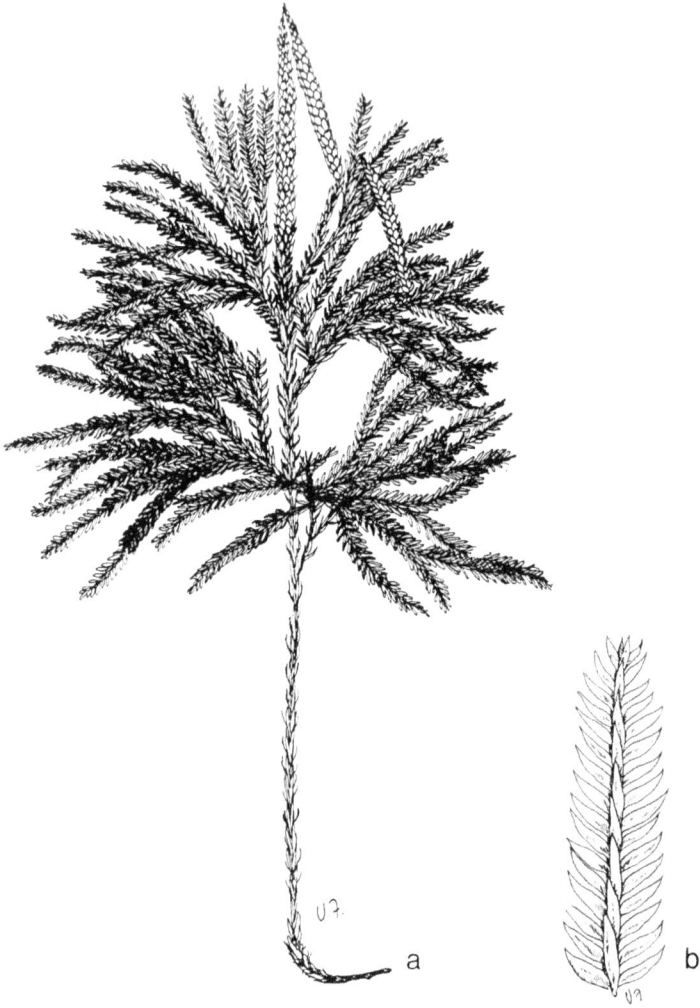

Fig. 5 *Lycopodium obscurum* var. *obscurum*; (a) habit, 1/2 ×; (b) portion of branch, 5×.

Lycopodiaceae

Similar to *L. dendroideum*, from which it may be distinguished by the strongly appressed to slightly divergent leaves on the lower portion of the aerial shoot. Leaves of the lateral branchlets arranged in 1 dorsal, 1 ventral, and 4 lateral ranks; leaves of ventral rank linear-attenuate to long triangular, smaller than leaves of other ranks; leaves of other ranks linear-acuminate to linear-acute.

Cytology: $n = 34$ (Löve and Löve 1976*).

Habitat: Woodlands.

Range: Nova Scotia to Ontario, Michigan, and Wisconsin, south to North Carolina, Tennessee, and Kentucky.

Remarks: Variety *obscurum* is flat-branched, having a single rank of leaves that run along the upper surface of a branchlet and reduced leaves along the ventral rank. This is a more southern and eastern species than the widespread *L. dendroideum*.

4.1. **Lycopodium obscurum** L. var. **isophyllum** Hickey
   equal-leaved ground-pine
Fig. 6, portion of branch. Map 6.

   Similar to var. *obscurum* in that leaves of lower portion of stem are strongly appressed to slightly divergent; leaves of branchlets are all of equal size and linear-attenuate; all leaves lie in planes tangential to the branchlet axis.

Habitat: Woodlands.

Range: Nova Scotia to Ontario and Minnesota, south to Tennessee and Kentucky.

Remarks: Because this variety has been recognized only recently (Hickey 1977), our knowledge of its distribution is still limited. Field workers can make a contribution when they become familiar with the three taxa, *L. dendroideum*, *L. obscurum*, and *L. obscurum* var. *isophyllum*.

Diphasiastrum group

   This group of species was treated in a monograph by Wilce (1965). Many of the species have flattened branches with 4 ranks of scale-like leaves (1 dorsal, 2 lateral, and 1 ventral), which suggest the name ground-cedar, but unfortunately this name has been used for more than one species in the group. The species to be considered here are the larger plants *L. complanatum*, *L. digitatum*, and *L. tristachyum*; and the smaller and more mat-like *L. alpinum*, *L. sitchense*, and *L. sabinifolium*. All have a chromosome number of $n = 23$ (Wilce 1965, but see Löve et al. 1977). The spores of all the species are reticulated (*L. clavatum* type, Wilce 1972) and the species apparently can hybridize with one another. Presumably, these hybrids produce fertile spores, because no meiotic irregularities occur during meiosis. However, this extrapolation has not been tested experimentally because of very low to nonexistent germination of spores of even nonhybrid taxa. Interspecific hybridization in this group is quite unlike that in *Equisetum* or the ferns, where hybrids usually show lack of chromosomal homology between species.

5. **Lycopodium complanatum** L.
   Diphasium complanatum (L.) Rothm.
   Diphasiastrum complanatum (L.) Holub
   flatbranch club-moss
Fig. 7 (*a*) habit; (*b*) portion of stobilus. Map 7.

Horizontal stems mostly below the surface of the ground; leaves distant, scale-like. Upright stems to 30 cm high or higher, with crowded or somewhat remotely forking branchlets. Branchlets flattened, often strongly constricted between yearly growths, 2.0–4.0 mm wide. Leaves 4-ranked; lateral leaves usually appressed; leaves of lower rank much reduced. Strobili mostly 1 or 2 on remotely bracted peduncles.
   Lycopodium complanatum is a familiar species in our boreal woods. Field characters are surficial to buried rhizomes, wide branches, and conspicuous annual constrictions. The irregular growth pattern often gives the plant a rather irregular or straggly look, in contrast to the extreme regularity of *L. digitatum*. The strobili are also irregular in number (1–4) per peduncle, and the naked peduncles seem very fine in relation to the size of the strobili.

Cytology: $n = 23$ (Hersey and Britton 1981*).

Habitat: Woodlands and clearings.

Range: Circumpolar; in North America from Greenland, Labrador, and Newfoundland to Alaska, south to New England, Michigan, Montana, Idaho, and Washington.

Remarks: There are several varieties to consider, and their disposition depends on whether one views the variation as belonging to the parental taxon or whether one considers the varieties to be, in reality, hybrids. Wilce (1965) equated var. *gartonis* Boivin with var. *elongatum* Vict., and they are now interpreted as *L. complanatum* × *tristachyum*. Stunted northern forms called var. *pseudoalpinum*

Fig. 6 *Lycopodium obscurum* var. *isophyllum*; portion of branch, 5 ×.

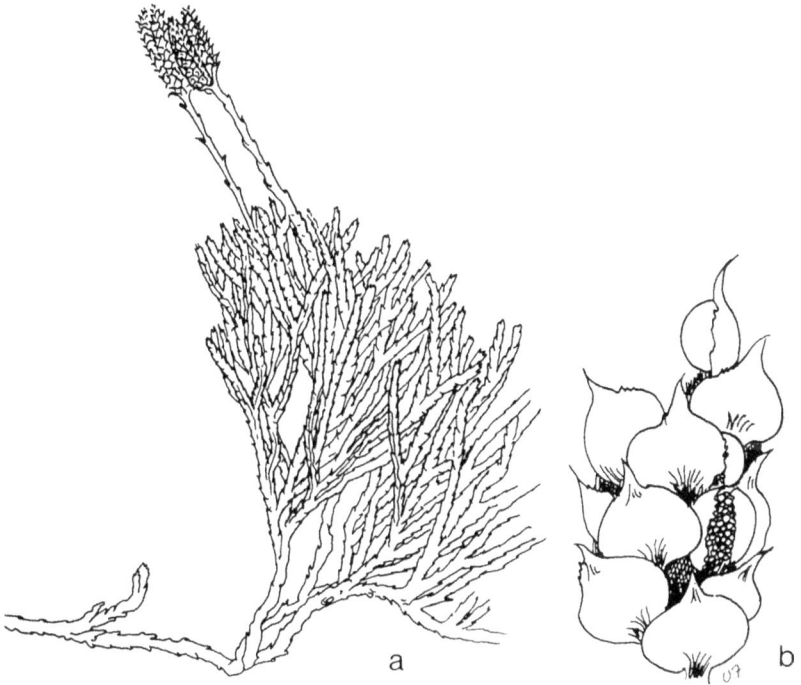

a

b

Fig. 7 *Lycopodium complanatum*; (a) habit, 2/3 × ; (b) portion of strobilus, 10 ×.

Farwell, var. *montellii* Kukkonen, and var. *canadense* Vict. are considered by us to be northern ecotypes.

6. **Lycopodium digitatum** A. Braun
   L. flabelliforme (Fern.) Blanch.
   L. complanatum L. var. *flabelliforme* Fern.
   L. complanatum L. var. *dillenianum* Doll
   Diphasiastrum digitatum (A. Braun) Holub
   Diphasium flabelliforme (Fern.) Rothm.
   crowfoot club-moss, running-pine
Fig. 8, habit. Map 8.

Stems horizontal, mostly on or near the surface of the ground; leaves distant, scale-like. Upright stems to 30 cm high or higher, with the branchlets of the branches arched and fan-like; constrictions between annual growth not present or only slightly evident. Branchlets 2.0–3.0 mm wide. Leaves 4-ranked; lateral leaves usually spreading; lower leaves much reduced. Strobili mostly 3 or 4 on remotely bracted peduncles; peduncle forking at one point.

Cytology: $n = 23$ (Hersey and Britton 1981*).

Habitat: Dry woods and clearings.

Range: Newfoundland to Ontario and Minnesota, south to New England, Kentucky, and Iowa.

Remarks: Linnaeus described *L. complanatum* in 1753, and by 1814 *L. tristachyum* was recognized. Many authors now believe that var. *flabelliforme* should be treated as a separate species, *L. digitatum*. Typical material is quite distinctive. The stems are on or very near the surface, the branchlets are very regular and fan-like, annual constrictions are lacking, and the strobili are usually in groups of four on long, naked peduncles.
   The species is characteristic of sandy woods and clearings in southeastern Canada, and is endemic in North America.

7. **Lycopodium tristachyum** Pursh
   Diphasiastrum tristachyum (Pursh) Holub
   ground-cedar
Fig. 9 (a) habit; (b) portion of strobilus. Map 9.

Horizontal stems usually deeply buried; leaves distant, scale-like. Upright stems to 30 cm high or higher. Sterile branches ascending to loosely divergent, flattened, 1.0–1.5 cm wide. Leaves 4-ranked, bluish green, lanceolate-subulate; lateral leaves appressed; lower leaves somewhat smaller. Strobili 2–6 on leafy-bracted peduncles.

Fig. 8 *Lycopodium digitatum*; habit, 1 × .

Fig. 9 *Lycopodium tristachyum*; (a) habit, 1/3 × ; (b) portion of strobilus, 6 × .

Lycopodiaceae

The vase-shaped and crowded branches, which are bluish green and have whitish wax on their underside, give a striking appearance to the sun forms. Shade forms are more diffusely branched, but the branchlets are still more rounded than those of *L. complanatum* and *L. digitatum*. Good field characters to observe are the annual constrictions along the branches, and ventral and lateral leaves of the same size and shape. The peduncles often branch and then branch again (2-forked) to give rise to 4 strobili.

Cytology: $n = 23$ (Hersey and Britton 1981*).

Habitat: Dry, sometimes sandy woods and clearings.

Range: Newfoundland to Ontario, Michigan, and Minnesota, south to West Virginia and Alabama.

Remarks: Variety *laurentianum* Vict. is considered by Wilce (1965) to be *L.* × *habereri* House.

8.   **Lycopodium alpinum** L.
     Diphasiastrum alpinum (L.) Holub
     alpine club-moss
Fig. 10 (*a*) habit; (*b*) leaves; (*c*) portion of strobilus. Map 10.

Stems elongate, horizontal, rooting at intervals, and bearing few leaves; leaves distant, yellow, bract-like. Erect stems dichotomously forked to 9 cm high. Sterile branchlets somewhat flattened. Leaves 4-ranked, dimorphic; dorsal leaves lanceolate-subulate, appressed, adnate for about half their length; lateral leaves deltoid-ovate to lanceolate, 4–5 mm long, adnate for about half their length, with the free part spreading and incurved at the tip; ventral leaf shorter and trowel-shaped. Strobili essentially sessile at the ends of branched leafy peduncle-like stems.
This northern and alpine species is often confused with *L. sitchense* and *L. sabinifolium*. The 4-ranked leaves, which are free from the stem for about half their length, are distinct from the 5-ranked rounded branches of *L. sitchense*. The ventral trowel-shaped leaves are unlike those of *L. sabinifolium*, which has pedunculate strobili.

Cytology: $n = 23$–24 probably 23 (Löve et al. 1977).

Habitat: Alpine and subalpine meadows and wooded alpine slopes.

Range: Circumpolar; in North America from Greenland to Alaska, south to Newfoundland, eastern Quebec, Michigan, Washington, and Montana.

Fig. 10 *Lycopodium alpinum*; (a) habit, 1 × ; (b) leaves, 12 × ; (c) portion of strobilus, 12 × .

Lycopodiaceae

9. **Lycopodium sitchense** Rupr.
   L. sabinifolium Willd. var. *sitchense* (Rupr.) Fern.
   Diphasiastrum sitchense (Rupr.) Holub
   Diphasium sitchense (Rupr.) Löve & Löve
   Sitka club-moss
Fig. 11 (*a*) habit; (*b*) portion of strobilus. Map 11.

Stems elongate, horizontal, rooting at intervals, bearing distant, yellowish, scale-like leaves. Erect stems dichotomously forked to 18 cm high. Sterile branchlets cylindrical. Leaves in 4 or, more often, 5 ranks, uniform, subulate, adnate for less than half their length, with the free parts usually incurved at the tips. Strobili sessile on leafy branches, not on naked peduncles.

Cytology: $n = 23$ (Löve and Löve 1976*).

Habitat: Alpine and subalpine barrens and wooded slopes.

Range: Greenland, Labrador, and Newfoundland to British Columbia and Alaska, south to Maine, New Hampshire, Montana, Washington, and Oregon.

10. **Lycopodium sabinifolium** Willd.
    Diphasiastrum sabinifolium (Willd.) Holub
    savin leaf club-moss
Fig. 12 (*a*) habit; (*b*) portion of strobilus. Map 12.

Stems elongate, horizontal, rooting at intervals, and bearing a few distant yellowish bract-like leaves. Erect stems dichotomously forked, up to 20 cm high. Sterile branchlets flattened. Leaves 4-ranked, linear-subulate, scarcely dimorphic; dorsal and ventral leaves appressed and slightly adnate; lateral leaves slightly larger and adnate for about half their length, with the free parts spreading and incurved at the tip. Strobili on leafy-bracted peduncles 1–8 cm long.

Cytology: $n = 23$ (Löve and Löve 1976*).

Habitat: Subalpine often dry and sandy woods and meadows.

Range: Labrador and Newfoundland to Algoma District, Ont., south to Pennsylvania and Michigan.

Remarks: Beitel (1979*a*, 1979*b*) stresses the indistinct strobilus base and the "scattered sporophylls and sporangia straggling down naked peduncles." The species is considered by Wilce (1965) and Beitel

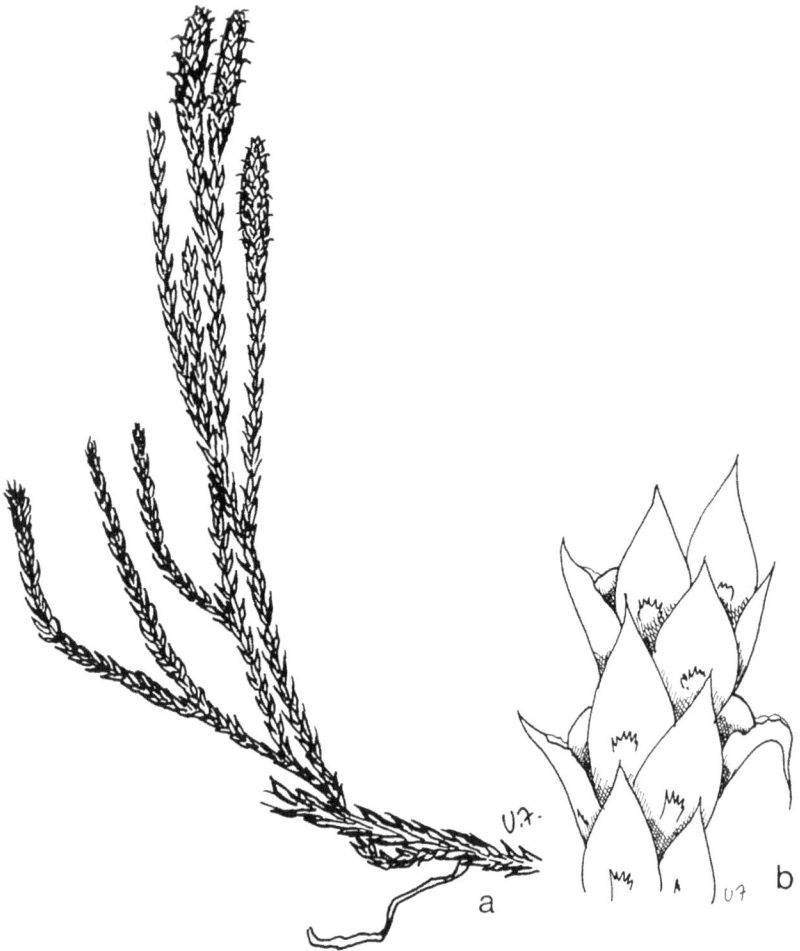

Fig. 11 *Lycopodium sitchense*; (a) habit, 1 × ; (b) portion of strobilus, 7 × .

Lycopodiaceae

a

Fig. 12 *Lycopodium sabinifolium*; (a) habit, 1 × ; (b) strobilus, 7 × .

(1979b) to be a hybrid between *L. sitchense* and *L. tristachyum*. Variants (perhaps segregants) resemble the parental species in appearance. The branchlets are flattened, with leaves in 4 ranks (compare with *L. sitchense*, which is 5-ranked and has round branchlets). In disturbed sites in Ontario, e.g., jack pine blowouts, one can usually find both *L. sabinifolium* and *L. sitchense* in the same location.

Lycopodiella group

The sporophyte is deciduous, except for the extreme tip of the rhizome. The plants are small and creeping, and the most striking attribute is the erect, fertile branches. The spores, known as the rugulate type (Wilce 1972) and having shallow, rolling ridges, are quite unlike the previous groups. The species have a chromosome number based on $n = 78$ (Löve et al. 1977). We have recognized only one species in this complex for Canada, although we know from Beitel (1979b) that the genus has been studied by Bruce (1975), who recognized additional species as well as a host of hybrids. Beitel (1979b) writes (quoting Bruce 1975) of two new tetraploid species (which are still undescribed) in the Great Lakes region, so that we can expect revisions to this group for Canada.

11. **Lycopodium inundatum** L. var. **inundatum**
   Lepidotis inundata (L.) C. Borner
   Lycopodiella inundata (L.) Holub
   bog club-moss
Fig. 13 (a) habit; (b) portion of strobilus. Map 13.

Stems horizontal or arching, forking, rooting at intervals; leaves linear-subulate, gradually long-acuminate, not adnate, spiralled in 8 or 10 ranks. Leaves on the underside of the stem twisted upwards. Fertile stems upright, ascending, or slightly incurved, with leaves similar to those of the sterile stems. Strobili single, sessile; strobilus (6–10 mm wide) definitely wider than stem.

Cytology: $n = 78$ (Löve and Löve 1976*).

Habitat: Acid bogs, shores, damp sandy banks, and disturbed situations.

Range: Labrador and Newfoundland to Ontario, south to Virginia, north-central United States; northern Saskatchewan; Alaskan Panhandle, south through British Columbia to Oregon and Idaho; Eurasia.

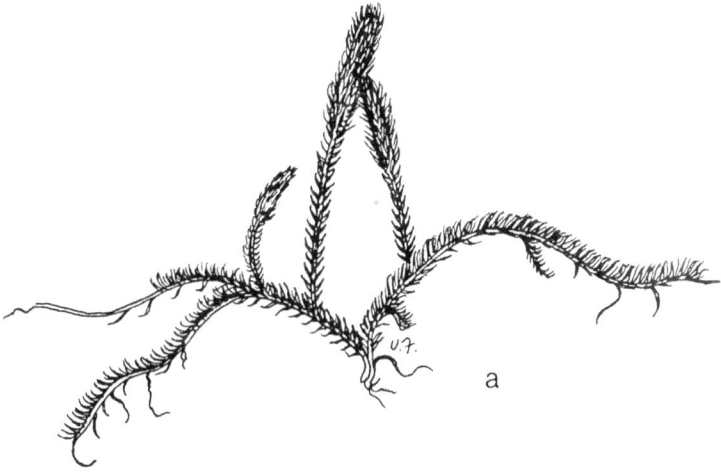

Fig. 13 *Lycopodium inundatum* var. *inundatum*; (a) habit, 2/3×; (b) portion of strobilus, 10×.

11.1 *Lycopodium inundatum* L. var. *bigelovii* Tuckerm.
Fig. 14, habit. Map 14.

Differs from var. *inundatum* by its taller (up to 35 cm) fertile stems and by its mostly ciliate-denticulate leaves. Peduncle leaves and sporophylls tightly appressed. Strobilus narrow, 3–4(5) mm wide.

Cytology: $n = 78$ (W.H. Wagner et al. 1970) for *L. appressum*.

Habitat: Wet shores, bogs, and savannas.

Range: Newfoundland and Nova Scotia to Florida and Texas.

Remarks: According to Beitel (1979b), following Bruce (1975), this taxon belongs in a separate species, *L. appressum* (Chapman) Lloyd & Underwood, the southern bog club-moss. *Lycopodium appressum* has a coastal and lowland distribution in the Gulf and Atlantic states, and Canadian plants would, in their view, be northern outliers. Beitel (1979b) includes Newfoundland in the distribution, and we are told there are two undescribed tetraploid species in this complex, plus several hybrids. The Canadian plants are in need of further study.

Huperzia group

The final group of lycopods to consider includes *L. lucidulum* and *L. selago*, which have leaves in many ranks and lack the specialized strobili that are characteristic of *L. clavatum*. The spores have small pits (foveolate) and are quite different in size and appearance from those of the other groups. The spores are triangular, with concave sides and truncate angles (Wilce 1972). Both *L. lucidulum* and *L. selago* s.l. are considered collective entities by some researchers (Beitel 1979b), and undoubtedly further segregate species will appear. The cytology of this group is difficult because of many meiotic irregularities (hybrid taxa?), small chromosomes, and large numbers of chromosomes. "Determinations" of chromosome number are often only estimates (Manton 1950).

12. *Lycopodium lucidulum* Michx.
 Huperzia selago (L.) Bernh. ssp. *lucidula* (Michx.) Löve & Löve
 shining club-moss
Fig. 15 (*a*) habit (*b*) portion of branch with sporangia. Map 15.

Stems ascending and sprawling, few-forked, to 40 cm long, leafy, rooting towards the base from among the brown marcescent leaves. Leaves mostly 6-ranked, 7–12 mm long, oblanceolate, spreading or deflexed, acuminate, sharply erose-serrulate near the apex,

Fig. 14 *Lycopodium inundatum* var. *bigelovii*; habit, 1 ×.

Fig. 15 *Lycopodium lucidulum*; (a) habit, 1×; (b) portion of branch with sporangia, 10×.

Lycopodiaceae

alternating in bands; shorter leaves appearing early in the season followed later by longer leaves. Stomates only on the lower surface. Sporangia in the axils of the shorter leaves. Gemmae or reproductive buds often borne in the upper leaf axils.

Lycopodium lucidulum may be distinguished from *L. selago* ssp. *patens* by its erose-serrulate rather than entire leaves and by the presence of stomates on the lower leaf surface only, rather than on both surfaces.

Cytology: $n = 67$ (Beitel and Wagner 1982*).

Habitat: Cool moist woods.

Range: Newfoundland to Ontario, Minnesota, and Iowa, south to South Carolina and Indiana.

Remarks: This is a characteristic species of rich boreal and hardwood forests in eastern Canada. The undulating outline to the branches and the toothed leaves of a dark and shiny green are good field characters. The species seems distinct, until one considers *L. lucidulum* var. *occidentale* or *L. selago* var. *patens*. The latter is considered under *L. selago*, and the former is a sporadic form (entire-leaved) in Canadian populations of *L. lucidulum*. Variety or forma *occidentale* as it occurs in Canada should not be confused with *L. porophilum* Lloyd & Underwood, even if listed in the synonymy of that species. *Lycopodium porophilum* is not yet known from Canada (compare with Wherry 1961), but is most frequent in central United States (Wisconsin, Iowa, Ohio, and Pennsylvania), growing on acidic sandstone cliffs and ledges. The color of the plant is markedly yellow green towards the base, it has some undulations in shoot outline, and the leaves are entire, linear-lanceolate, with parallel sides. Stomata occur on both surfaces of the leaves. *Lycopodium porophilum* is reported to hybridize frequently with *L. lucidulum*, and the hybrids can rapidly reproduce by gemmae.

13.   ***Lycopodium selago*** L. ssp. ***selago***
    L. selago L. var. *appressum* Desv.
    Huperzia selago (L.) Bernh.
    mountain club-moss
Fig. 16 (*a*) habit; (*b*) portion of strobilus. Map 16 (s.l.).

Horizontal stems short, leafy, rooting from among the marcescent leaves; erect stems to 20 cm high or higher, branched several times, usually near the base. Leaves yellow-green, crowded, 8- to 10-ranked, 3–8 mm in length, ovate-lanceolate, entire or nearly so, acuminate, usually hollow at the base, and with stomates on both surfaces. Sporangia in the axils of leaves produced early in the season followed later by sterile leaves, thus appearing in bands. Gemmae or reproductive buds often borne in the upper leaf axils.

Cytology: $n = 132$ (Löve and Löve 1966a).

Habitat: Arctic tundra species, south in the mountains, on barrens, in bogs, and in cold woods.

Range: Circumpolar; in North America from Greenland to Alaska, south to Virginia, Michigan, Wisconsin, Montana, and Washington.

Remarks: According to Beitel (1979b), *Lycopodium selago* s.l. is a complex of species, hybrids, and environmental forms. Because all the Canadian material has not been compared and is inadequately known at this time, we have fallen back on an older and simpler interpretation of these plants, recognizing *L. selago* ssp. *selago*, ssp. *patens*, and ssp. *miyoshianum*. Beitel (1979b) would refer ssp. *patens* to a "catch-all" taxon, i.e., a complex of a species, hybrids, and environmental forms. Some specimens are perhaps hybrids of ssp. *selago* and *L. lucidulum*. Undoubtedly, one can expect to see quite different treatments of this species in the future. For the present, we are following Calder and Taylor (1968) until new research is published and has been carefully evaluated.

13.1 *Lycopodium selago* L. ssp. *patens* (Beauv.) Calder & Taylor

Differs from ssp. *selago* in its longer leaves, 8–12 mm, lance-attenuate, reflexed or strongly divergent, and by its stems to 30 cm long or longer.

Habitat: Cold woods and rocky situations.

Range: Newfoundland to Manitoba and Wisconsin, south to New England; British Columbia and Alaska.

13.2 *Lycopodium selago* L. ssp. *miyoshianum* (Makino) Calder & Taylor
Fig. 17 (a) habit; (b) portion of strobilus.

Differs from ssp. *selago* by its leaves dark green, thin and flexuous, narrowly lanceolate, about 6 mm long, strongly imbricated but not appressed, and by its usually longer stems.

Habitat: Mountain slopes.

Range: British Columbia through Alaska to Japan, Korea, and China.

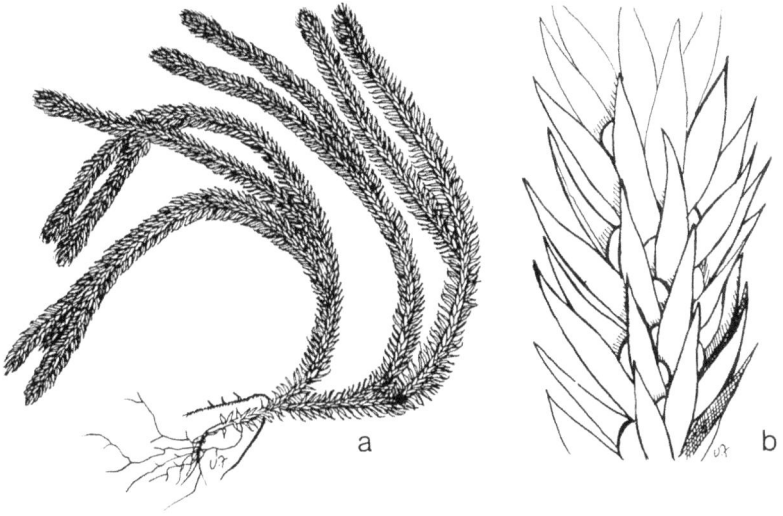

Fig. 16 *Lycopodium selago* ssp. *selago*; (a) habit, 1/2 ×; (b) portion of strobilus, 10 ×.

Fig. 17 *Lycopodium selago* ssp. *miyoshianum*; (a) habit, ½ ×; (b) portion of strobilus, 3 ×.

Lycopodiaceae 47

The occurrence of hybrids in *Lycopodium* is not subject to direct experimental testing because of our inability to germinate and grow spores in large numbers. A further difficulty is the varying concepts of species within the genus. Some researchers would deny that hybrids exist, whereas others believe that hybrids are quite frequent. Until there is greater unanimity concerning the characteristics of the basic species in our flora, the characteristics of the hybrids between these species (or even their existence) will be uncertain. Accordingly, we have dismissed from consideration hybrids in all the groups except those in section *Complanata* or genus *Diphasiastrum* Holub. As mentioned previously, they are unusual hybrids for pteridophytes, in that few if any meiotic irregularities or lack of chromosomal homology are observed (Hersey and Britton 1981), and it is assumed that the spores are viable. The most frequently cited hybrids involve the triangle of *L. complanatum*, *L. digitatum*, and *L. tristachyum*. The most convincing ones are those whose parents differ rather widely in morphology.

Wilce (1965) cited nine Canadian specimens of *Lycopodium complanatum* × *digitatum* from Quebec and Ontario, but considered it rare. The parents, however, do not appear very different, so that hybrids are difficult to distinguish.

Wilce (1965) cited 24 specimens of *Lycopodium complanatum* × *tristachyum* (*Diphasiastrum* × *zeilleri* (Rouy) Holub) for Canada, from Northwest Territories, Saskatchewan, Manitoba, Ontario, Quebec, Prince Edward Island, Newfoundland, and Labrador. It is supposedly more common in Minnesota than is *L. tristachyum*. A number of our collections have come from the Thunder Bay District, Ont.

The extremes of the species are distinctive, so that the hybrid *L. digitatum* × *tristachyum* (*Lycopodium* × *habereri* House) seems reasonably clear-cut. Wilce (1965) reported 16 collections for Canada from Ontario, Quebec, and New Brunswick. The cytology was studied by Hersey and Britton (1981).

Lycopodium alpinum × *complanatum* (*L.* × *issleri* (Rouy) Lawalrée) is known from a few localities in Europe, as well as from Maine in North America (Wilce 1965). It should be looked for where the ranges of the parents overlap.

A specimen of *Lycopodium alpinum* × *sitchense* from the Mealy Mountains in southern Labrador has been seen. Wilce (1965) cited two specimens for North America, one from Oregon and the other from Washington.

According to Wilce (1965) the species *L. sabinifolium* arose from a cross of *L. sitchense* with *L. tristachyum*. *L. sabinifolium* is extremely variable, and variants approach the parents in morphology. Accordingly, it is not possible to detect hybrids of *L. sabinifolium* with either *L. sitchense* or *L. tristachyum*.

2. SELAGINELLACEAE   spikemoss family

1. *Selaginella* Beauv.   spikemoss

Low, creeping plants with branching stems and few fine roots.
Leaves simple, imbricated, in 4 or 6 rows, with or without a bristle tip.
Some sporangia containing macrospores, others microspores, borne in
the axils of the leaf-like sporophylls of a terminal cone.
The genus *Selaginella*, in which several hundred species are
included, is the only genus in the family Selaginellaceae. These
species are widely distributed and are mostly tropical. About 37
species occur in North America north of the Mexican border. A few
species are cultivated as greenhouse plants, and one, *S. lepidophylla*,
the resurrection plant, is imported as a novelty. Members of the *S.
rupestris* section that grow in semiarid and subalpine regions are
adapted to being almost completely desiccated and revive a few hours
after moisture becomes available.

A.   Leaves flat, not bristle-tipped.
    B.   Leaves dimorphic; sporophylls and leaves eciliate . . . .
    . . . . 2. **S. apoda**
    B.   Leaves uniform; sporophylls and leaves ciliate . . . .
    . . . . 1. **S. selaginoides**
A.   Leaves grooved on the back, bristle-tipped.
    C.   Leaves abruptly adnate to the stem and differing in color
        from it . . . . 3. **S. wallacei**
    C.   Leaves decurrent on the sides of the stem.
        D.   Bristles of leaves about 1 mm long . . . . . 5. **S. densa**
        D.   Bristles shorter.
            E.   Epiphytic; stems lax and freely branching; cones
                hardly differentiated . . . . 4. **S. oregana**
            E.   Terrestrial; densely matted; cones distinct
                F.   Leaves tapering to the setae . . . .
                . . . . 6. **S. rupestris**
                F.   Leaves truncate at the apex, then setate . . .
                . . . . 7. **S. sibirica**

1.   *Selaginella selaginoides* (L.) Link
Fig. 18 (*a*) habit; (*b*) portion of strobilus. Map 17.

Plants delicate, branching, forming small mats. Leaves uniform,
2–4 mm long, spreading-ascending, acute, ciliate. Fertile branches
upright, with lower leaves similar to those of the stem but becoming
larger upwards to form the sporophylls of a subcylindric spike.

The uniform, ciliate leaves and upright fertile branches readily separate *S. selaginoides* from the only other herbaceous-leaved species in our area, *S. apoda*.

Cytology: $2n = 18$ (Löve and Löve 1976*).

Habitat: Moist banks and shores, bogs, and boggy woods.

Range: Circumpolar; in North America from Greenland and Labrador to Alaska, south to New England, Michigan, Minnesota, and southern British Columbia.

Remarks: This widespread circumpolar species, which occurs northward nearly to the limit of trees, is often partly buried in mosses or muskeg, and thus is easily overlooked. It is rare in Manitoba (White and Johnson 1980) and Saskatchewan (Maher et al. 1979).

2. **Selaginella apoda** (L.) Fern.
    S. eclipes Buck
Fig. 19 (*a*) habit; (*b*) portion of the branch bearing sporangia. Map 18.

Plants delicate, freely branching, matted. Leaves membranous, 4-ranked, the lateral 2 rows bluntish, oblong to oval, and spreading; dorsal and ventral leaves pointed, smaller, and appressed. Spikes sessile; sporophylls eciliate, similar to the foliage leaves.
The pale or whitish green eciliate heterophyllous leaves readily set this species off from all others in our area.

Cytology: $2n = 18$ (Löve and Löve 1976*).

Habitat: Wet woods, swamps, bogs, and shores.

Range: Southwestern Quebec and Ontario to Wisconsin, south to Florida and Texas.

Remarks: Buck (1977) described a new species, *S. eclipes*, with a range north of, but essentially adjacent to, the range he ascribed to *S. apoda*. The characters used to differentiate *S. eclipes* from *S. apoda* largely overlapped those of the latter, and the author therefore suggested that it might prove to be better placed at a subspecific level. Because this publication is not an appropriate vehicle for making such a transfer, we have included *S. eclipes* in the synonymy of *S. apoda*. A map depicting the distribution of *S. eclipes* and *S. apoda* in the restricted sense is given by Buck (1977). Alston (1955) reported *S. apoda* as occurring west to British Columbia, but in Canada the species is found only in southwestern Quebec and Ontario.

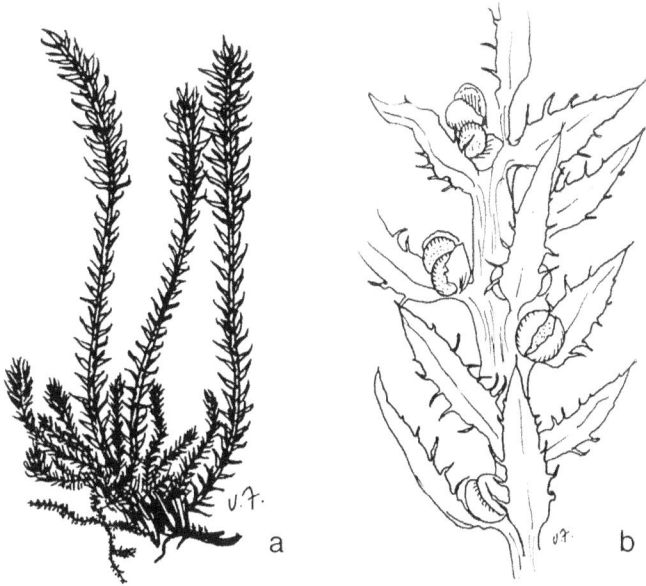

Fig. 18 *Selaginella selaginoides*; (a) habit, 1 × ; (b) portion of strobilus, 8 ×.

Fig. 19 *Selaginella apoda;* (a) habit, 1 × ; (b) portion of branch bearing sporangia, 2 ×.

Selaginellaceae

3. *Selaginella wallacei* Hieron.
   S. montanensis Hieron.
   Fig. 20 (*a*) habit; (*b*) leaves; (*c*) portion of strobilis. Map 19.

Main stems prostrate, sparsely rooted, forming loose mats. Ascending branches numerous. Leaves abruptly adnate to the stem, tightly appressed, more or less glaucous, oblong-linear, more or less obtuse at the apex, grooved on the back, ciliate, about 3 mm long, including the approximately 0.5 mm long scabrous seta. Sporophylls ovate-deltoid, shorter than the leaves, ciliate. Setae nearly smooth.
   S. wallacei may be distinguished from *S. densa* var. *scopulorum*, the taxon it sometimes most closely resembles, by its remote branches and its abruptly adnate rather than decurrent leaf bases.

Cytology: none.

Habitat: Open and shaded rocky slopes.

Range: Southern British Columbia and adjacent mountain slopes of Alberta, south to northern California.

Remarks: Specimens from moist, shady situations have long stems that form loose mats and somewhat distant leaves, whereas specimens from drier, more open situations are more compact and the leaves are closer together. The species is rare in Alberta (Argus and White 1978).

4. *Selaginella oregana* D.C. Eat.
   Fig. 21 (*a*) habit; (*b*) portion of strobilus. Map 20.

Stems long, lax, and freely branching. Leaves bright green, loosely imbricate, ovate-triangular, adnate for about half their length, eciliate or slightly ciliate towards the tip; setae short, green to whitish. Spikes sessile, inconspicuous; sporophylls ovate, long-acuminate, eciliate towards the tip.
   According to R.M. Tryon (1955) *S. oregana* is the only species in the *S. rupestris* complex that is commonly an epiphyte. The long pendant branches are characteristic of the species.

Cytology: None.

Habitat: Usually epiphytic on such trees as *Acer macrophyllum*.

Range: West coast of Vancouver Island, British Columbia, south in coastal United States to northern California.

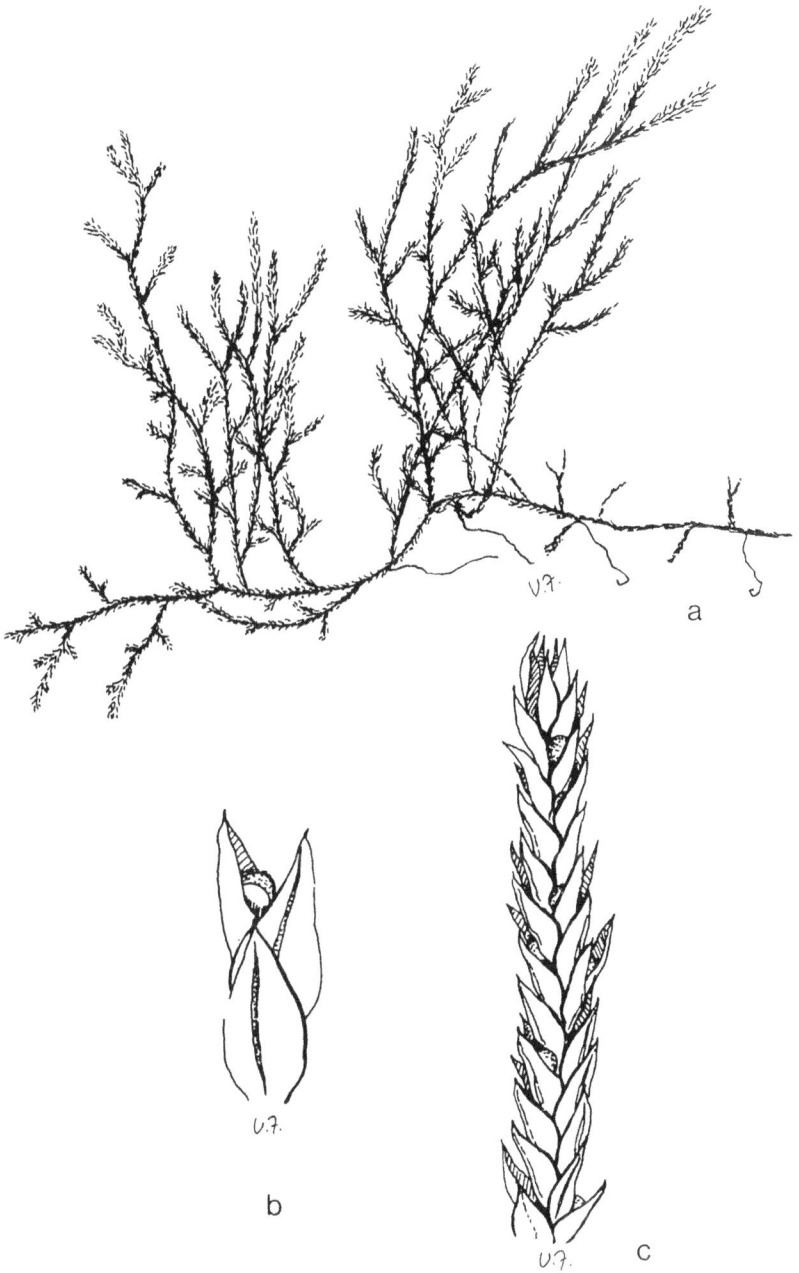

Fig. 20 *Selaginella wallacei*; (a) habit, 1 ×; (b) leaves, 10 ×; (c) portion of strobilus,
5 ×.

Selaginellaceae

Remarks: *Selaginella oregana* was recently collected at Barclay Sound and Power River on the west coast of Vancouver Island; an early collection by Scouler, labeled Observatory Inlet, is most likely mislabeled as to locality.

5. **Selaginella densa** Rydb.
Fig. 22 (a) habit; (b) leaf; (c) portion of strobilus. Map 21.

Stems forming dense cushion mats. Leaves decurrent on the sides of the stem, with under leaves longer than upper leaves on the same portion of the stem; leaves grooved on the back, ciliate, about 3 mm long, including the approximately 1-mm-long scabrous seta. Setae often forming a distinct brush at the tips of the branches. Sporophylls ovate-deltoid, apiculate, ciliate.

On the Canadian prairies both *S. densa* and *S. rupestris* occur. The former can usually be distinguished by the longer (about 1 mm long) setae.

Cytology: $2n = 18$ (Löve and Löve 1976*).

Habitat: Dry prairies.

Range: Southwestern Manitoba to southeastern British Columbia, south to New Mexico and Arizona.

5.1 **Selaginella densa** Rydb. var. **scopulorum** (Maxon) Tryon
S. scopulorum Maxon
Fig. 23 (a) habit; (b) leaf; (c) portion of strobilus. Map 22.

Differs from var. *densa* by its sporophylls eciliate in the upper part and by the slightly shorter setae of the leaves.

Habitat: Alpine rocky slopes and ridges.

Range: Southwestern Alberta and southern British Columbia, south to Texas, Arizona, and northern California.

Remarks: R.M. Tryon (1955) deemed *S. densa* to be one of the most complex species in the *S. rupestris* complex. He upheld three varieties that freely intergrade, var. *densa*, var. *scopulorum* (Maxon) Tryon, and var. *standleyi* (Maxon) Tryon. The last variety, which has the apex of leaves predominantly or entirely truncate in profile, was recorded from southwestern Alberta and British Columbia by R.M. Tryon (1955), but we have been unable to assign any specimens there.

Selaginellaceae

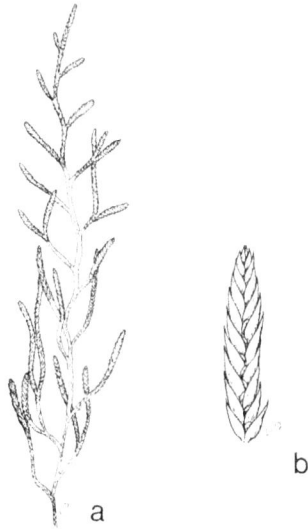

Fig. 21 *Selaginella oregana*; (a) habit, 1/2 × ; (b) portion of strobilus, 2 1/2 ×.

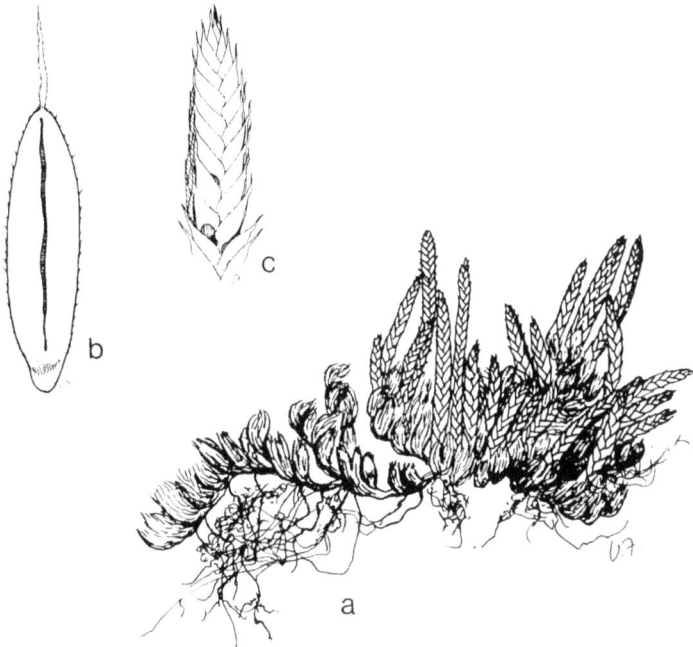

Fig. 22 *Selaginella densa*; (a) habit, 1 3/5 × ; (b) leaf, 21 × ; (c) portion of strobilus, 3 3/5 ×.

Selaginellaceae 55

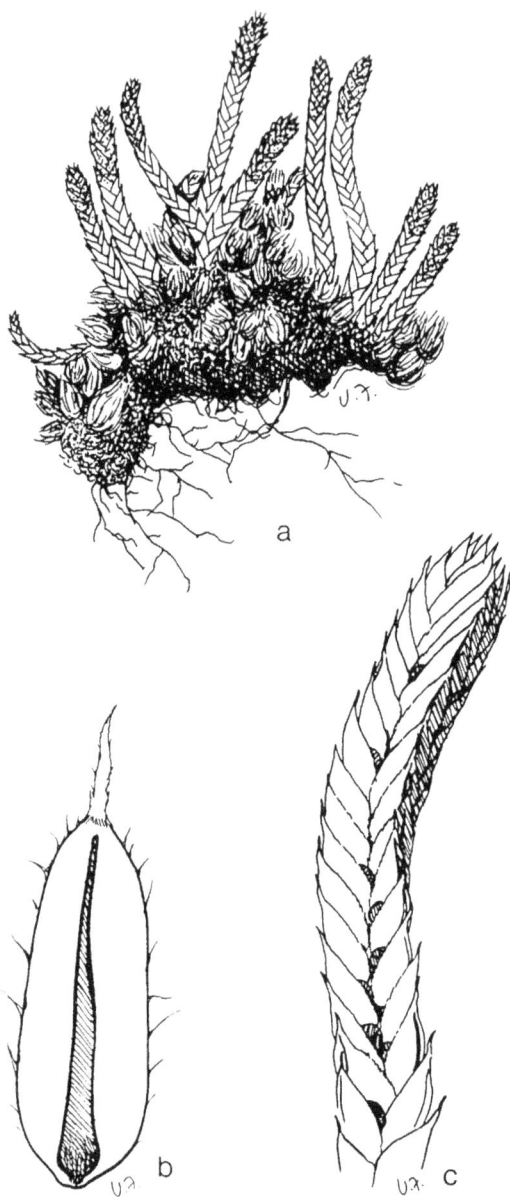

Fig. 23 *Selaginella densa* var. *scopulorum*; (a) habit, 2 1/2 ×; (b) leaf, 40 ×;
(c) portion of strobilus, 7 ×.

6.   **Selaginella rupestris** (L.) Spring
Fig. 24 (a) habit; (b) leaves; (c) portion of strobilus. Map 23.

Prostrate stems forming open mats. Leaves decurrent on the sides of the stem; leaves linear-lanceolate, about 2.8 mm long (including the approximately 0.7-mm-long scabrous seta), grooved on the back, ciliate. Sporophylls narrowly ovate-deltoid, apiculate, ciliate, about as long as the leaves.
Selaginella rupestris and its allies were studied in detail by R.M. Tryon (1955); *S. rupestris* most closely resembles *S. densa* var. *densa*, from which it can most easily be distinguished by its radially symmetrical leafy stem.

Cytology: $2n = 18$ (Löve and Löve 1976*).

Habitat: Sand dunes and open or shaded, dry, often igneous rocky bluffs.

Range: Greenland, Quebec, and New Brunswick to northern Alberta, south to Georgia, Michigan, Kansas, and Oklahoma.

Remarks: According to R.M. Tryon (1955), *S. rupestris* is the only species that is certainly apogamous; however, some populations have four megaspores in a sporangium, and because microsporangia are present in the strobilis, the plants are presumably sexual. The species is rare in Nova Scotia (Maher et al. 1978).

7.   **Selaginella sibirica** (Milde) Hieron.
Fig. 25 (a) habit; (b) leaves; (c) portion of strobilus. Map 24.

Stems forming small intricate mats. Leaves densely appressed-ascending, decurrent on the sides of the stem; leaves linear-ligulate, about 2.7 mm long (including the approximately 0.5-mm-long scabrous seta), grooved on the back, subtruncate to truncate at the apex. Sporophylls broadly ovate-deltoid, short-apiculate, ciliate, shorter than the leaves.
Selaginella sibirica is perhaps closest in aspect to some forms of *S. densa*, from which it can usually be distinguished by the milk-white rather than lutescent setae and the intricate rather than discrete branches.

Cytology: $2n = 18$ (Zhukova and Petrovsky 1972); $2n = 20$ (Johnson and Packer 1968).

Habitat: Dry exposed rocks and ridges.

Fig. 24 *Selaginella rupestris*; (a) habit, 1 ×; (b) leaves, 10 ×; (c) portion of strobilus, 10 ×.

Selaginellaceae

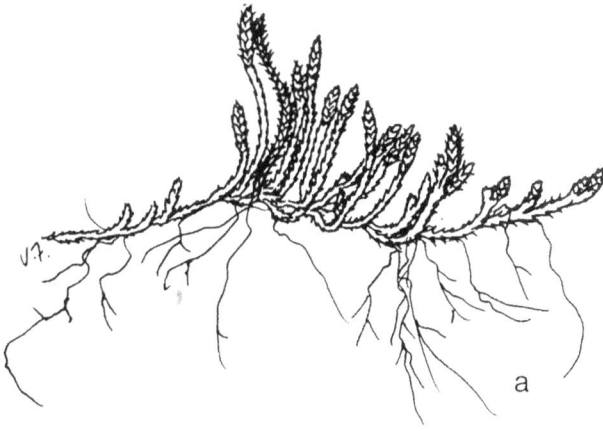

Fig. 25 *Selaginella sibirica*; (a) habit, 1×; (b) leaves, 12×; (c) portion of strobilus, 12×.

Range: Amphi-Beringian; in North America, Alaska to northwestern District of Mackenzie.

Remarks: R.M. Tryon (1955) noted that although the North American material is relatively uniform, there is a phase in Asia that has longer tawny rather than milk-white setae.

3. ISOETACEAE   quillwort family

1. *Isoetes* L.   quillwort

Perennial usually aquatic herbs. Leaves superficially grass-like, few to numerous from a lobed, corm-like rhizome. Sporangia in a hollow at the expanded base of the leaves and more or less covered by the thin edges of the hollow (velum), and with a small ligule situated above. Spores dimorphic, numerous, variously ornamented; megaspores (female) borne in megasporangia; microspores (male) borne in microsporangia.

This is a technically difficult poorly known genus, with perhaps 100 species in the world. Pfeiffer (1922) discussed, in monographs, all the species (approximately 60 at the time) in the genus, and Kott (1980*b*) studied eight species in northeastern North America. Even with a battery of modern techniques, such as scanning electron microscopy (SEM), cytology, and chemical analysis (chromatography) (Kott and Britton 1982*a*), the species are difficult to identify and their phylogeny is in doubt. It is possible that a controlled hybridization program might be informative.

Simple vegetative characters such as length of leaves, rigidity of leaves, and their color and shape, for example, are extremely variable, depending greatly on habitat. Most systems of classification depend almost entirely on megaspore and microspore size and ornamentation, which results in the necessity of microscopy. The average field pteridologist is not likely to be challenged by these difficult plants and should perhaps be satisfied with a determination to genus. To distinguish quillworts from other aquatic plants such as *Eleocharis*, which often look somewhat similar, one should first apply the "thumb test." After pulling up the plant, one can feel the swollen base or corm when it is squeezed gently between the thumb and forefinger. When the sporophylls (leaves) are removed singly, the sporangia at their bases are evident.

Cytology has proven useful for classification. The basic chromosome number ($\times$) is 11, and diploids ($2\times$) to decaploids ($10\times$) are known in nature. Because spore size is correlated with ploidy level, careful measurement of a sample of spores is often useful for identification of dried material.

In eastern Canada there are seven species to consider. Two are wide-ranging and frequent. *Isoetes echinospora* ($2\times$) with small spiny megaspores is found usually in shallow water, and *I. macrospora* ($10\times$) with large ridged megaspores is usually in deep water. The other species are much less widespread, and indeed some are considered rare.

In western Canada, there are at least six species. Three are wide-ranging and frequent: *I. echinospora*, with small spiny megaspores and almost smooth microspores; *I. maritima* ($4\times$), with larger spores and spiny or papillate microspores; and *I. occidentalis*

(6×), which sometimes has been confused with *I. lacustris* of Europe. The last is 10× and is undoubtedly closer to *I. macrospora* than to *I. occidentalis*. The other three species are more localized in their distribution and include the interesting terrestrial species *I. nuttallii* (2×) and the amphibious species *I. howellii (2×)*. It is interesting to note that although 12 Canadian species are treated here, there is no species occurring in Canada with megaspores that have the regular, neat honeycombs of *I. engelmannii*, which occurs in the eastern United States as far north as New York State, and of *I. japonica*, which occurs in Japan.

A. Plants terrestrial; corms more or less 3-lobed; leaves trigonous; megaspores smooth to spongy fibrillar .... 9. *I. nuttallii*
A. Plants aquatic; corms 2-lobed; leaves usually rounded.
    B. Megaspores with sparse or dense spines.
        C. Spines on megaspores long and acute, not reduced in size near the equator; microspores smooth or with very fine thread-like spines; across Canada and abundant .... 1. *I. echinospora*
        C. Spines on megaspores blunt, sometimes confluent into ridges, reduced to small tubercles near equator; microspores echinate with coarse pronounced spines; British Columbia and Alberta .... 2. *I. maritima*
    B. Megaspores not spiny, but with various types of ridges.
        D. Megaspores with scattered or more or less confluent low ridges or wrinkles.
            E. Amphibious; hyaline wing-margins extending 1–5 cm above the sporangium; ligule elongated triangular .... 10. *I. howellii*
            E. Submerged; hyaline wing-margins not extending more than 1 cm above the sporangium; ligule cordate .... 11. *I. bolanderi*
        D. Megaspores with more or less connected, distinct ridges.
            F. Megaspores with short, closely set, meandering but not anastomosing ridges or mounds; ridges and mounds minutely spiny .... 3. *I. eatonii*
            F. Megaspores with ridges variously textured, branched, or anastomosing.
                G. Megaspores with ridges rounded, smooth, and with a smooth unornamented band encircling the distal side of the equatorial ridge.
                    H. Microspores usually smooth to slightly low-papillate, 37–45 µm in length .... 6. *I. hieroglyphica*
                    H. Microspores roughly echinate, 25-31 µm in length .... 5. *I. acadiensis*

Isoetaceae

G. Megaspores with anastomosing or branched ridges that have rough or sharp crests; equatorial zone variously ornamented.
  I. Megaspores usually averaging over 600 μm; leaves usually dark green and stiff .... 8. *I. macrospora*
  I. Megaspores usually averaging less than 600 μm.
    J. Ligule cordate; megaspores cream-colored ....
       .... 12. *I. occidentalis*
    J. Ligule elongate; megaspores white.
      K. Megaspores with ridges on the distal face forming reticulations; microspores roughened to smoothish . . .
         .... 7. *I. tuckermanii*
      K. Megaspores with close ridges not forming reticulations; microspores papillate
         .... 4. *I. riparia*

1. **Isoetes echinospora** Dur.
   I. muricata Dur.
   I. echinospora Dur. var. *muricata* (Dur.) Engelm.
   I. braunii Dur.
   I. echinospora Dur. var. *braunii* Engelm.
   I. muricata Dur. var. *braunii* (Engelm.) Reed
   I. echinospora Dur. ssp. *muricata* (Dur.) Boivin var.
      savilei Boivin
Fig. 26 (a) habit; (b) ventral side of leaf base showing sporangia and ligule; (c) megaspore. Map 25.

Corm 2-lobed. Leaves 7–25 or more, usually erect, fine and soft, bright green to yellowish green. Sporangium to 10 mm long and 3 mm wide, unspotted or spotted. Velum covering one-quarter to three-quarters of sporangium. Ligule deltoid to elongate, up to 2.5 mm long. Megaspores spherical, white, averaging 480 μm (350–550) in diameter, covered with sparse to dense spines; microspores kidney-shaped, averaging 26 μm (23–32) in length, usually smooth or with fine thread-like spines under SEM.

Cytology: $2n = 22$ (Kott and Britton 1980*; Britton and Ceska unpublished*).

Habitat: Shallow water up to about 1 m in depth in ponds, lakes, and slow-moving rivers.

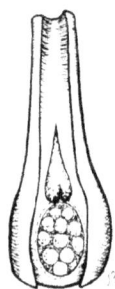

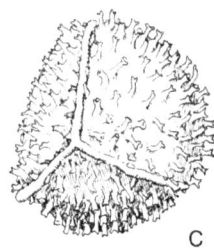

Fig. 26 *Isoetes echinospora*; (a) habit, 3/4 × ; (b) ventral side of leaf base showing sporangia and ligule, 4 × ; (c) megaspore, 60 × .

Isoetaceae

Range: Labrador and Newfoundland to Alaska, south to Pennsylvania, Wisconsin, Colorado, and California.

Remarks: *Isoetes echinospora* is Canada's most abundant and widespread species. It is often found in shallow water in sand or gravel. The sharp straight leaves of some specimens are seen growing intermixed with *Lobelia dortmanna* and *Eriocaulon*. The Canadian species is sometimes considered distinct from the European *I. echinospora* and is then called *I. muricata*. A similar species, *I. asiatica*, is known from northern Japan and Sakhalin Island.

2. **Isoetes maritima** Underw.
   I. macounii A.A. Eat.
   I. echinospora var. *maritima* (Underw.) A.A. Eat.
   I. beringensis Komarov
Fig. 27, megaspore. Map 26.

Corm 2-lobed. Leaves 8–15, erect, rigid, or somewhat recurved, dark green, 2–5 cm long, 1.5 mm wide. Sporangia oval, 4 mm long, 2.5 mm wide, covered one-third to one-half by the velum. Ligule small, inconspicuous. Megaspores spherical, white, 490–670 µm in diameter, covered with spines that are rather blunt and sometimes confluent into ridges or plates; spines reduced in size near the equator. Microspores kidney-shaped, white, 30–36 µm in length, rough with short sharp spines under SEM; light microscopy suggests papillose or reticulated surface.

Cytology: $2n = 44$ (Britton and Ceska unpublished[*]).

Habitat: Shallow water to about 1 m in lakes and estuaries.

Range: Alaska, British Columbia, Alberta, and south to Washington.

Remarks: Historically, *Isoetes maritima* was a plant related to tidal flats because it was found by Macoun near Port Alberni, B.C., and treated by Hultén (1968) as a coastal species. It is now considered present in interior lakes also, as well as east of the Rockies near Jasper.

3. **Isoetes eatonii** Dodge
   I. gravesii A.A. Eat.
Fig. 28, megaspore. Map 27.

Corm 2-lobed. Leaves 12–100, 8–45 cm long or longer, erect, usually fine and soft, bright green to yellowish green. Sporangium to 12 mm long and 5 mm wide, unspotted to tan-colored. Velum covering one-sixth to one-quarter of sporangium. Ligule 3.5 mm long, elongate.

Megaspores flattish, white, averaging 400 µm (320–530) in diameter with closely set, short, meandering, spiny ridges or mounds. Microspores rounded, averaging 23 µm (22–25) in length, roughened to smooth.

Cytology: $2n = 22$ (Kott and Britton 1980*).

Habitat: Ponds and slow rivers to 1 m in depth.

Range: In Canada, apparently isolated in the Severn River, Muskoka District, and Simcoe County, Ont. (Kott and Bobbette 1980); in the United States occurring in the New England states, New Jersey, Pennsylvania, and New York.

4. **Isoetes riparia** Engelm.
   I. echinospora Dur. var. *robusta* Engelm.
   I. braunii Dur. f. *robusta* (Engelm.) Reed
   I. canadensis (Engelm.) A.A. Eat.
Fig. 29, megaspore. Map 28.

Corm 2-lobed. Leaves 5–35, 6–35 cm long or longer, usually erect, fine, and lax, bright green to yellow-green. Sporangium to 7 mm long and 4 mm wide, unspotted or with horizontal streaks. Velum covering one-quarter of sporangium. Ligule elongate to 3 mm long. Megaspores spherical, averaging 540 µm (430–680) in diameter, with widely or closely set long and branching or short and broken ridges. Microspores kidney-shaped, averaging 31 µm (24–35) in length, granular-textured and usually with spaced spine-tipped tubercules.

Cytology: $2n = 44$ (Kott and Britton 1980*).

Habitat: River shores, creeks, and tidal mud flats.

Range: Southern Quebec and southeastern Ontario southward through Maine, Vermont, and eastern New York.

Remarks: The megaspores have characteristics that resemble both those of *I. echinospora* and those of *I. macrospora*. At times, the megaspores appear eroded, with rough projections that one could mistake for spines, and at other times the broken ridges approach the sculpture of *I. macrospora*.

5. **Isoetes acadiensis** Kott
Fig. 30, megaspore. Map 29.

Corm 2-lobed. Leaves 9–35 or more, 5–21 cm long, mostly recurved, dark green and sometimes tinged with red. Sporangium to 5 mm long and 3 mm wide, unspotted or with a few brown spots. Velum covering one-sixth to one-third of sporangium. Ligule elongate, to 3 mm in length. Megaspores spherical, 400–570 µm in diameter, with smooth rounded reticulating or branching ridges. Microspores kidney-shaped, 25–30 µm in length, roughly echinate.

Cytology: $2n = 44$ (Kott 1981).

Habitat: Shallow water along borders of lakes, ponds, and rivers.

Range: Newfoundland, Nova Scotia, and New Brunswick to Maine, Massachusetts, and New Hampshire.

Remarks: This recently described species (Kott 1981) appears to have a distribution and ecology similar to *I. tuckermanii*. It was formerly included in *I. hieroglyphica* because of the similarity of megaspore sculpturing.

6. **Isoetes hieroglyphica** A.A. Eat.
    I. macrospora Dur. f. *hieroglyphica* (A.A. Eat.) Pfeiffer
Fig. 31, megaspore. Map 30.

Corm 2-lobed; leaves 7–15 or more, 5–11 cm long, erect or recurved. Sporangium to 5 mm long and 3 mm wide, usually unspotted. Velum covering one-third of sporangium. Megaspores spherical, white, averaging 635 µm (580–700) in diameter, with low rounded ridges forming a network. Microspores kidney-shaped, averaging 40 µm (37–45) in length, smoothish to low papillate.

Cytology: Unknown.

Habitat: Lakes.

Range: Southwestern Quebec and adjacent New Brunswick, Maine, and Wisconsin. Few collections are known, and knowledge of distribution is incomplete.

7. **Isoetes tuckermanii** A. Br.
Fig. 32, megaspore. Map 31.

Corm 2-lobed. Leaves 10–45 or more, 4–25 cm long or occasionally longer, erect or recurved, soft and fine, bright green, or sometimes yellowish green. Sporangium to 5 mm long and 3 mm wide, usually unspotted. Velum covering one-quarter or less of sporangium. Ligule elongate, to 2 mm long. Megaspores spherical, white, averaging 518 μm (400–650) in diameter, with rough-crested ridges forming a honeycomb. Microspores kidney-shaped, averaging 27 μm (24–33) in length, tuberculate to almost smooth.

Cytology: $2n = 44$ (Kott and Britton 1980*).

Habitat: Shallow water of estuaries, slow-moving streams, lakes, and ponds.

Range: Newfoundland, Nova Scotia, New Brunswick, and the St. Lawrence estuary region of Quebec, south through the New England states at least to Maryland.

Remarks: The soft and fine recurved leaves and the Atlantic coastal plain distribution are typical of this tetraploid species. It is sometimes difficult to distinguish it from *I. macrospora*. Without cytology, the easiest way to separate *I. macrospora* from *I. tuckermanii* is to measure carefully about 20 microspores with the use of a microscope.

8. **Isoetes macrospora** Dur.
    I. heterospora A.A. Eat.
Fig. 33, megaspore. Map 32.

Corm 2-lobed. Leaves few to 70 or more, 3–17 cm long, stiff and erect or with recurving tips, dark green. Sporangium to 5 mm long and 4 mm wide, usually unspotted. Velum covering one-sixth to one-quarter of sporangium. Ligule deltoid, to 2 mm long. Megaspores spherical, white, averaging 640 μm (400–800) in diameter, with ridges that form honeycomb-like areas. Microspores kidney-shaped, averaging 42 μm (32–50) in length, with the surface having evenly spaced blunt or rounded papillae.
The usually large, coarse, and stiff-leaved plants from water as deep as 6 m are field aids for recognition of this eastern species.

Cytology: $2n = 110$ (Kott and Britton 1980*).

Habitat: Usually in deep water of oligotrophic lakes in the Precambrian Shield.

Range: Newfoundland, Nova Scotia, Quebec, and Ontario to Minnesota, south through the Appalachian region of the United States to Virginia.

9. **Isoetes nuttallii** A. Br.
  I. suksdorfii Baker
Fig. 34, megaspore. Map 33.

Corms more or less 3-lobed. Leaves up to 60, 7–17 cm long, 3-angled, slender with conspicuous hyaline margins towards the base. Sporangium conspicuous, about 5 mm long and 1.5 mm wide. Velum completely covering the sporangium. Ligule small, triangular. Megaspores 400–500 μm in diameter, densely fibrillar spongy or even smooth. Microspores 28–31 μm long, spiny tuberculate.

Cytology: $2n = 22$ (Britton and Ceska unpublished*).

Habitat: Usually terrestrial, on springy but not regularly inundated terrain.

Range: Southern Vancouver Island, British Columbia, to California.

Remarks: This species has most of its distribution to the south of Canada. It is associated with the limited zone where the madrona, or *Arbutus*, tree grows in Canada. Growth starts in the fall, and the plant is dormant by early summer in hot, dry weather.

10. **Isoetes howellii** Engelm.
  I. melanopoda Gay & Dur. var. *californica* A.A. Eat.
Fig. 35, megaspore. Map 34.

Corm 2-lobed. Leaves 5–28, up to 30 cm long, slender but tough, with membranous margins at the base above the sporangia. Sporangia about 6 mm long. Velum covering about one-third of sporangium. Ligule narrow, elongated-triangular. Megaspores about 475 μm (420–610) in diameter, with inconspicuous anastomosing wrinkles or slightly tuberculate ridges. Microspores about 27 μm (25–30) in length, coarsely and roughly very spinulose.

Cytology: $2n = 22$ (Britton and Ceska unpublished*).

Habitat: Muddy shores and wet depressions, both in and out of water.

Range: In Canada known only from the vicinity of Kamloops, B.C.; in the United States from Oregon to California, east to Montana and Idaho.

Remarks: This species, like *I. nuttallii*, has only the fringe of its distribution in Canada. It is closely related to, and perhaps conspecific with, the wide-ranging species *I. melanopoda* Gay & Dur. (Taylor et al. 1975).

Isoetaceae   69

11. *Isoetes bolanderi* Engelm.
Fig. 36, megaspore. Map 35.

Corm 2-lobed. Leaves up to 20 in number, up to 15 cm long, slender and soft. Sporangium about 4 mm long. Velum covering about one-third of sporangium. Ligule small, cordate. Megaspores white, sometimes bluish, about 370 μm (350–390) in diameter, with very low tuberculate ridges or only wrinkles. Microspores about 27 μm (25–30) in length, obscurely fine spinulose.

Aids for identification are the slender and soft leaves, which are light yellow green, and a centre of distribution largely in the alpine lakes of Colorado, Montana, and Wyoming.

Cytology: $2n = 22$ (Britton and Ceska unpublished*).

Habitat: Lakes and ponds in often deep water, often at high altitudes.

Range: Southwestern British Columbia and Waterton Lakes, Alta., south to California, Wyoming, and Arizona.

Remarks: At this time, *I. bolanderi* is known from very few localities in Canada close to the United States border.

12. *Isoetes occidentalis* Henderson
   I. lacustris L. var. *paupercula* Engelm.
   I. paupercula (Engelm.) A.A. Eat.
   I. piperi A.A. Eat.
   I. flettii (A.A. Eat.) Pfeiffer
Fig. 37, megaspore. Map 36.

Corm 2-lobed. Leaves 10–30 or more, 5–20 cm long, more or less rigid, dark green. Sporangium almost orbicular, 5–6 mm in diameter. Velum covering one-quarter to one-third of the sporangium. Ligule short-triangular. Megaspores 500–700 μm in diameter, cream-colored or white, with sharp ridges and crests, and sometimes tuberculate or almost smooth. Microspores 36–43 μm long, papillose.

Cytology: $2n = 66$ (Britton and Ceska unpublished*).

Habitat: Ponds and lakes.

Range: Coastal Alaska, British Columbia, south to California and Colorado, at low elevations. Frequent on Vancouver Island and in lakes around the Fraser Valley.

Isoetaceae

27

28

29

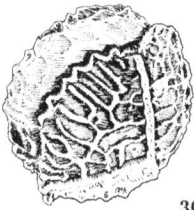

30

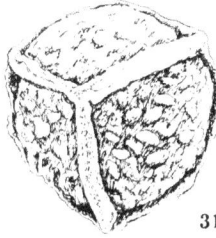

31

32

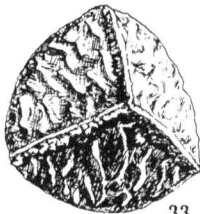

33

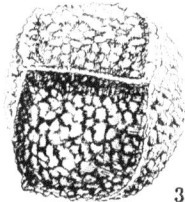

34

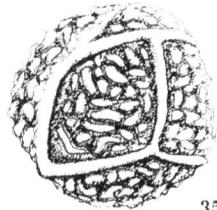

35

Fig. 27   *Isoetes maritima*; megaspore, 45 × .

Fig. 28   *Isoetes eatonii*; megaspore, 75 × .

Fig. 29   *Isoetes riparia*; megaspore, 50 × .

Fig. 30   *Isoetes acadiensis*; megaspore, 55 × .

Fig. 31   *Isoetes hieroglyphica*; megaspore, 45 × .

Fig. 32   *Isoetes tuckermanii*; megaspore, 50 × .

Fig. 33   *Isoetes macrospora*; megaspore, 45 × .

Fig. 34   *Isoetes nuttallii*; megaspore, 60 × .

Fig. 35   *Isoetes howellii*; megaspore, 60 × .

Fig. 36   *Isoetes bolanderi*; megaspore, 75 × .

Fig. 37   *Isoetes occidentalis*; megaspore, 50 × .

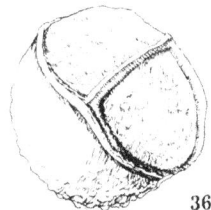

36

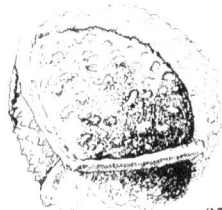

37

Isoetaceae

Remarks: The rigid, dark green leaves, often with a reddish base, have suggested to some previous workers an association with the European *I. lacustris*. *Isoetes occidentalis* is a hexaploid, not a decaploid, and the megaspores are very variable. The almost smooth, chalk white, fragile (easily cracked) megaspores of some collections are distinctive, but unfortunately they can range to almost spiny megaspores (as seen in a variant once known as *I. flettii*) or they can have rounded protuberances (as in the *I. piperi* type). The large microspores with characteristic papillae are more uniform.

4. EQUISETACEAE horsetail family

1. *Equisetum* L. horsetail

Rhizomatous perennials. Stems rush-like, jointed, sometimes hollow, branched or unbranched; internodes of stems commonly ridged longitudinally, with stomata in rows or bands in the grooves and with ridges bearing siliceous tubercules or bands. Leaves small, whorled, fused into nodal sheaths. Spores green, spherical, wrapped with 4 elaters, and borne in sporangia on sporophylls in cones. Cones terminal on vegetative stems, or occasionally on branches, or in some species on specialized precocious shoots that lack chlorophyll.

The genus *Equisetum* is the only genus in the family Equisetaceae. It is mostly cool north-temperate in distribution. According to Hauke (1978), there are 15 species, eight in the subgenus *Equisetum*, and seven in the subgenus *Hippochaete*. Ten species are known to occur in Canada, all but one of which, *E. laevigatum*, occur also outside North America. All species have the same chromosome number, $n = 108$. Sterile hybrids, a few of them widespread, occur within the two subgenera.

A.   Stems unbranched.
    B.   Fertile stems green.
        C.   Stomata in bands or scattered in the grooves; stems annual; cones not apiculate.
            D.   Central cavity four-fifths the diameter of the stem; sheaths with 15–20 dark brown teeth . . . . 1. *E. fluviatile*
            D.   Central cavity about one-sixth the diameter of the stem; sheaths with 10 or fewer white-margined teeth . . . . 2. *E. palustre*
        C.   Stomata in two lines in each groove; stems perennial or mostly annual (*E. laevigatum*); cones apiculate or obtuse (*E. laevigatum*).
            E.   Sheaths with 3 (4) teeth; stems lacking a central cavity . . . . 9. *E. scirpoides*
            E.   Sheaths with 4 or more teeth; stems with a central cavity.
                F.   Teeth few, not articulated at the base, persistent . . . . 10. *E. variegatum*
                F.   Teeth numerous, articulated at the base.
                    G.   Stems annual, soft; sheaths with black bands at the apex only . . . . 8. *E. laevigatum*
                    G.   Stems perennial, firm; sheaths becoming black-banded at the base and apex . . . . 7. *E. hyemale* ssp. *affine*
    B.   Fertile stems not green.

H. Coning stems fleshy, lacking stomata, withering after sporulation.
    I. Sheaths with more than 14 teeth, cones 4–8 cm long .... 3. *E. telmateia*
    I. Sheaths with less than 14 teeth; cones 2-4 cm long .... 4. *E. arvense*
H. Coning stems not fleshy, with stomata, and becoming green and branched after sporulation.
    J. Sheaths chestnut brown, flaring upwards, with teeth cohering in several broad lobes; branches usually branched again .... 5. *E. sylvaticum*
    J. Sheaths green, rather tight, with teeth white-margined, free or nearly so; branches usually unbranched .... 6. *E. pratense*
A. Stems branched.
  K. Sterile stems 0.5–3.0 m tall (British Columbia) .... 3. *E. telmateia*
  K. Sterile stems to 0.6 m tall (widespread).
    L. Fertile and sterile stems similar, green; first internode of the primary branches (if present) equaling or mostly shorter than the stem sheath; coning in summer.
      M. Central cavity about one-sixth the diameter of the stem .... 2. *E. palustre*
      M. Central cavity about four-fifths the diameter of the stem .... 1. *E. fluviatile*
    L. Fertile and sterile stems not alike; first internode of the primary branches considerably longer than the stem sheath; coning in spring.
      N. Stem sheath teeth chestnut brown, papery .... 5. *E. sylvaticum*
      N. Stem teeth dark, stiff.
        O. Branches spreading, with teeth of their sheaths deltoid .... 6. *E. pratense*
        O. Branches ascending, with teeth of their sheaths lance-attenuate .... 4. *E. arvense*

1. **Equisetum fluviatile** L.
  E. limosum L.
  water horsetail
Fig. 38 (*a*) fertile branch; (*b*) sterile branch; (*c*) node. Map 37.

Stems up to 1 m long or longer, but usually shorter, 3–8 mm thick, annual, single, but often forming dense stands from branching smooth light brown rhizomes. Central cavity four-fifths or more the diameter of the stem; vallecular cavities absent; 10–30 smooth ridges present; stomata in a broad band in each groove; sheaths tightly appressed, with 15–20 teeth; teeth dark brown, narrow, acuminate,

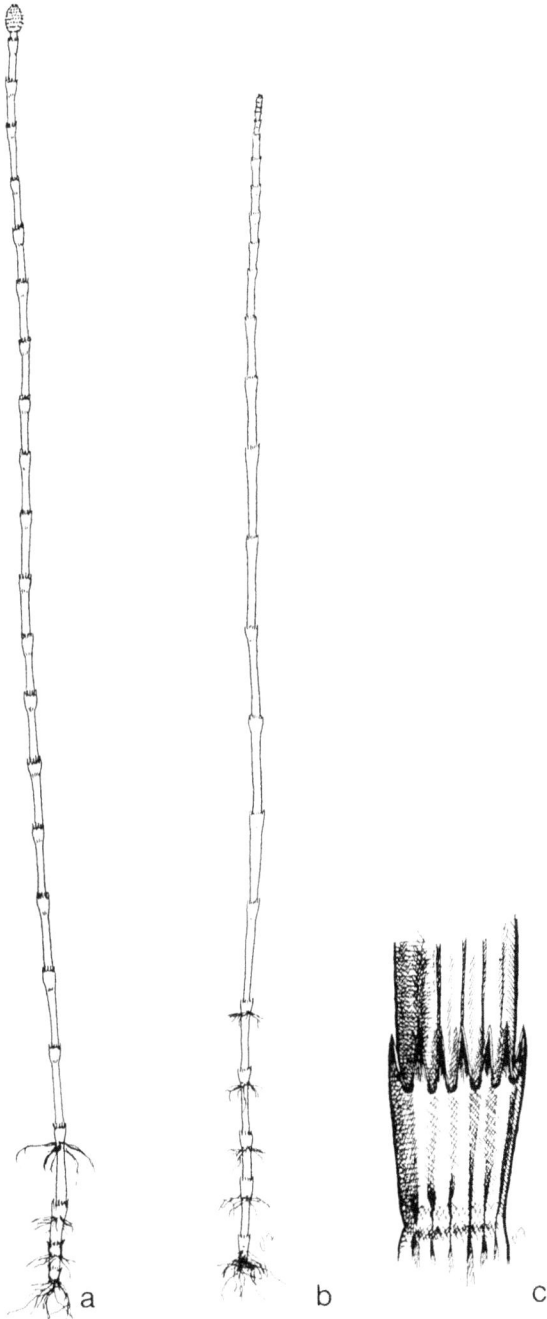

Fig. 38 *Equisetum fluviatile*; (a) fertile stem, 1/4 ×; (b) sterile stem, 1/4 ×; (c) node, 4 ×.

Equisetaceae   75

persistent. Plant unbranched, or with branches occurring sporadically, or verticillate. Branches up to 15 cm long, hollow, with 4-6 ridges, with the first internode shorter than the stem sheath and with the teeth narrowly pointed. Cones up to 2.5 cm long, yellow to brown, obtuse, peduncled, deciduous, shedding spores from May to August.

The water horsetail may be distinguished from all other species of *Equisetum* in Canada by the soft annual stems in which the central cavity is about four-fifths the diameter of the stem. The stems collapse readily when squeezed, because of the thin walls.

Cytology: n = 108 (Löve and Löve 1976*).

Habitat: Quiet, shallow water of rivers and lakes, wet shores, swales, and ditches.

Range: Circumpolar; in North America from Labrador to Alaska, south to New England, Virginia, Indiana, Wyoming, and Oregon.

Remarks: The form with simple stems or with merely a few scattered branches has been described as f. *linaeanum* (Döll) Broun. This is what Linnaeus, who believed it to be a distinct species, called *E. limosum* (Fernald 1950). The species often forms extensive stands in shallow, slow-moving water.

2.   **Equisetum palustre** L.
     marsh horsetail
Fig. 39 (*a*) sterile stem; (*b*) fertile stem; (*c*) node. Map 38.

Stems annual, 20-80 cm long, 1-3 mm thick, erect, solitary or clustered, and growing from shiny, black to brown, occasionally tuber-bearing rhizomes. Central cavity one-sixth to one-third the diameter of the stem; vallecular cavities about the same size as the central cavity and alternating with the 5-10 prominently angled smooth or rough ridges; stomata in a single wide band in the valley; sheaths green with the teeth long, narrow, black, scarious-margined. Branches (sometimes few to none) spreading in regular whorls from the middle nodes, with the first internode shorter than the subtending stem sheath; sheaths with 5-6 teeth similar to the stem teeth but with less obvious scarious margins. Cones 1-3.5 cm long, not apiculate, deciduous, peduncled at the end of the main stems. Spores shed from June to August.

Points by which this species may be distinguished from the somewhat similar *E. arvense* are discussed under that species. The small central cavity causes the stem to feel firm when squeezed.

Cytology: n = 108 (Taylor and Mulligan 1968*).

Fig. 39 *Equisetum palustre*; (a) sterile stem, 1/3 × ; (b) fertile stem, 1/3 × ; (c) node, 3 × .

Habitat: Wet woods and meadows, shores, and shallow waters.

Range: Circumpolar; in North America from Newfoundland to Alaska south to New York, Minnesota, Idaho, and California.

Remarks: Marsh horsetail usually occurs in wetter situations than does field horsetail, with which it might be confused. It has been reported as being poisonous to horses in Europe (Bottarelli 1968; Richter 1961).

3. **Equisetum telmateia** Ehrh. ssp. *braunii* (Milde) Hauke
   giant horsetail
Fig. 40 (a) fertile stem; (b) sterile stem. Map 39.

   Stems annual, of two kinds, sterile and fertile, both erect, mostly solitary from felted tuber-bearing rhizomes. Sterile stems to 2 m long or longer, 0.5–2.0 cm thick; central cavity two-thirds to three-quarters the diameter of the stem, and prominent vallecular cavities alternating with 14–30 somewhat scabrous ridges; internodes whitish, lacking stomata; sheaths pale below, dark above, with two-keeled teeth; teeth long-attenuate, broadly hyaline-margined, united in groups of two or three. Branches whorled, solid, with 4 or rarely 5 grooved scabrous ridges; stomata in bands on each side of the valleys. Fertile stems unbranched, lacking chlorophyll, generally shorter than the sterile stems but thicker and fleshy, with longer sheaths and longer cohering teeth; fertile stems normally withering and dying after spores are shed but occasionally persisting and becoming branched. Cones to 7 cm or more long, shedding spores in April and May.
   Giant horsetail can usually be recognized easily by its size alone; small plants can be distinguished from *E. arvense* by the larger, looser sheaths, which have 2-ribbed teeth.

Cytology: $n = 108$ (ssp. *telmateia*, Sorsa 1965).

Habitat: Swamps and low wet places by lakes and streams.

Range: Subspecies *telmateia* is found in Europe, North Africa, and western Asia; ssp. *braunii* occurs along the Pacific coast of North America from Kodiak Island, Alaska, to California; disjunct in Keewenaw County, Mich., but this inland station remains to be rediscovered.

Remarks: *Equisetum telmateia* ssp. *braunii* is usually found near the Pacific coast, but it has been collected at Penticton and observed at Kelowna in the Okanagan Valley, B.C. The species was collected by

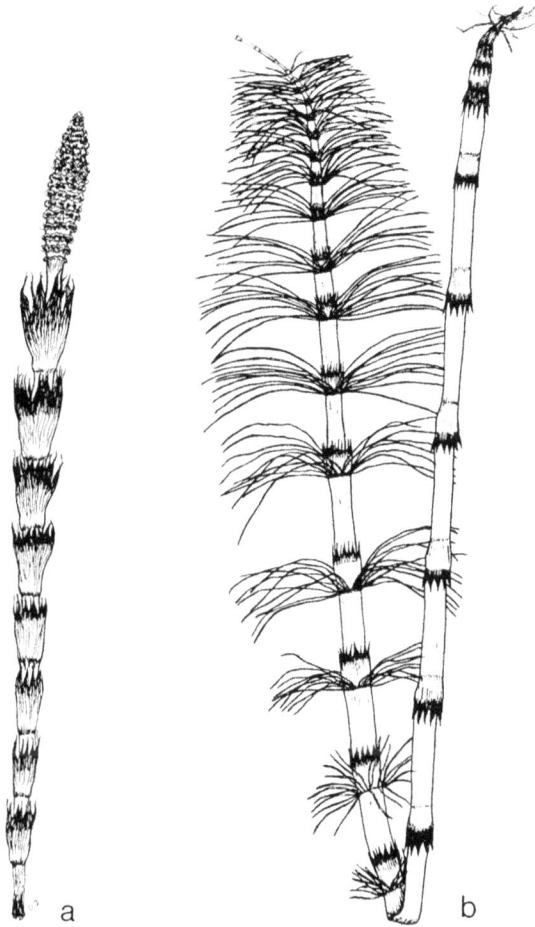

Fig. 40 *Equisetum telmateia* ssp. *braunii*; (a) fertile stem, 1/3 × ; (b) sterile stem, 1/3 × .

Farwell far inland on the Keewenaw Peninsula of northern Michigan in 1880 and 1895 (Billington 1952). It has not been found there since and may have been extirpated.

4. **Equisetum arvense** L.
   E. boreale Bongard
   E. calderi Boivin
   field horsetail

Fig. 41 (a) fertile stem; (b) sterile stem; (c) habit of reduced arctic form; (d) branch node; (e) root nodules. Map 40.

Stems of two kinds, sterile and fertile, annual, growing from dark brown to black, hairy, occasionally tuber-bearing rhizomes. Sterile stems upright to 50 cm long or longer, to prostrate or diffusely branched, 1.5–5 mm thick; central cavity one-third to two-thirds the diameter of the stem, and large vallecular cavities alternating with the 4–14 ridges; silica in dots on the ridges; stomata in 2 broad bands in the valleys; sheaths with 4–14 short narrow dark scarious-margined teeth and occasionally cohering in pairs. Branches solid, whorled, spreading or ascending, mostly unbranched, 3- or 4-angled; teeth lance-attenuate; first internode longer than the subtending stem sheath. Fertile stems lacking chlorophyll, precocious and fleshy, withering and dying after spores are shed, generally shorter than the sterile stems; sheaths 0.5–2.5 mm long, with 8–12 brown, scarious-margined, persistent, distinct or partly united teeth; terminal cone long-peduncled, not apiculate. Spores shedding from late March to mid May or later, depending on latitude, altitude, and season.

Sterile stems of *E. arvense* are perhaps most frequently confused with *E. pratense* and *E. palustre*. *Equisetum pratense* is more delicate in aspect, and the stems are whitish green; also, the teeth are deltoid rather than lance-attenuate as in *E. arvense*. In *E. arvense* the first internode of the branches is longer than the subtending teeth, whereas in *E. palustre* the first internode is shorter than the subtending teeth. This species is very variable and many forms have been described; these forms, however, are the result of exposure, peculiar ecology, season, or damage, and they do not warrant taxonomic status (Hauke 1966).

Cytology: $n = 108$ (Löve and Löve 1976*).

Habitat: Damp open woods, low open ground and meadows, roadside fill, and embankments, often where the surface is dry and sandy; but even in apparently xeric situations the rhizome system can be found penetrating saturated soil. In arctic situations it may grow on shattered limestone or in pockets of soil with permafrost close to the surface.

Range: Circumpolar; in North America from Greenland to Alaska, south to Georgia, Alabama, Texas, and California.

Equisetaceae

Fig. 41 *Equisetum arvense*; (a) fertile stem, 1/2 × ; (b) sterile stem, 1/2 × ; (c) habit of reduced arctic form, 1/2 × ; (d) branch node, 1/2 × ; (e) root nodules, 1/2 × .

Equisetaceae

Remarks: This species has been studied in great detail by Hauke (1966, 1978) and others. Cody and Wagner (1981) published a review of field horsetail in the series *Biology of Canadian Weeds*. The plant is a serious weed in low-lying pastures and in some crops, and has been reported as poisonous to livestock in Canada (Gussow 1912). Others (Pohl 1955; Rapp 1954) have more recently reviewed the reports of toxicity in the genus *Equisetum*. Field horsetail is difficult to eradicate by cultivation because of the deeply buried rhizome that continues to send up new branches.

5. **Equisetum sylvaticum** L.
   wood horsetail
Fig. 42 (a) sterile and fertile stems; (b) node; (c) portion of strobilus. Map 41.

Stems annual, of two kinds, sterile and fertile, both erect, mostly solitary, from shiny, light brown, smooth, hairy, occasionally tuber-bearing rhizomes. Sterile stems up to 70 cm in length and 1.5–3 mm thick; central cavity one-half to two-thirds the diameter of the stem, and prominent vallecular cavities alternating with the 10–18 ridges; silica tubercules in 2 rows on the ridges; stomata borne in 2 bands in the valleys; sheaths loosely inflated, with the reddish brown papery teeth persistent, usually united into 3 or 4 groups; branches whorled, arched, appearing lacy from secondary branches; branches with 3 or 4 (rarely 5) ridges, with stomata in a single line on either side of the valleys, and with teeth narrow, pointed, and spreading. Fertile stems at first unbranched and lacking chlorophyll, precocious and fleshy, becoming green and branched after the spores have been released; sheaths and teeth usually larger than in sterile stems. Peduncled cones to 3 cm in length, blunt, deciduous, shedding spores in April and May, or later at higher latitudes.

The wood horsetail is readily recognized by its secondary branches, which give the plant a lacy appearance, and by the loosely inflated, reddish brown sheaths.

Cytology: $n = 108$ (Löve and Löve 1976*).

Habitat: Moist open woods, wet banks, swamps, and meadows.

Range: Circumpolar; in North America from southern Greenland and Labrador to Alaska, south to Washington, Montana, Michigan, and Virginia.

Remarks: This is a very variable species in which many varieties and forms, which have little taxonomic significance, have been described (Hauke 1978).

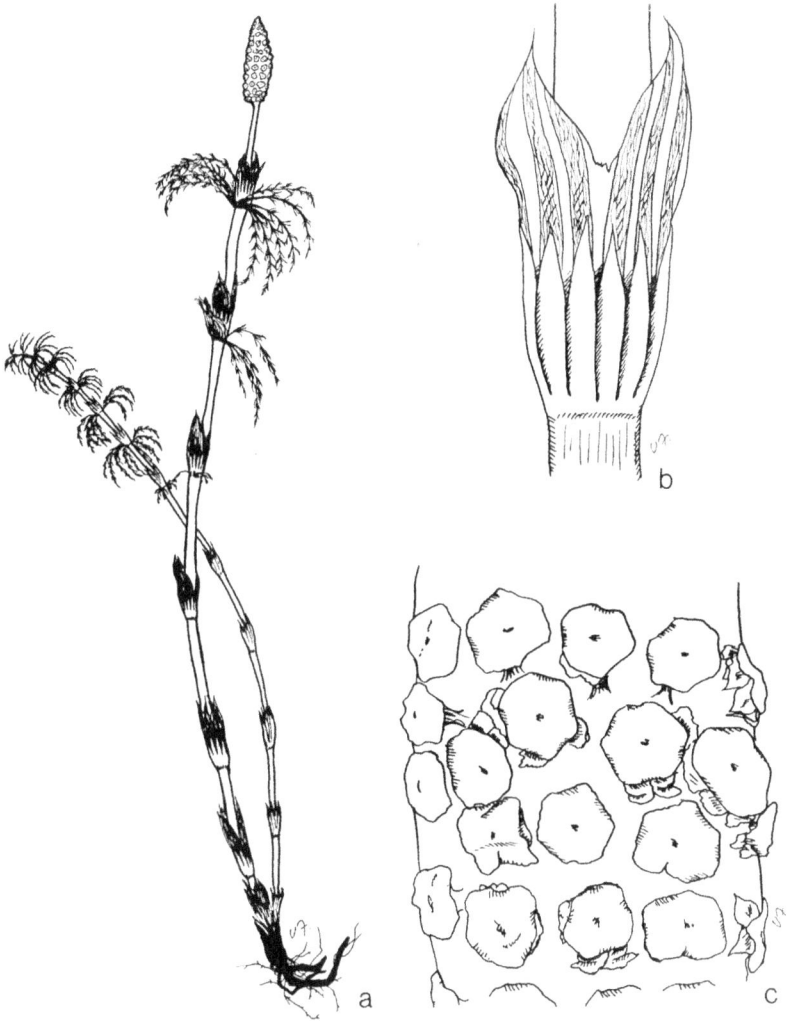

Fig. 42 *Equisetum sylvaticum*; (*a*) sterile and fertile stems, 1/2×; (*b*) node, 5×; (*c*) portion of strobilus, 5×.

6. *Equisetum pratense* Ehrh.
    meadow horsetail
Fig. 43 (*a*) sterile and fertile stems; (*b*) node. Map 42.

Stems annual, of two kinds, sterile and fertile, both upright, mostly solitary from dull black rhizomes. Sterile stems whitish green, to 50 cm in length and 1–3 mm thick; central cavity from one-sixth to one-third the diameter of the stem, and small vallecular cavities alternating with the 8–18 ridges; silica spicules long, thin, in 3 rows on the ridges of the middle and upper internodes; stomata in 2 bands in the valleys; sheaths pale, with the teeth narrow, persistent, white-margined, and dark-centred; branches whorled, horizontal to drooping, with 3 ridges; deltoid teeth slightly incurved, with thin white margins. Fertile stems apparently not common, at first unbranched and lacking chlorophyll, precocious and fleshy, becoming green and branched after the spores are shed; sheaths and teeth about twice as long as those of the sterile stems; peduncled cones to 2.5 cm in length, blunt, deciduous, shedding spores from late April to early July.

This species is perhaps most easily confused with *E. arvense*, the field horsetail, from which it can be separated by its more delicate aspect, the whitish green stems that possess thin silica spicules on the ridges of the middle and upper internodes, and the deltoid, rather than lance-attenuate, teeth on the branches.

Cytology: $n = 108$ (Löve and Löve 1961).

Habitat: Moist woods or meadows, in sun or partial shade.

Range: Circumpolar; in North America from Newfoundland and Labrador to Alaska, south to Montana, Michigan, and New York.

Remarks: *Equisetum pratense* is apparently more common in northwestern Canada, where it sometimes forms dense stands on open wooded floodplains of rivers and streams.

7. *Equisetum hyemale* L. ssp. *affine* (Engelm.) Stone
    E. prealtum Raf.
    Hippochaete hyemalis (L.) Bruhin ssp. *affinis* (Engelm.) Holub
    scouring-rush
Fig. 44 (*a*) sterile and fertile stems; (*b*) node. Map 43.

Stems perennial, up to 1.2 m long or longer, but much shorter throughout most of the Canadian range, 0.3–1.0 cm thick, upright, usually unbranched, single or several together from a thick, dark brown, dull, rough rhizome. Central cavity three-quarters or more the diameter of the stem; small vallecular cavities alternating with 14–50

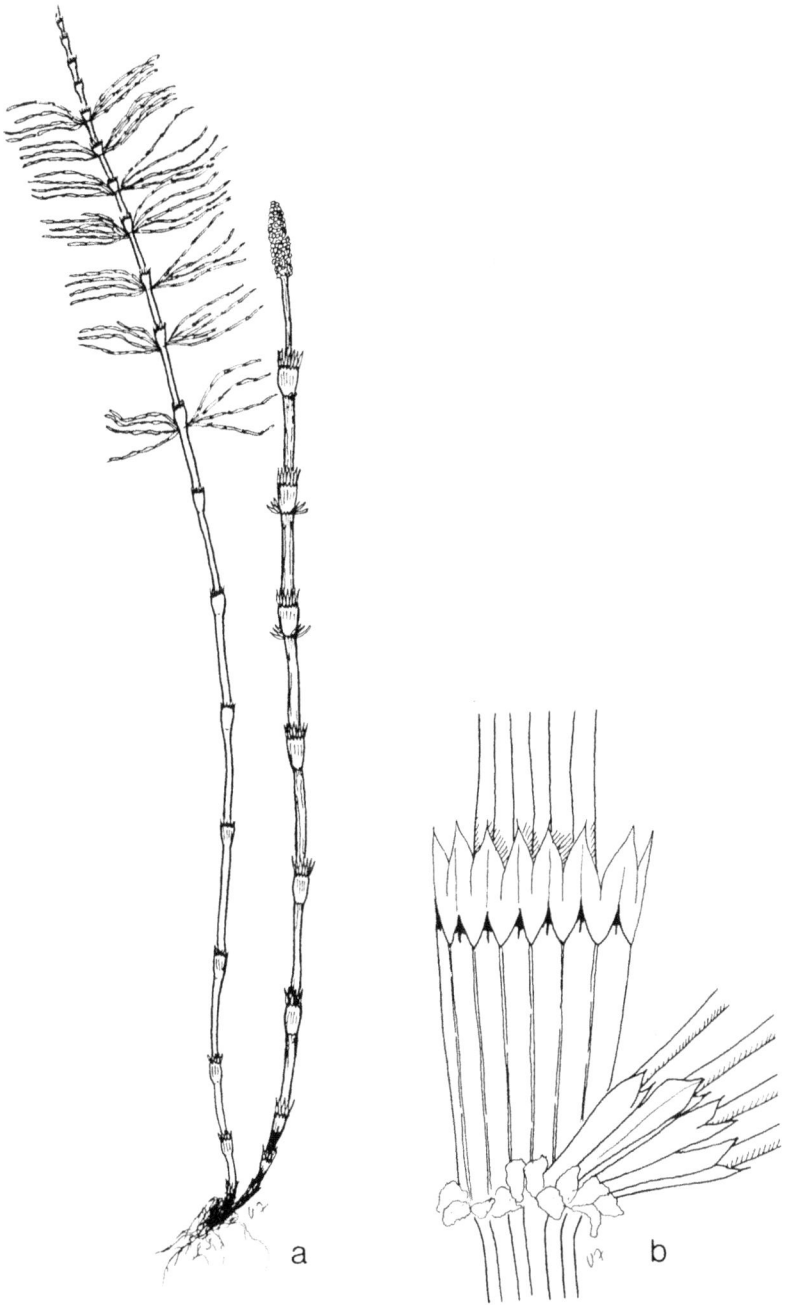

Fig. 43 *Equisetum pratense*; (a) sterile and fertile stems, 1/2 × ; (b) node, 5 × .

Equisetaceae

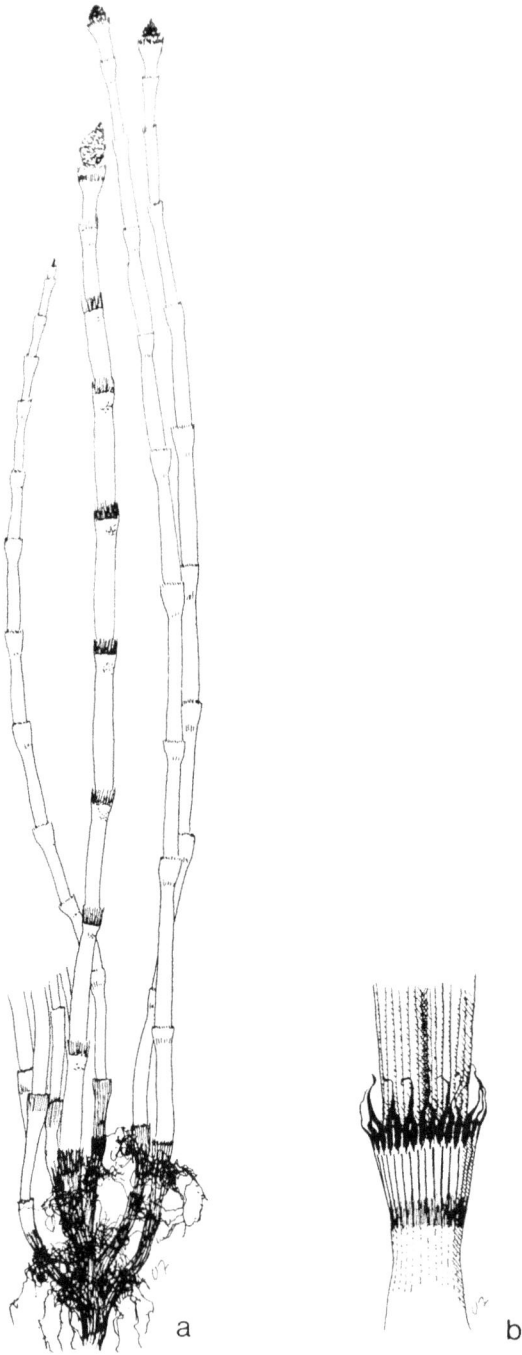

Fig. 44 *Equisetum hyemale* ssp. *affine*; (a) sterile and fertile stems, 1/2 ×; (b) node, 2 ×.

Equisetaceae

ridges. Ridges broad, flat, or rounded, with prominent cross-bands to double rows of tubercules; stomata in 2 lines, one on each side of the grooves; sheaths constricted at the base, the same color as the stem when young, but soon developing dark bands at the base and summit, with the part between base and summit white or ashy gray; teeth lanceolate, usually promptly deciduous, dark brown with broad scarious margins. Cones to 2 cm in length when expanded, yellow to black, apiculate, short-peduncled, shedding spores from June to September, or persisting unopened till the following spring.

The scouring-rush is perhaps most easily confused with the smooth scouring-rush, from which it can usually be distinguished by its perennial rather than annual stems and by its sheaths, which are only slightly expanded upward and are black-banded at the base and apex, rather than flaring, with the black bands occurring at the apex only.

Cytology: $n = 108$ (Cody and Mulligan 1982*).

Habitat: Sandy and gravelly river terraces, lakeshores, old fields, railway embankments, and roadsides.

Range: *Equisetum hyemale* s. l. is circumpolar, the ssp. *affine* North American, from Newfoundland to southern Alaska, south to Texas and New Mexico.

Remarks: This species may form extensive, dense stands on sandy slopes; Muenscher (1955) considered it to be a weed. Early Canadian settlers used the rough, silica-encrusted stems of scouring-rush to clean pots and pans, hence the common name; the stems have also been used for honing the reeds of woodwind musical instruments. Hauke (1963) treated the subgenus *Hippochaete*, to which this and the following species belong, in a monograph.

8.   ***Equisetum laevigatum*** A. Br.
     E. kansanum Schaffn.
     Hippochaete laevigata (A. Br.) Farwell
     smooth scouring-rush
Fig. 45 (a) fertile stems; (b) node. Map 44.

Stems annual, up to 1 m or more in length, but usually shorter in our range, 2–7 mm thick, soft, upright, usually unbranched, single or several together from a thick, dull, rough, dark brown rhizome. Central cavity two-thirds to three-quarters the diameter of the stem; small vallecular cavities alternating with 14–26 ridges; ridges rounded, smooth or with cross-bands of silica; stomata in two lines, one on each side of the groove; sheaths constricted at the base, flaring towards the top, the same color as the stem except for a narrow black

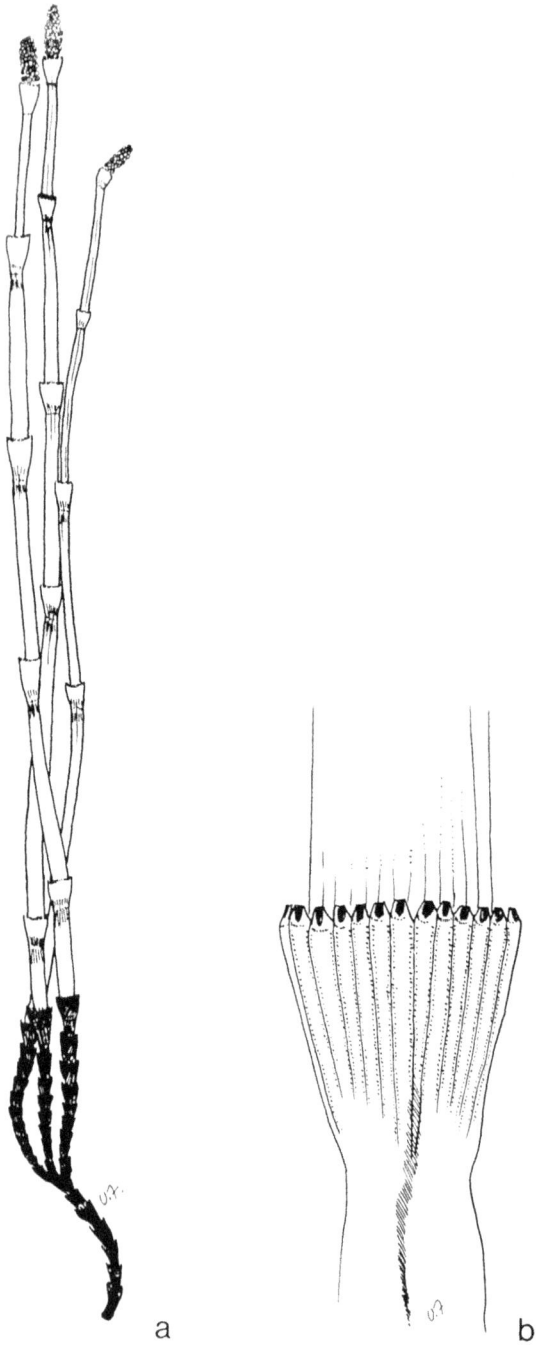

Fig. 45 *Equisetum laevigatum*; (a) fertile stems, 1/2 × ; (b) node, 5 × .

Equisetaceae

band at the apex, or in old stems the lower becoming girdled with brown; teeth lanceolate-subulate, promptly deciduous, dark with scarious margins. Branches resulting from injury smaller and rougher than stems, retaining their white teeth. Cones up to 2 cm in length when expanded, yellow to brown, blunt or only slightly apiculate, short-peduncled, shedding spores from mid May to July.

See under *E. hyemale* for characters used to differentiate that species from *E. laevigatum*.

Cytology: $n = 108$ (Löve and Löve 1976*).

Habitat: Moist or dry sandy river terraces and banks, meadows, and prairies.

Range: North American, from southern Ontario to southern interior British Columbia, south to northern Mexico.

Remarks: Hauke (1961) has summarized the problems related to the various forms of this taxon and the hybrid between it and *E. hyemale* ssp. *affine* (*E.* × *ferrissii*). *Equisetum laevigatum* is the only species of *Equisetum* endemic in North America.

9. **Equisetum scirpoides** Michx.
   Hippochaete scirpoides (Michx.) Farwell
   dwarf scouring-rush
Fig. 46 (*a*) habit; (*b*) strobilus; (*c*) node. Map 45.

Stems perennial, 3–20 cm long or longer, slender, 0.5–1.0 mm thick, usually unbranched, ascending or prostrate, arched-recurving and flexuous, caespitose from fine branching rhizomes. Centre of stem solid, with 3 or rarely 4 vallecular cavities alternating with the deeply grooved ridges; stomata in single lines on either side of the ridge; silica rosettes in lines on the crests of the ridges; sheaths green below, black above, loose, with 3 or rarely 4 deltoid teeth; teeth scarious-margined, subpersistent, but their subulate tips usually soon breaking off. Cones small, 2–3 mm long, apiculate, black, shedding spores in July or August, or persisting unopened until the following spring.

E. scirpoides can be separated from *E. variegatum*, with which it might be confused, by its usually 3 rather than 4 teeth, and by its flexuous stems, which are solid rather than hollow.

Cytology: $n = 108$ (Löve and Löve 1976*).

Habitat: Tundra, mossy places and woods, the stems often partly buried in humus.

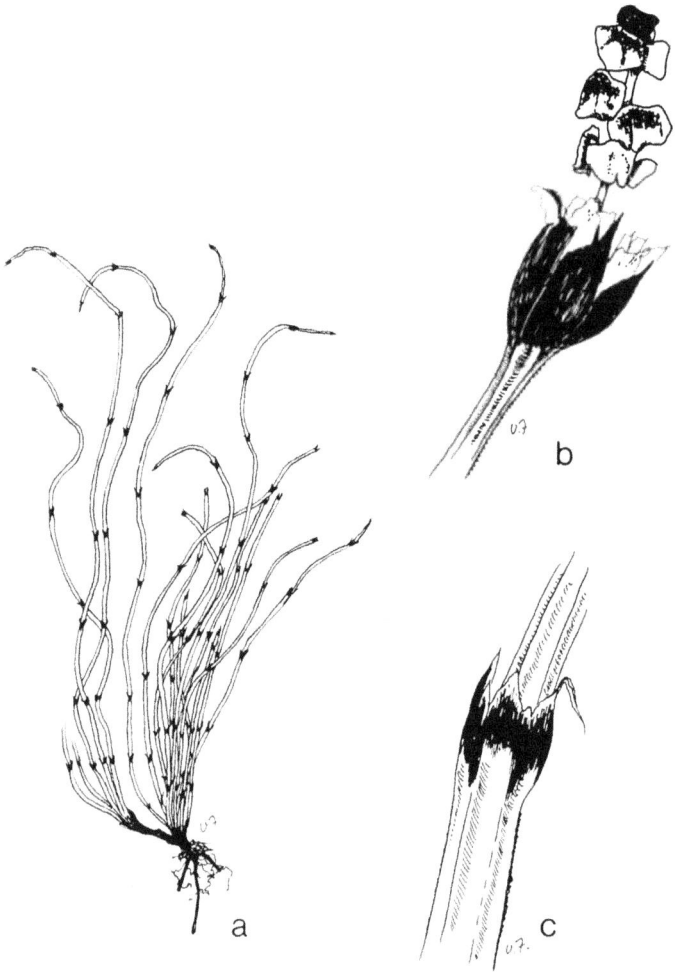

Fig. 46 *Equisetum scirpoides*; (a) habit, 1/2 × ; (b) strobilus, 5 × ; (c) node, 5 × .

Range: Circumpolar; in North America from Labrador to Alaska, south to New England, Washington, and Illinois.

Remarks: This is a common species in the northern boreal forest, but is often overlooked because it is more or less buried in mosses and humus.

10. **Equisetum variegatum** Schleich. ssp. **variegatum**
    Hippochaete variegata (Schleich.) Bruhin
    variegated horsetail
Fig. 47 (a) habit; (b) node. Map 46.

Stems 6–50 cm long, 0.5–3.0 mm thick, evergreen, usually unbranched, ascending, tufted from smooth branching rhizomes. Central cavity one-third to two-thirds the diameter of the stem; large vallecular cavities alternating with 3–12 furrowed ridges; stomata in single lines on each side of the ridge; silica tubercules in two lines on ridges, separated by the furrow; sheaths green at the base, black above, slightly spreading; teeth lanceolate to lance-deltoid, obtuse, persistent, with or without filiform tips, and with a brown central portion and wide white margins. Cones 5–10 mm long, apiculate, shedding spores in July or August or more often persisting unopened until the following spring.

The variegated horsetail may be distinguished from the dwarf scouring-rush by its straighter, more upright stems and by the characters mentioned under the latter species.

Cytology: $n = 108$ (Löve and Löve 1976*).

Habitat: Tundra, moist sand, river banks, and meadows.

Range: Circumpolar; in North America from Greenland to Alaska, south to Oregon, Utah, Michigan, New York, and New England.

Remarks: This circumpolar species, like *E. arvense*, is found north of the treeline as far north as northern Ellesmere Island, where it is quite reduced in size. *Equisetum variegatum* ssp. *variegatum* is rare in Nova Scotia (Maher et al. 1978).

10.1 **Equisetum variegatum** Schleich. ssp. **alaskanum** (A.A. Eat.)
     Hultén
Fig. 48 (a) habit; (b) node. Map 47.

Can be distinguished by its more robust stature and by its teeth incurved, completely black, or at most narrowly white-margined, with the black covering part or most of the sheath.

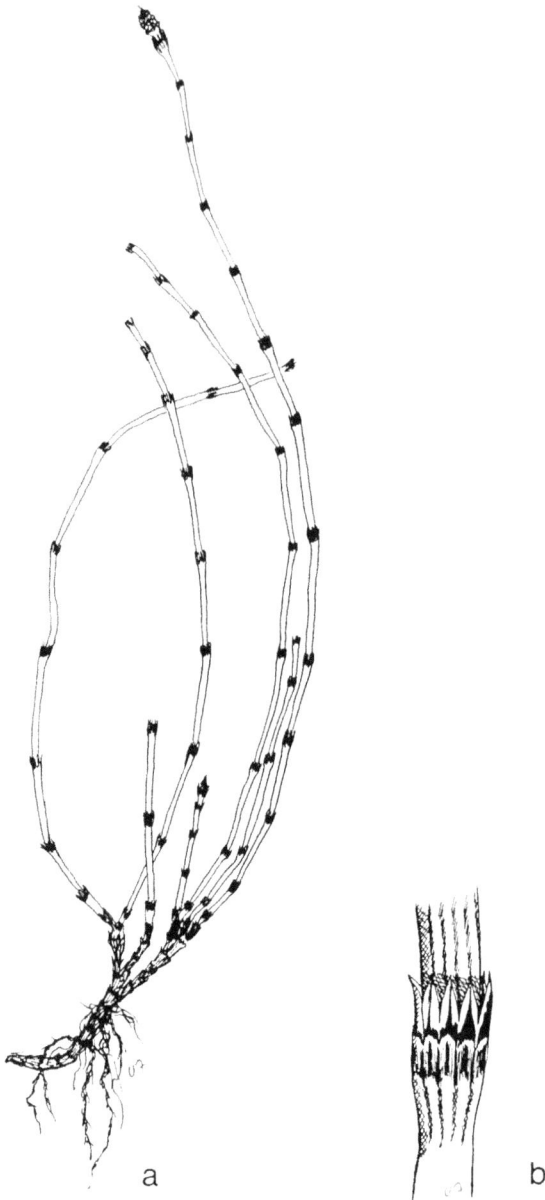

Fig. 47 *Equisetum variegatum* ssp. *variegatum*; (a) habit, 1/2 × ; (b) node, 4 × .

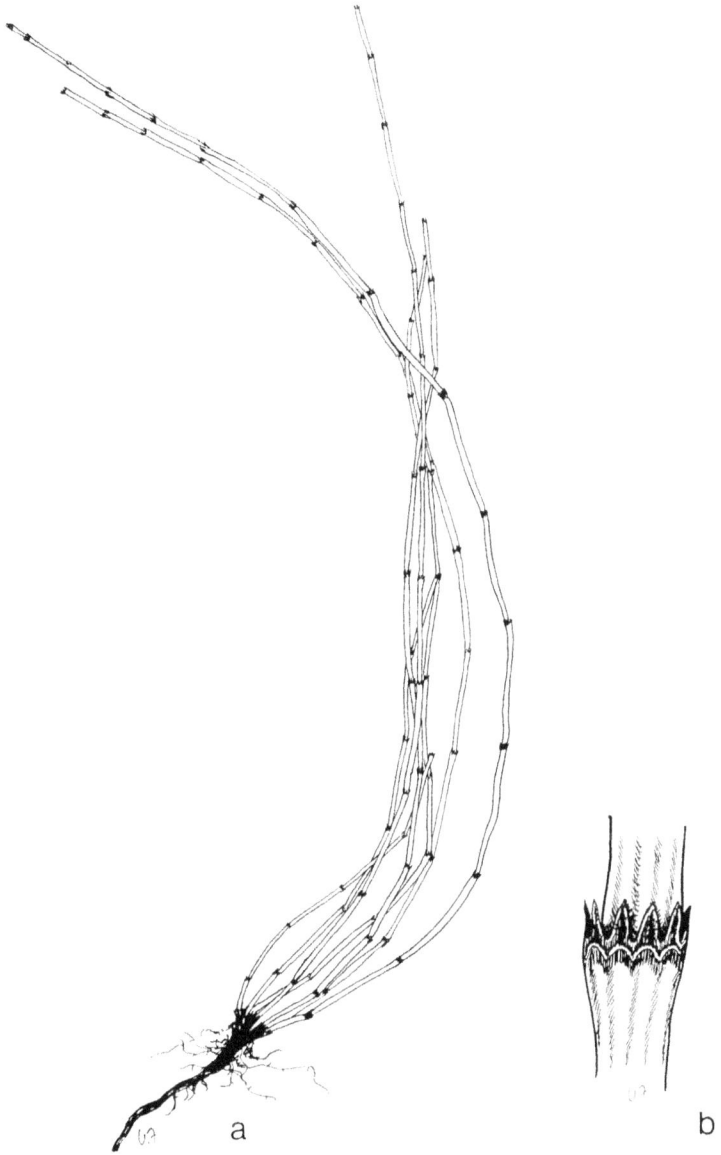

Fig. 48 *Equisetum variegatum* ssp. *alaskanum*; (a) habit, 1/2 × ; (b) node, 5 × .

Habitat: Found in habitats similar to those of ssp. *variegatum*.

Range: Alaska through southwestern Yukon, south along the British Columbia coast to Vancouver Island.

Hybrids of *Equisetum*

Hybrids may occur spontaneously between various species in subgenus *Equisetum* and in subgenus *Hippochaete*. They may be recognized on the basis of their intermediacy between the parents and by the presence of abortive spores. Some are found in nature much more frequently than are others. Such hybrids will spread vegetatively by fragmentation of the rhizomes. All the hybrids that follow have been produced under experimental conditions in petri dishes, but not all are known to occur in Canada (Duckett 1979).

E. arvense × *fluviatile* (*E.* × *litorale* Kuhlewein) occurs frequently in southern Quebec and southern Ontario, and rarely in British Columbia.

E. arvense × *telmateia* (*E.* × *dubium* Dostal) occurs rarely in Czechoslovakia.

E. arvense × *palustre* (*E.* × *rothmaleri* C.N. Page) occurs rarely in Scotland.

E. arvense × *pratense* (*E.* × *suecicum* Rothm.) occurs in Europe.

E. fluviatile × *palustre* occurs in Scotland.

E. palustre × *telmateia* (*E.* × *font-queri* Rothm.) occurs rarely in British Columbia.

E. pratense × *sylvaticum* (*E.* × *mildanum* Rothm.) occurs in Europe.

E. hyemale ssp. *affine* × *laevigatum* (*E.* × *ferrissii* Clute, *E. hyemale* var. *intermedium* A.A. Eat.) occurs occasionally from Quebec to British Columbia.

E. hyemale ssp. *affine* × *variegatum* (*E.* × *trachyodon* A. Braun, *E. hyemale* var. *jesupi* (A.A. Eat.) Vict., *E. variegatum* var. *jesupi* A.A. Eat.) occurs frequently from western Newfoundland to British Columbia.

E. laevigatum × *variegatum* (*E.* × *nelsonii* (A.A. Eat.) Schaffn.) is common in the Great Lakes region, especially adjacent to the Lakes. Manitoulin Island was given special mention by Hauke (1963).

E. scirpoides × *variegatum* (*E.* × *arcticum* Rothm.) may be represented by a specimen in DAO from the Richardson Mountains, District of Mackenzie.

5. OPHIOGLOSSACEAE   adder's-tongue family

Herbs perennial, more or less succulent. Rhizome short, bearing one or more stalked or sessile fronds and a fertile spike or panicle. Sporangia naked, bivalvate, producing thick-walled spores. Gametophyte subterranean, usually without chlorophyll and associated with an endophytic mycorrhiza.

A. Sporangia cohering in a simple spike; fronds simple, entire, usually one; veins reticulate .... 1. *Ophioglossum*
A. Sporangia separate in a pinnate, a compound, or rarely, a simple spike; sterile segments of fronds not simple; veins free ....
.... 2. *Botrychium*

1. *Ophioglossum* L.   adder's-tongue

1. ***Ophioglossum vulgatum*** L. var. ***pseudopodum*** (Blake) Farw. adder's-tongue
Fig. 49 (*a*) fronds; (*b*) sporangia. Map 48.

Sporophyte bearing usually 1 frond, 15–25 cm long, from an erect rootstock. Sterile segment sessile, glabrous, entire, attached near the middle, varying in shape from broadly lanceolate, to ovate, to oblanceolate, 4.0–9.5 cm long, 1.5–3.0 cm wide. Fertile segment a simple stalked spike bearing two rows of cohering sporangia.
This is a very distinctive plant, if one can find a specimen. They tend to be hidden in grass or under sensitive ferns or are associated with plants that have leaves that are confusingly similar in general shape, e.g., plantains, trout lilies, or even the orchid *Pogonia*. Quite often the search involves a hands-and-knees approach, near ground level.

Cytology: $n = 480$ (Mulligan and Cody 1969*).

Habitat: Moist humus-rich depressions, wet meadows, and sometimes grassy hillsides and high, dry, sunny locations, or even bogs.

Range: *Ophioglossum vulgatum* s.l. is circumpolar; var. *pseudopodum*, occurs from Nova Scotia to Ontario, Vancouver Island, and Washington, south to Virginia, Arizona, and Mexico; var. *pycnostichum* Fern. is found in the United States from Michigan (Wagner 1971) eastward to Ohio, and south to Florida and Alabama (Cranfill 1980); and var. *alaskanum* (E.G. Britt.) Christens is known only from Kodiak Island, Alaska.

Fig. 49 *Ophioglossum vulgatum* var. *pseudopodum*; (a) fronds, 1/2 ×
(b) sporangia, 2 ×.

Ophioglossaceae

Remarks: The extremely high chromosome numbers in *Ophioglossum* have been noted by many biologists. *Ophioglossum reticulatum* from India has $n = 630$, and consequently every body cell has 1260 chromosomes. However, number of chromosomes and amounts of DNA are only two factors to consider; *Selaginella apoda* or *S. densa*, which have $n = 9$, seem to have as many morphological characters as *O. vulgatum*, which has $n = 480$.

Löve recognizes var. *pseudopodum* as one species, *O. pusillum* Raf., and var. *pycnostichum* as another species, *O. pycnostichum* (Fern.) Löve & Löve. The latter has $n = $ ca. 630 (Löve and Löve 1976). We find it difficult to evaluate the chromosome count of $2n = $ ca. 1260 for a collection of *O. pycnostichum* from Prince Edward Island, near Souris, because we have not seen any herbarium material of var. *pycnostichum* from Canada. The matter is further confused by the fact that var. *pycnostichum* is noted for spores that are smaller (44 μm) (Wagner 1971; Cranfill 1980) and chromosomes that are fewer in number ($n = 250$–260) (Cranfill 1980) than those of var. *pseudopodum* (spores, 50-54 μm) (Wagner 1971) (chromosomes, $n = 480$) (see above).

2. *Botrychium* Sw.   moonwort, grape fern

Sporophyte bearing 1 to several fronds from an erect unbranched rootstock. Roots thick and fleshy. Sterile segment sessile or stalked, in Canadian species pinnately or palmately once to many times decompound; venation dichotomous, open. Fertile segment stalked; spike simple (rarely) to pinnately compound. Sporangia naked and distinct, borne laterally on its branches.

The genus *Botrychium* is a small one of approximately 40 species. There are about 20 species in North America, and we have recognized 13 in Canada's flora. The only really common and familiar species in Canada is *B. virginianum*, the rattlesnake fern, which is in a separate section of the genus; some researchers have even placed it in a separate genus, e.g., *Osmundopteris*, *Japanobotrychium*, or *Botrypus*. It has a chromosome number of $n = 92$ and is not based on $x = 45$, as are Canada's other species.

The genus was treated in a monograph by Clausen (1938) and has been extensively studied by W.H. Wagner at the University of Michigan for over 30 years. Clausen's approach to taxonomy is quite different from Wagner's. Clausen attempted to choose a central type and then group the variants as subspecies or varieties around this type. Wagner, on the other hand, has attempted to separate out the genetically distinct species and downplay the environmental forms. The genus is one that has resisted the usual biosystematic approach. Many of the species are basic diploids, and hybrids are considered extremely rare. Experimental crosses are not yet feasible, because the spores can be germinated only in low numbers and no plants have been raised to maturity. Spore sizes have been useful diagnostically, but spore morphology has not, and the chemotaxonomists have yet to produce a diagnostic scheme for us. The plants are even difficult to transplant, and we are certain that crop scientists must be amused at our inability to separate out environmental components either by controlled growth conditions or by field-plot techniques.

W.H. Wagner (1960b) made use of mass collections, relying on the dictum that several species of *Botrychium* often grow at the same location. Unfortunately, this has left us with two divergent viewpoints, one represented by workers who are impressed with the large range of variation exhibited by *one* species, including variants of different ages, and the other by workers who stress the large number of species at one location. Herbarium material can also be frustrating. It is not unusual to have six or more small plants on a sheet that the collector has identified as one species (e.g., *B. lunaria*), but no notes of the area covered by the collector or of his impressions regarding which specimen was the most common type present. Various annotators subsequently identified each of the plants as *B. minganense, B. simplex,* or *B. matricariifolium,* for example – several species for one sheet. W.H. Wagner and Lord (1956) and W.H. Wagner (1960b) have carefully examined *Botrychium* species for new taxonomic characters, such as vernation, color of the roots, and folding of the leaf, among others, but the nature of the plants is such that judgments are still more subjective than one would wish. We are lacking clear objective criteria for specific identification. Some workers continue to emphasize differences and others similarities. Perhaps we can all agree with Wherry (1961) when he writes about the small species, "The search for these tiny plants, and the correct naming of them after collection, is recommended to every fern student as a rewarding undertaking." We trust fern students will not become frustrated by our inability to describe clearly the limits of variation of each taxon!

A. Sterile blades usually large, up to 20 cm wide or wider at the base, deltoid, tripinnate; lamina usually thin and papery . . . . . .
. . . . 1. *B. virginianum*
A. Sterile blades various, usually leathery or fleshy.
  B. Blade evergreen; plants appearing in summer and spores maturing in autumn.
    C. Segments of blade all of about the same size and shape, ovate, obovate, rhomboid, or oblong; chief terminal segments not elongate.
      D. Sterile blades very fleshy; margins slightly hyaline; ultimate segments somewhat acute or obtuse and either crenate or entire . . . .
. . . . 5. *B. multifidum*
      D. Sterile blades membranous; margins not hyaline; ultimate segments somewhat acute, usually serrate, rarely entire . . . . 6. *B. rugulosum*
    C. Segments of blade not all of the same size and shape; chief terminal divisions usually elongate and little divided.

E. Chief terminal divisions of blade broad and rounded; blades normally winter green .... 4. *B. oneidense*

E. Chief terminal divisions of blade usually narrower and acute or somewhat acute; blades becoming bronze-colored in late autumn.

    F. Divisions of blade deeply and finely lacerate or divided .... 2. *B. dissectum*

    F. Divisions of blade not deeply lacerate or divided .... 3. *B. obliquum*

B. Blade deciduous; plants appearing in spring and spores soon maturing.

    G. Sterile blade deltoid, usually sessile .... 13. *B. lanceolatum*

    G. Sterile blade oblong or ovate, sessile or stalked.

        H. Blades variously divided, usually bipinnate in mature specimens.

            I. Blade either palmately or pinnately divided, ovate or broadly ovate-oblong, usually sessile .... 12. *B. boreale*

            I. Blade usually pinnately, but sometimes ternately divided, commonly oblong, usually sessile .... 11. *B. matricariifolium*

        H. Blades simple or once pinnate; basal divisions sometimes divided again, thus appearing ternate.

            J. Sterile blade short, less than 4 cm long, stalked; divisions often remote, but at times imbricate, with at most 3 pairs, obliquely ovate to obovate to oblong, seldom flabellate, often dissimilar in shape .... 10. *B. simplex*

            J. Sterile blade 5–20 cm long, often imbricate; divisions at times remote, over 3 pairs (to 10), shapes various.

                K. Pinnae 3–6 pairs, bluish green, fan-shaped to lunate, standing straight out from the rachis, and abruptly reduced to irregular segments at blade tip. ... . 7. *B. lunaria*

                K. Pinnae 6–10 pairs, yellowish green, rhomboid to fan-shaped, ascending towards the rachis, and uniformly reduced at the blade tip.

                    L. Blade short, oblong-ovate, inserted above the middle; small divisions flabellate, often

L. Blade much longer than wide, inserted below the middle, entire to incised; plants to 20 cm tall . . . . 8. *B. minganense*

1. **Botrychium virginianum** (L.) Sw. var. *virginianum*
    rattlesnake fern
Fig. 50 (*a*) frond; (*b*) pinnule. Map 49.

Fronds erect, 50 cm long or longer, glabrous or nearly so, deciduous. Blades broadly deltoid, sessile, attached above the middle, bipinnate to tripinnate; ultimate segments oblong-lanceolate, toothed, membranous or slightly fleshy. Fertile segment pinnately compound. This is a large, fairly common species that has a broad distribution across Canada. The highly divided blade and characteristic fertile segment make identification simple.

Cytology: $n = 92$ (Britton 1953*; Löve and Löve 1976). This number is a departure from the $x = 45$ found in the other species.

Habitat: Usually in dry to somewhat moist deciduous woodlands, but occasionally in wet cedar woods and boggy areas.

Range: Newfoundland to British Columbia, south to Florida and California.

Remarks: The common name comes from the fanciful similarity in appearance of the cluster of unopened sporangia to the rattles on the tail of a rattlesnake.

1.1 **Botrychium virginianum** (L.) Sw. var. *europaeum* Ångstr.
    B. virginianum (L.) Sw. var. *laurentianum* Butters
Fig. 51, frond. Map 50.

Similar to var. *virginianum* but usually smaller and stiffer. Blade leathery; ultimate segments less toothed, often crowded, and overlapping.

Habitat: Thickets and damp often coniferous woods.

Range: More northern than var. *virginianum*; Labrador to Alaska, south to the northern United States; northern Eurasia.

Remarks: This variety would seem to be somewhat clinal and may not merit taxonomic rank; compare with *Dryopteris fragrans* var. *remotiuscula* and *Woodsia alpina* (*W. bellii*).

Ophioglossaceae

Fig. 50 *Botrychium virginianum* var. *virginianum*; (a) frond, 1/3 ×; (b) pinnule, 2 1/2 ×.

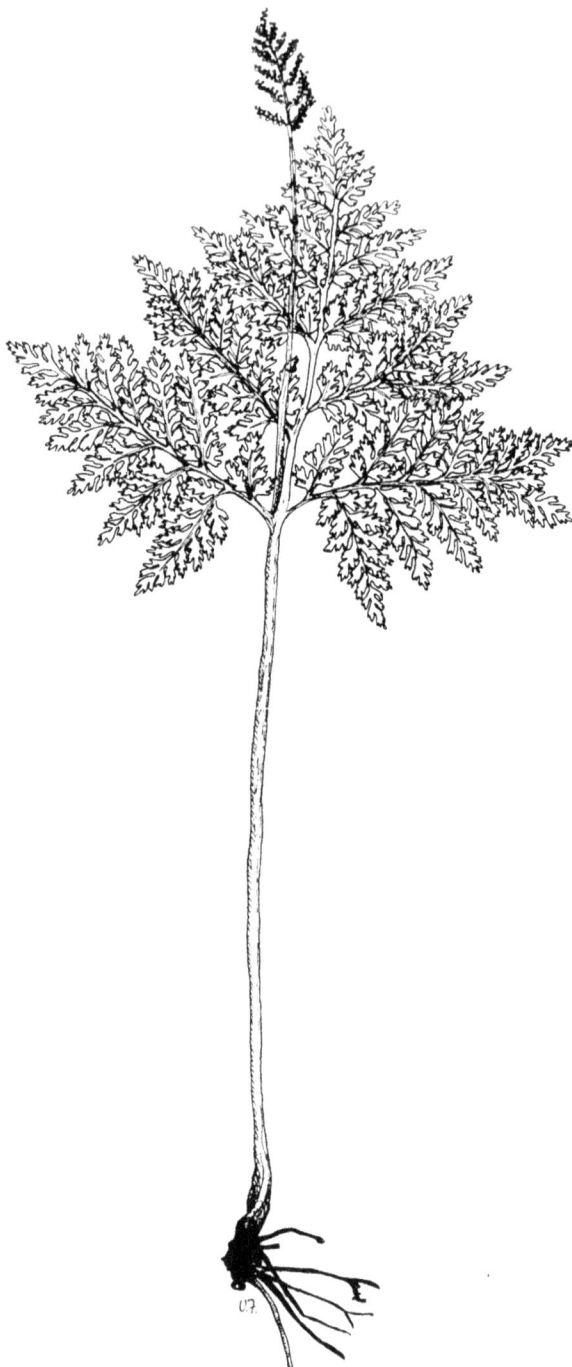

Fig. 51 *Botrychium virginianum* var. *europaeum*; frond, 1/2 ×.

Ophioglossaceae

The variety is, however, quite striking when one compares the far northern plants with those in the south. It is rare in the District of Mackenzie (Cody 1979) and the Yukon (Douglas et al. 1981).

Species 2-6 comprise a group of five species that are sometimes called the "fall botrychiums." W.H. Wagner (1959, 1960*b*, 1962) and W.H. Wagner and Rawlings (1962) have considered this group of *Botrychium* species in the subgenus *Sceptridium* in some detail. Wagner recognizes four species in this group for northeastern United States and adjacent Canada, as follows: *Botrychium dissectum* (including f. *obliquum*), *B. multifidum*, *B. oneidense*, and *B. rugulosum*. We are following Wherry (1961) in recognizing five species, whereas earlier workers, e.g., Clausen (1938) have recognized as few as two species, *B. dissectum* and *B. multifidum*. All our species are basic diploids with $x = n = 45$.

2. **Botrychium dissectum** Spreng.
   cut-leaved grape fern
Fig. 52, frond. Map 51.

Fronds up to 27 cm long; stem and blade less coriaceous than *B. multifidum*. Blades long-petioled, triangular, ternate, attached at or near the base; ultimate divisions of blade cut in linear segments; segments more or less notched at the apex. Fertile segment paniculate. Spores mature from September to November.

Cytology: $n = 45$ (Britton 1953*; W.H. Wagner 1960*b*).

Habitat: Sterile hilltops, dry pastures, dry woodlands, and grassy banks.

Range: New Brunswick to Ontario and Minnesota, south to North Carolina, Tennessee, and Missouri.

Remarks: Extremely dissected and lacerated plants appear almost skeletonized and are very distinctive. The fronds are often bronze or even turn reddish in late fall. This species occurs in southeastern Canada.

3. **Botrychium obliquum** Muhl.
   B. dissectum Spreng. var. *obliquum* (Muhl.) Clute
   B. dissectum Spreng. f. *obliquum* (Muhl.) Fern.
Fig. 53, frond. Map 52.

Fronds 30 cm long or longer; stem and blade somewhat coriaceous. Blades triangular, ternate, attached at or near the base;

Fig. 52 *Botrychium dissectum*; frond, 1/2 ×.

Ophioglossaceae

Fig. 53 *Botrychium obliquum*; frond, 1/2 ×.

ultimate segments of blade elongate, somewhat acute, somewhat divided below; margins of segments entire or inconspicuously crenate. Fertile segment paniculate. Spores mature from September to November.

Cytology: n = 45 (Britton 1953*).

Habitat: Sterile fields, dry pastures, meadows, thickets, dry woodlands, and rich swampy woods.

Range: Nova Scotia to Ontario, Wisconsin, and Iowa, south to South Carolina, Georgia, and Louisiana.

Remarks: This taxon, which often grows intermixed with *B. dissectum*, is considered a mere form by W.H. Wagner (1961). For many field workers, it is much more easily identified than the others and may occur in large numbers. The genetic relationships between *B. obliquum* and *B. dissectum* are still obscure (see also comments by Wherry 1961).

4. **Botrychium oneidense** (Gilbert) House
   B. dissectum Spreng. var. *oneidense* (Gilbert) Farw.
   B. dissectum Spreng. f. *oneidense* (Gilbert) Clute
Fig. 54 (a) sterile frond; (b) fertile frond; (c) sporangia. Map 53.

Fronds 40 cm long or longer; stem and blade somewhat coriaceous. Blades triangular, ternately decompound, little divided, attached at or near the base; chief terminal segments of blade broadly ovate and obtuse. Fertile segment paniculate. Spores mature in September and October.

Cytology: n = 45 (W.H. Wagner 1955)

Habitat: Rich moist woodland.

Range: New Brunswick to Ontario and Minnesota, south to North Carolina, Ohio, and Indiana.

Remarks: This species has had a checkered career. It has been variously treated as a variety, a form, or even a hybrid of both *B. dissectum* and *B. multifidum*. It was studied in some detail by W.H. Wagner (1961) and is considered a good species. The broader, more rounded divisions and the more shady habitat are characteristic of the species.
   Botrychium oneidense appears to be a rare plant in Canadian flora. We have seen very few good colonies of it. Most plants are without fertile segments, and many of the collections seem to be occasional plants that are selected from larger colonies of *B. obliquum*. Some specimens of *B. multifidum* also closely resemble *B. oneidense*.

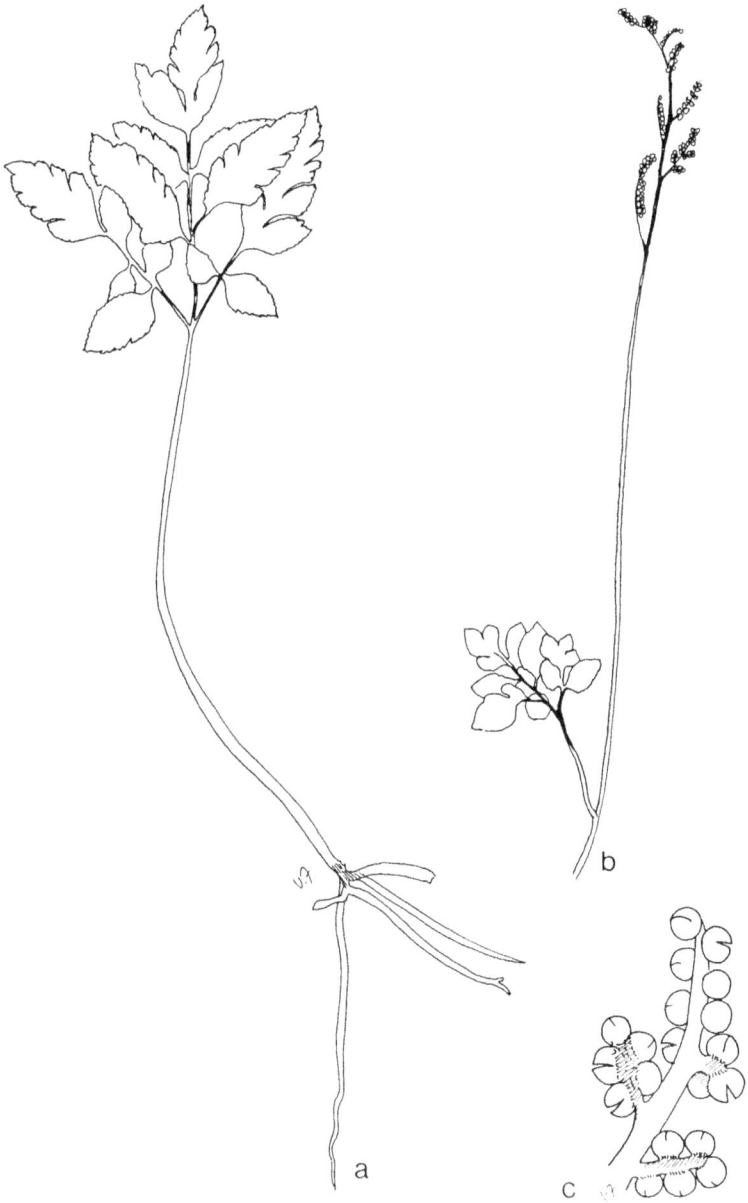

Fig. 54 *Botrychium oneidense*; (a) sterile frond, 3/5×; (b) fertile frond, 3/5×; (c) sporangia, 5×.

5. ***Botrychium multifidum*** (Gmel.) Rupr.
   leathery grape fern
Fig. 55, frond. Map 54 (s.l.).

Fronds up to 20 cm long; stem and blade coriaceous. Blade evergreen, long-petioled, ternate, attached near the base of the plant; ultimate segments of blade crowded, sometimes imbricate, varying from flabellate to ovate, but more or less the same size, obtuse or somewhat acute. Fertile segment paniculate. Spores mature in August and September.

Cytology: $n = 45$ (W.H. Wagner 1955, 1960*b*; Taylor and Mulligan 1968*).

Habitat: Grassy hillsides, sterile fields, exposed meadows, and sandy open places.

Range: Labrador and Newfoundland to British Columbia and southern District of Mackenzie, south to New York, Minnesota, and Wisconsin; northern Eurasia.

Remarks: The species is rare in the District of Mackenzie (Cody 1979) and Saskatchewan (Maher et al. 1979).

5.1 ***Botrychium multifidum*** (Gmel.) Rupr. var. ***intermedium***
   (D.C. Eat.) Farw.
   B. silaifolium Presl
Fig. 56, frond.

Similar to var. *multifidum*, but taller. Blades much larger, up to 20 cm wide and 15 cm long; ultimate segments of blade not so closely crowded, varying from oblong to obovate or ovate and usually somewhat crenate. Spores mature in August and September.
These larger plants with many divisions of the blade are strikingly distinctive in their bright yellow green coloration.

Habitat: Open fields, pastures, dry hillsides, borders of woods, and sandy places.

Range: Southern Labrador and Nova Scotia to British Columbia and Alaska, south to New York, Pennsylvania, Michigan, Montana, and California.

6. ***Botrychium rugulosum*** W.H. Wagner
   Botrychium ternatum Am. auth.
Fig. 57 frond. Map 55.

Ophioglossaceae

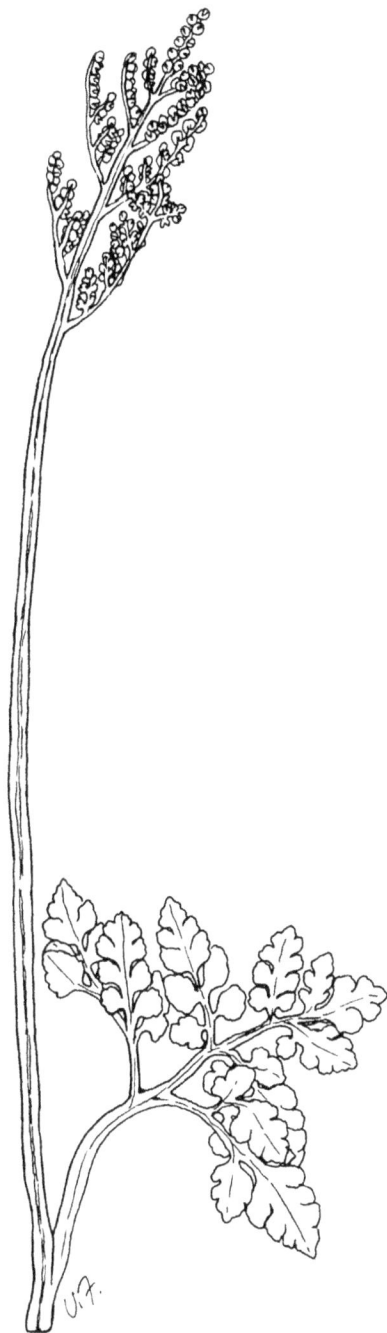

Fig. 55 *Botrychium multifidum*; frond, 1 × .

Ophioglossaceae

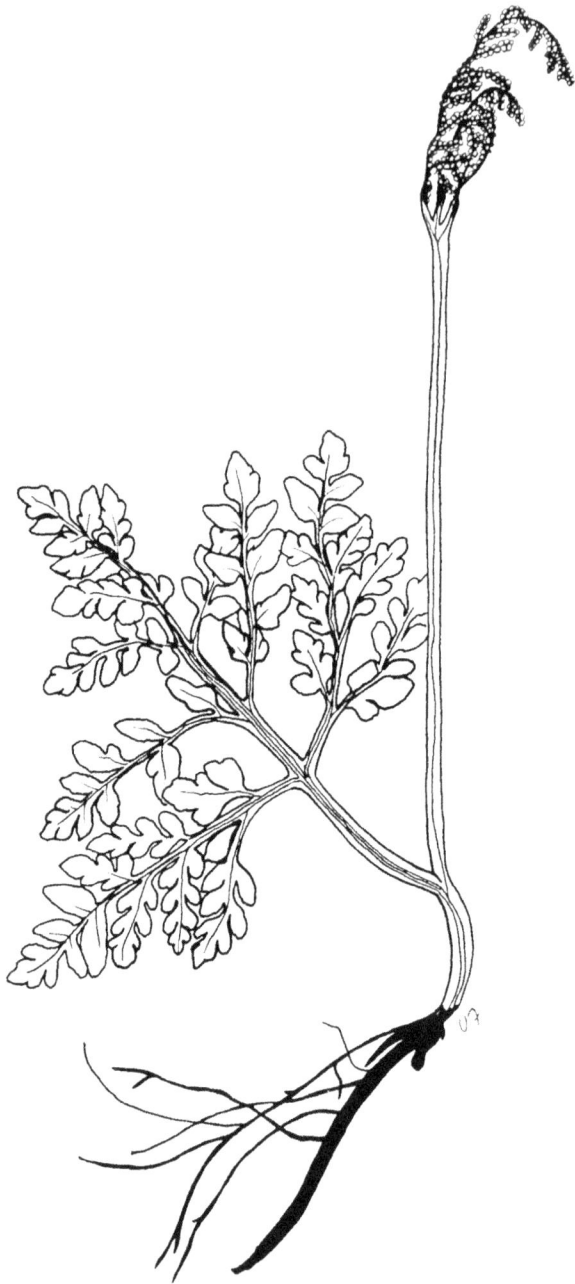

Fig. 56 *Botrychium multifidum* var. *intermedium*; frond, 1/2 ×.

Ophioglossaceae

Fronds 25 cm long or longer, thin and membranous. Blades inserted at the base, ternate, with the three major divisions stalked; ultimate segments of blade all about the same size, ovate to oblong, acutish, serrate or entire, and concave in the living state. Fertile segment paniculate. Spores mature from August to October.

Cytology: $n = 45$ (W.H. Wagner 1960b).

Habitat: Swampy woods, brushy fields, and wooded stream banks.

Range: Western Quebec and Ontario south and west to New York, Michigan, Indiana, and Iowa; Asia.

Remarks: W.H. Wagner (W.H. Wagner and F.S. Wagner 1982b) has recently separated this taxon from the closely related Japanese B. ternatum (Thumb.) Swartz. Botrychium rugulosum has been sought in Canada at some length. We have found no good colonies of the species, although we can occasionally find a few plants that match Wagner's descriptions (W.H. Wagner 1959, 1960b). We appreciate his confirming identifications of these collections. We consider the species to be extremely rare in Canada's flora.

The remaining species of this genus produce their fertile spikes in spring and early summer.

7.   **Botrychium lunaria** (L.) Sw.
     moonwort
Fig. 58, frond. Map 56.

Fronds up to 25 cm long, somewhat leathery. Blades more or less oblong, sessile, inserted at or below the middle, pinnate; segments of blade opposite, flabellate and often overlapping, with margins entire or somewhat incised. Fertile segment racemose or paniculate. Spores mature from June to August.

Cytology: $n = 45$ (W.H. Wagner and Lord 1956; Löve and Löve 1976*).

Habitat: Open, turfy, or gravelly slopes, shores, and meadows, usually on basic soils.

Range: Circumpolar; in boreal North America from Greenland, Labrador, and Newfoundland to British Columbia and Alaska, south to Maine, Michigan, Wyoming, Colorado, and California.

Remarks: The moonwort has a long and illustrious history in early herbals. We are told that the "seeds" can make one invisible or can be

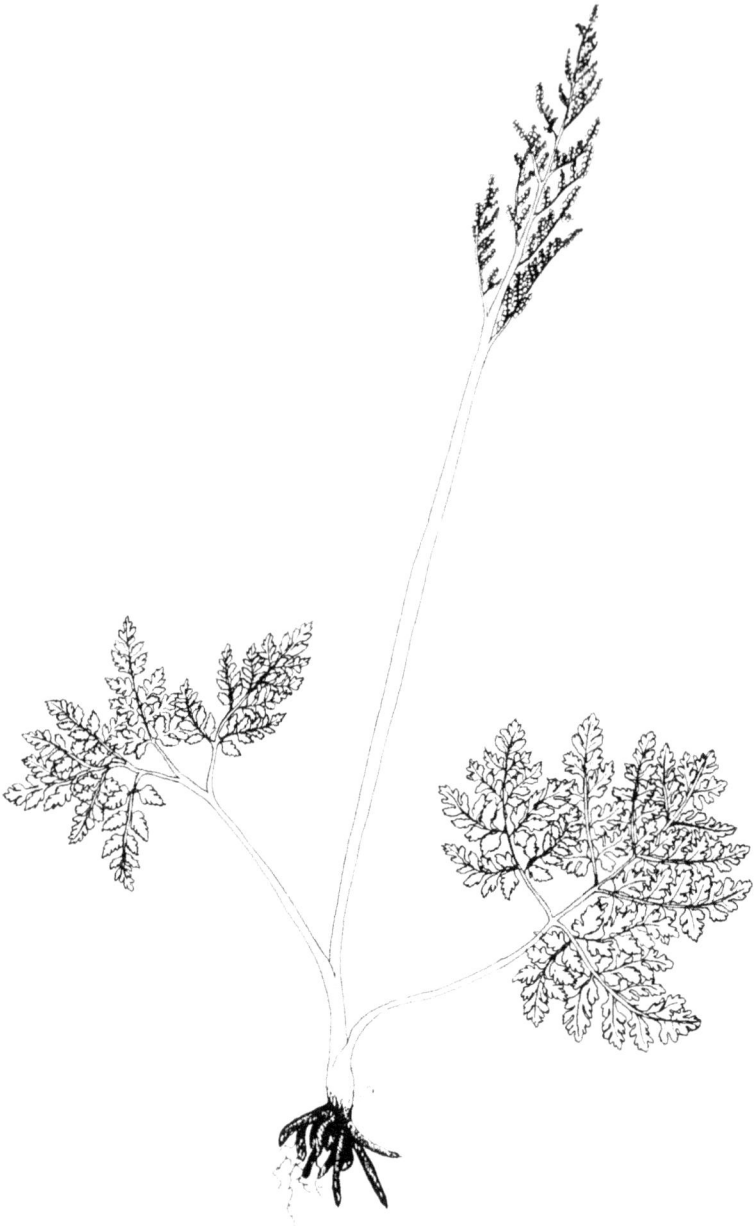

Fig. 57 *Botrychium rugulosum*; frond, 1/2 ×.

Ophioglossaceae

used to unlock doors. It is more common northward in Canada and is an early quest for the field worker to see and photograph. The fully mature plants are distinctive, but the species should be identified with care when it is compared with *B. minganense* (W.H. Wagner and Lord 1956). Moonwort is rare in the District of Mackenzie (Cody 1979), Nova Scotia (Maher et al. 1978), and Saskatchewan (Maher et al. 1979).

8.   ***Botrychium minganense*** Vict.
Botrychium lunaria (L.) Sw. var. *minganense* (Vict.) Dole
Fig. 59, frond. Map 57.

Fronds up to 30 cm long, somewhat membranous. Blades narrowly oblong, sessile or nearly so, inserted below the middle, pinnate or occasionally pinnate-pinnatifid at the base; segments of blades opposite, obovate, rhomboidal or oblong, frequently incised, remote. Fertile segment paniculate. Spores mature in July and August.

Botrychium minganense can be distinguished from *B. lunaria* by its yellowish green hue and by its trough-shaped sterile segments, which are ascending rather than at right angles to the stalk and which rarely overlap with each other. (W.H. Wagner and Lord 1956).

Cytology: $n = 90$ (W.H. Wagner and Lord 1956; Löve and Löve 1976).

Habitat: Marly meadows and open alpine areas.

Range: Southern Labrador to Hudson Bay, Alaska, and British Columbia south to New York, Michigan, Montana, and California.

Remarks: This tetraploid species of uncertain origin, other than as a derivative of *B. lunaria*, is usually present at *B. lunaria* localities. It has been given various ranks, from form to species, and was originally described by one of Canada's eminent botanists, Frère Marie-Victorin, from material collected on the Mingan Islands in the lower St. Lawrence River. It is rare in the District of Mackenzie (Cody 1979).

9.   ***Botrychium dusenii*** (Christ) Alst.
Fig. 60, frond. Map 58.

Fronds up to 13 cm long; stem and blade deciduous. Blades short, oblong-ovate, inserted above the middle, sessile or nearly so, pinnate; segments of blade usually remote, about as wide as long, often spatulate and distinctly petioled, sometimes cuneate, often crenate. Fertile segment pinnate, or the lowermost pinnae again pinnate. Spores mature from July to early September.

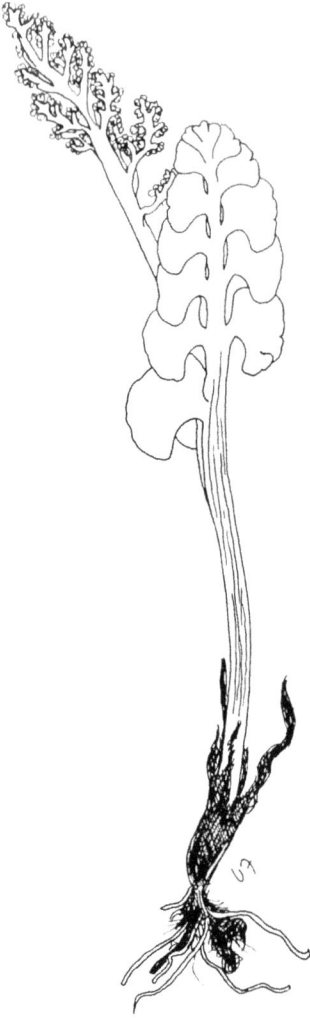

Fig. 58 *Botrychium lunaria*; frond, 2/3 ×.

Fig. 59 *Botrychium minganense*; frond, 4/5 ×.

Ophioglossaceae

Cytology: $n = 45$ (W.H. Wagner and Lord 1956). From material that had previously been considered to be western *B. minganense.*

Habitat: Mountains and western plains.

Range: District of Mackenzie, Alberta, and British Columbia, south to California and Arizona; South America.

Remarks: W.H. Wagner, after a trip to a scientific meeting in Edmonton, said that the limited amount of western material available of *B. minganense, B. lunaria, B. simplex,* and *B. matricariifolium* should be investigated and that a species thought to be restricted to South America, *B. dusenii* (Reeves 1977), closely matched specimens from western Canada. W.H. Wagner took up his own challenge and has recently reported (W.H. Wagner and F.S. Wagner 1981) further studies on western moonworts. He says, "Plants we previously identified as the South American *B. dusenii* (Christ) Alston from various western states may prove to be variations of *B. crenulatum.*" We are also told, "More detailed reports on all of these plants will be made in the future." Accordingly, our treatment here is tentative, pending further study of these interesting, diminutive plants.

10. **Botrychium simplex** E. Hitchc. var. **simplex**
    least grape fern

Fronds up to 16 cm long, rather fleshy. Blades simple, lobed or pinnately divided, inserted at the base or towards the middle; segments of blade oblong, rhomboid or reniform, and usually overlapping, with the basal segments occasionally pinnatifid. Fertile segment simple or compound. Spores mature in late May and June.

Cytology: $n = 45$ (W.H. Wagner 1955).

Habitat: Pastures, meadows, lakeshores, and gravelly slopes.

Range: Newfoundland to Ontario and British Columbia, south to Pennsylvania, Wisconsin, Colorado, and California; Europe and Japan.

Remarks: The least grape fern is indeed a small species and is therefore often overlooked in the field. Unfortunately, young or small individuals of other species often resemble this one. Tryon et al. (1953) say that to distinguish this species from the very small *B. matricariifolium* may require measuring the size of spores—*B. simplex* (35–50 µm) and *B. matricariifolium* (25–35 µm). Individuals with a few sporangia on the edge of the blade occur quite often. This is also true for *B. minganense*. This species is rare in Nova Scotia (Maher et al. 1978).

10.1 **Botrychium simplex** E. Hitchc. var. **tenebrosum** (A.A. Eat.) Clausen
Fig. 61, frond. Map 59 (s.l.).

Similar to var. *simplex*, but usually very slender and taller. Blade not sessile, inserted above the middle, simple or lobed; lobes subopposite, obovate-oblong.

Habitat: Deep woods and the borders of swamps.

Range: New Brunswick to British Columbia, south to Pennsylvania, Michigan, Minnesota, and Washington.

Remarks: W.H. Wagner (1960b) has suggested that this is a "questionable" variety. It does however, look quite distinct from var. *simplex*, being long-attenuate, with a high insertion of the very small blade. The habitat of this variety, which is in deep cedar swamps, seems quite unlike that of var. *simplex*, which is open pastures and meadows.

11. **Botrychium matricariaefolium** A. Br.
      matricary grape fern
Fig. 62, frond. Map 60.

Fronds 28 cm long or longer, membranous to fleshy. Blades narrowly deltoid to ovate, short-stalked, inserted above the middle, pinnatifid to bipinnate-pinnatifid; segments of blade blunt and usually toothed. Fertile segment paniculate. Spores mature in June and July.
This species is somewhat larger than *B. simplex* and seems to occur more frequently than some of the other species. The shape of the blade is variable (deltoid to ovate) but it is stalked, and the toothed segments are distinctive (compare *lanceolatum*).

Cytology: $n = 90$ (W.H. Wagner 1955; W.H. Wagner and Lord 1956; Löve and Löve 1976*) This is a tetraploid species of uncertain origin.

Habitat: In acid soil in old sandy and sterile fields, dry wooded slopes, rocky woods, moist cedar woods, and rich swamps.

Range: Southern Labrador and Newfoundland to southern British Columbia, south to Pennsylvania, Minnesota, and Idaho.

Ophioglossaceae

Fig. 60   *Botrychium dusenii*; frond, 1 × .

Fig. 61   *Botrychium simplex* var. *tenebrosum*; frond, 1 × .

Ophioglossaceae

Fig. 62 *Botrychium matricariifolium*; frond, 1/2 ×.

Ophioglossaceae

Remarks: The species is distinctive for a lesser *Botrychium*, although specimens with deltoid blades and a more fleshy nature (found in northern, open sites) mimic *B. boreale*, and the very small plants look like *B. simplex* var. *simplex*. *Botrychium matricariifolium* is rare in Manitoba (White and Johnson 1980) and Saskatchewan (Maher et al. 1979).

12. *Botrychium boreale* Milde ssp. *boreale*
Fig. 63, frond. Map 61.

Fronds erect, up to 26 cm long, stout and fleshy. Blades pinnately divided, sessile, inserted above the middle; primary divisions of blades palmately lobed or crenate, acute at the apex. Fertile segment simple or paniculate. Spores mature in June and July.

Cytology: $n = 90$ (W.H. Wagner 1963, without variety).

Habitat: Grassy and rocky slopes and alpine meadows.

Range: Circumpolar, but in North America known only in Greenland and Alaska.

12.1 *Botrychium boreale* Milde ssp. *obtusilobum* (Rupr.) Clausen
B. boreale Milde var. *obtusilobum* (Rupr.) Broun
Fig. 64 (*a*) frond; (*b*) sporangia. Map 62.

Similar to ssp. *boreale*, but with the divisions oblong and obtuse at the apex.

Habitat: Grassy and rocky alpine slopes and meadows, extending down into open deciduous and evergreen woodland.

Range: Western Alberta and interior British Columbia, south to Washington and Montana, and north through the Yukon and Alaska to eastern Siberia.

Remarks: This is the western counterpart of the eastern *B. matricariifolium*, but usually lacks the short blade stalk, is more often deltoid, and has the fleshy characteristics of so many of the smaller *Botrychium* species (Reeves 1977). It is rare in the District of Mackenzie (Cody 1979), the Yukon (Douglas et al. 1981), and Alberta (Argus and White 1978).

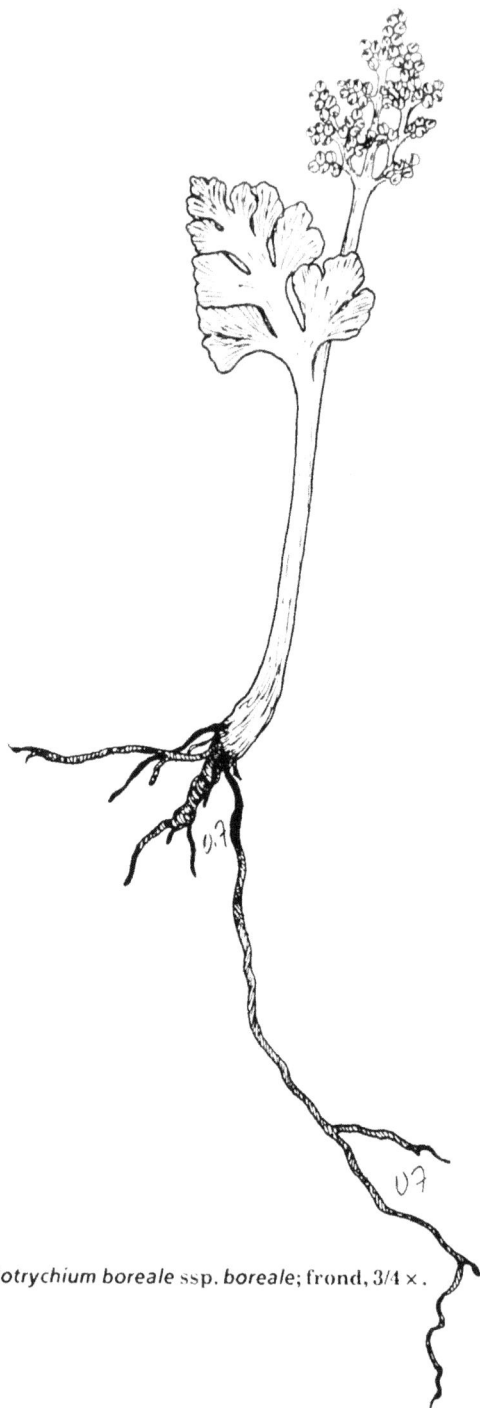

Fig. 63 *Botrychium boreale* ssp. *boreale*; frond, 3/4 × .

Ophioglossaceae

Fig. 64   *Botrychium boreale* ssp. *obtusilobum*; (a) frond, 1/2 × ; (b) sporangia, 7 ×

13. *Botrychium lanceolatum* (Gmel.) Ångstr. var. *lanceolatum*
     lance-leaved grape fern
Fig. 65, frond. Map 63.

Fronds 20 cm long or longer, stout and fleshy. Blades sessile, broadly deltoid, inserted near the summit; segments of blades lanceolate, pinnatifid. Fertile segment paniculate. Spores mature in July and August.
The sessile triangular blade with sharp toothing (but not as acute as in taxon 13.1) is characteristic of the species (Reeves 1977). See also remarks under *B. matricariifolium*.

Cytology: $n = 45$ (W.H. Wagner 1963; Löve and Löve 1976).

Habitat: Alpine meadows, dry slopes, sandy open places, and swampy forests.

Range: Greenland, south through Newfoundland, eastern Quebec, and northern Maine; Aleutian Islands, south through British Columbia to Washington, Wyoming, and Colorado; Eurasia.

Remarks: This species is rare in the Yukon (Douglas et al. 1981) and Saskatchewan (Maher et al. 1979).

13.1 *Botrychium lanceolatum* (Gmel.) Ångstr. var.
      angustisegmentum Pease & Moore
      B. angustisegmentum (Pease & Moore) Fernald
Fig. 66, frond. Map 64.

Differs from var. *lanceolatum* in being lax and membranous, with the divisions of the blade narrow and more acute.

Cytology: $n = 45$ (W.H. Wagner 1955; Löve and Löve 1976*).

Habitat: Shaded woodlands, edges of swamps, along streams, and occasionally in open fields.

Range: Newfoundland to southern Ontario and Michigan, south to Pennsylvania and Ohio.

Remarks: This more southern variety was sufficiently distinct for Fernald to give it specific status. The blade is much thinner or less fleshy than that of var. *lanceolatum*, and the divisions are narrower and very sharp-pointed. It is a dark but bright green, rather than the bluish green of *B. matricariifolium*. It is a rare plant in Ontario, and some of the records we have for the urbanized south of that province now represent extirpated plants. *B. lanceolatum* var. *angustisegmentum* is rare in Nova Scotia (Maher et al. 1978) and Ontario (Argus and White 1977).

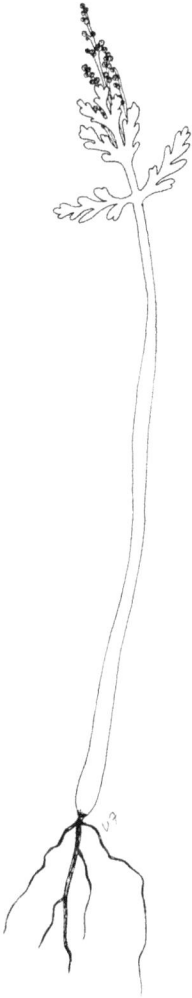

Fig. 65 *Botrychium lanceolatum*
var. *lanceolatum*; frond, 1/2 ×.

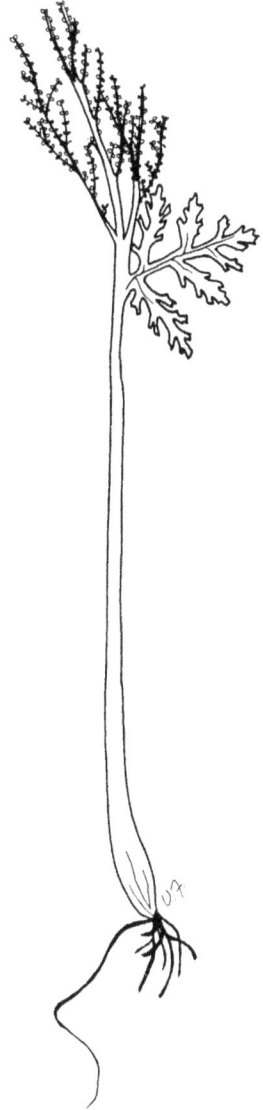

Fig. 66 *Botrychium lanceolatum*
var. *angustisegmentum*; frond, 2/3 ×.

6. OSMUNDACEAE   flowering fern family

1. *Osmunda* L.   flowering fern

Tall ferns of marshy places, frequently in large clumps; fertile fronds surrounded by sterile ones. Rootstocks creeping or suberect. Fronds with stipes winged at the base. Blades with free, usually forked veins extending to the margins. Sporangia naked, large, globose, bivalved, borne on modified, contracted pinnae.

The genus *Osmunda*, which contains 12 species, is one of three genera in the family Osmundaceae. The species occur from Canada's temperate region south to tropical swamps. We have three species in Canada.

A. Fronds bipinnate, some of them fertile at the tip; pinnules finely toothed .... 1. *O. regalis* var. *spectabilis*
A. Fronds pinnate; sterile pinnae deeply pinnatifid; lobes usually entire.
    B. Fertile fronds with fertile pinnae near the middle; no tufts of wool at the base of the pinnae of sterile fronds ....
    .... 2. *O. claytoniana*
    B. Fertile and sterile fronds separate; pinnae of sterile fronds with a tuft of wool in the axils .... 3. *O. cinnamomea*

1. *Osmunda regalis* L. var. *spectabilis* (Willd.) Gray
royal fern
Fig. 67 (a) frond; (b) sporangia. Map 65.

Fronds up to 1 m long and 25 cm wide, bipinnate. Pinnules oblong to lance-oblong, up to 6 cm long, sessile, subentire to finely toothed, rounded to the base; fertile pinnules contracted, borne at the tip of the frond.

Forma *anomala* (Farw.) Harris has the normally fertile part of the frond intermixed with sterile pinnae, and some of the normally sterile pinnae more or less fertile.

The royal fern may be recognized by the clumps of large doubly pinnate fronds. The fertile pinnules are contracted at the tips of the fertile fronds.

Cytology: $n = 22$ (Cody and Mulligan 1982*).

Habitat: Swamps, low lying woods, wet marshy meadows, and cedar bogs.

Fig. 67 *Osmunda regalis* var. *spectabilis*; (a) frond, 1/3 × ; (b) sporangia, 3 × .

Range: Variety *spectabilis* is found in eastern North America, Newfoundland to Rainy River District, Ont. (not to Saskatchewan as given in some manuals), south to Florida, Alabama, Mississippi, Louisiana, and Texas; var. *regalis* occurs in Eurasia.

Remarks: Three other subspecies are known to occur in Europe and Asia. All have the same chromosome number, as do all the members of the family Osmundaceae counted to date.
The range "to Sask." given in Fernald (1950) is probably based on the reports in Macoun (1890) "on Muskeg Island, Lake Winnipeg (J.M. Macoun). Through Canada westward to the Saskatchewan (Eaton)." No specimens have been found to substantiate these early reports, and no recent collections have been seen from west of the Rainy River District, Ont.

2. **Osmunda claytoniana** L.
interrupted fern
Fig. 68 (a) frond; (b) sporangia. Map 66.

Fronds up to 1.2 m long, 15–25 cm wide or wider, pinnate-pinnatifid. Sterile pinnae oblong-lanceolate; pinnules elliptic-oblong to oblong-oval, blunt; lower pinnules 1.3–1.8 cm long; young pinnae and rachis with a rusty wool, promptly glabrous. Fertile frond with 3–5 pairs of dark brown contracted fertile pinnae situated at about the middle.
The contracted fertile pinnae in the middle of the fertile fronds give the interrupted fern its name.

Cytology: $n = 22$ (Britton 1964*).

Habitat: Moist wooded slopes, swamp margins, and open thickets.

Range: Eastern North America, Newfoundland to southeastern Manitoba, south to Georgia, Kentucky, and Arkansas.

Remarks: Another subspecies, ssp. *vestita* (Wahl.) Löve and Löve, occurs in Asia. It has the same chromosome number as ssp. *claytoniana*.

3. **Osmunda cinnamomea** L.
cinnamon fern
Fig. 69 (a) fertile frond; (b) sterile frond; (c) sporangia. Map 67.

Sterile fronds up to 1.2 m long, 15–20 cm wide or wider, similar to *O. claytoniana*, but with a tuft of wool at the base of each of the linear-lanceolate pinnae. Fertile fronds shorter than the sterile, the

Fig. 68 *Osmunda claytoniana*; (a) frond, 1/4 × ; (b) sporangia, 4 ×.

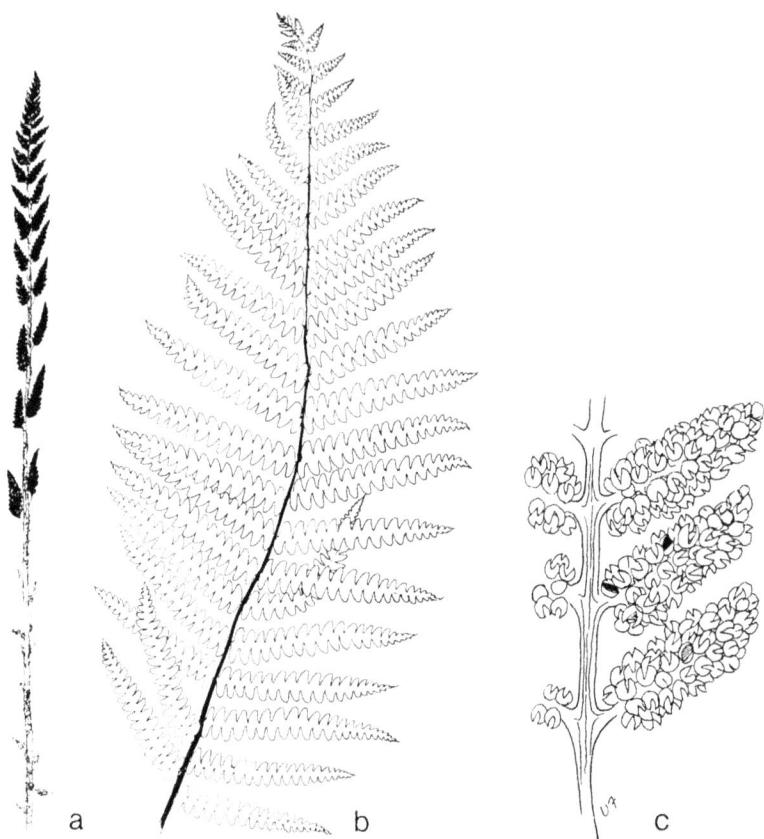

Fig. 69 *Osmunda cinnamomea*; (*a*) fertile frond, 1/4 ×; (*b*) sterile frond, 1/4 ×; (*c*) sporangia, 2 ×.

dark brown pinnae contracted, withering after the spores are cast. Immature sterile and fertile fronds covered with a thick rusty wool, still partly present on the fertile fronds even at maturity.

Forma *frondosa* (T. & G.) Britt. has the fertile frond partly leafy and the fertile and sterile pinnae variously mixed.

The cinnamon fern can quickly be recognized by the densely woolly cinnamon-colored fertile fronds, which quickly shrivel after the spores are shed.

Cytology: $n = 22$ (Britton 1964*).

Habitat: Low ground, thickets, and wet marshy woods.

Range: In eastern North America from Newfoundland to Rainy River District, Ont., south to the Gulf States and New Mexico.

Remarks: An Asiatic subspecies, ssp. *asiatica* (Fern.) Hultén, is found in eastern Asia.   Both subspecies have the same chromosome number.

Hybrids of *Osmunda*

In spite of the fact that our species of *Osmunda* frequently occur side by side, only one hybrid, *O.* × *ruggii* Tryon (*O. claytoniana* × *regalis*) has yet been found, and that only twice (W.H. Wagner et al. 1978). This hybrid is not known in Canada.

7. SCHIZAEACEAE

1. *Schizaea* Sm.

1. **Schizaea pusilla** Pursh
curly-grass
Fig. 70 (*a*) fronds; (*b*) fertile pinnae. Map 68.

Fronds dimorphic, forming dense tufts. Sterile fronds slenderly linear, spiraling and curling, 8 cm or more in length. Fertile fronds erect, very slender, up to 12 cm in length, with 3–8 tiny crowded pairs of pinnae at the tip folded together and thus looking one-sided. Sporangia pear-shaped, bivalvate, in a double row along the vein on the back of the pinnae.

This species is so insignificant that only the keenest observers can find its grass-like fronds, which are hidden among other vegetation. The tiny infolded pinnae at the tip of the fertile fronds are unlike any other fern in Canada.

Cytology: $n = 103$ (W.H. Wagner 1963).

Habitat: Damp peaty and sandy depressions, sphagnum bogs, and low, mossy, open woods and crevices along shores, both on tablelands and lowlands; easily overlooked among other vegetation.

Range: Newfoundland and Nova Scotia and on the Pine Barrens of New Jersey. There is a specimen in the University of Toronto Herbarium collected by E.A. Moxley reputedly from Sauble Beach, Bruce County, Ont. However, various botanists have searched for it there without success, and the locality is questionable. The site is now a cottage area, and so if the plant did grow there it has been extirpated.

Remarks: *Schizaea* is one of four genera of the family Schizaeaceae. The approximately 160 species in the family are mostly tropical in distribution. The climbing fern, *Lygodium palmatum*, which is rare from Florida to New Hampshire in the eastern United States, is one of this family.

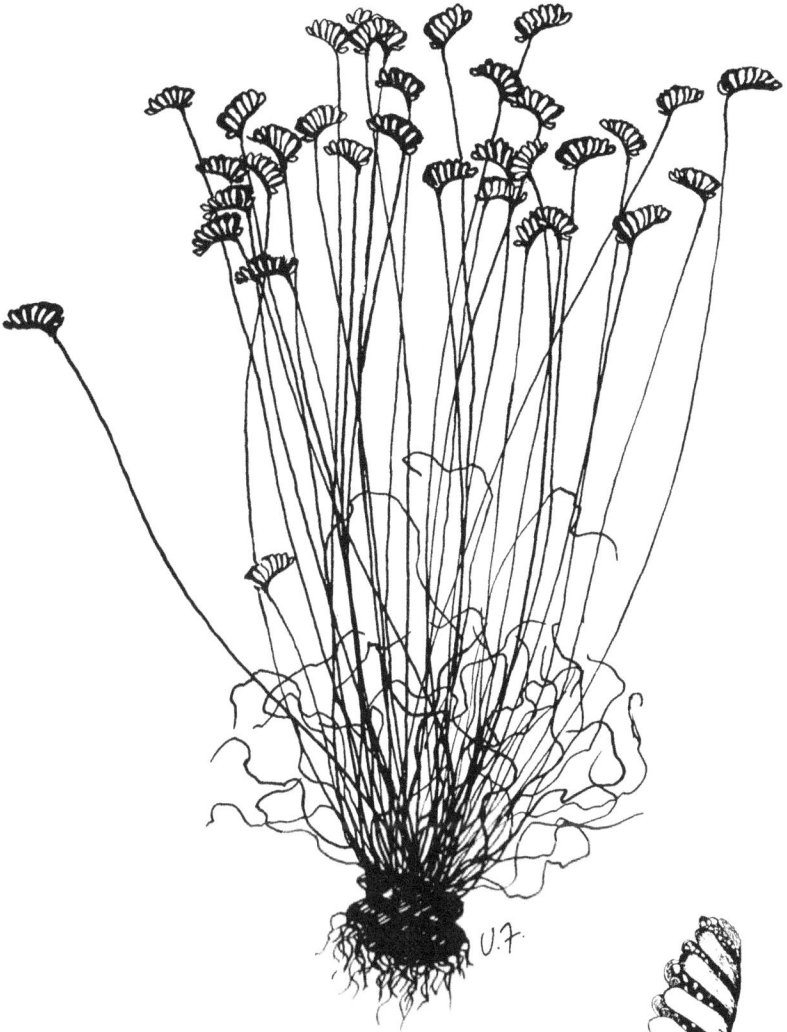

Fig. 70 *Schizaea pusilla*; (a) fronds, 1 × ; (b) fertile pinnae, 4 × .

8. HYMENOPHYLLACEAE filmy fern family

1. *Mecodium* Copeland filmy fern

1. *Mecodium wrightii* (van den Bosch) Copeland
Hymenophyllum wrightii van den Bosch
Fig. 71 (a) fronds; (b) pinnule with sporangia. Map 69.

Fronds up to 7 cm long, arising along the extensively creeping and branched thread-like rhizome. Stipes thread-like, blackish, with a tuft of hair-like scales at the base and narrowly decurrent from the basal pinnae. Blades 3–5 cm long, pinnate; pinnae pinnatifid through 3 or 4 dichotomous branchings; ultimate segments few, linear, blunt, very delicate, pale green and almost translucent so that the dark dichotomously branched veins are very conspicuous. Sori at the ends of the veinlets. Indusia bivalvate to the base.

This delicate fern could be easily passed over in its moist, mossy habitat. The dark, dichotomously branched veins are conspicuous in the pale green almost translucent pinnae.

Cytology: $x = n = 27(?)$ (Manton and Vida 1968); 42(?) (Tatuno and Takei 1969).

Habitat: Forming mats on shaded cliff faces, boulders, and bases of trees.

Range: In North America from the Alaskan Panhandle, south to Vancouver Island, and apparently of very local occurrence; also in Japan and Korea.

Remarks: This is the only filmy fern known in Canada. It was first found at Dawson Inlet on the west coast of Graham Island in the Queen Charlotte Islands by H. Persson while he was collecting mosses. Its disjunct distribution in North America, from the main range in Japan and Korea, was discussed by Iwatsuki (1961). W. Schofield, also while searching for mosses, has since discovered the gametophyte on nearby Chaatl Island and on the British Columbia mainland near Prince Rupert. The gametophytes bear deciduous marginal cell-masses, which allow the gametophyte to reproduce asexually. Thus they may be found in quantity where the sporophyte generation is absent. A detailed description of the habitat is given by Calder and Taylor (1968). Taylor (1967) described the gametophyte and reported its occurrence in British Columbia, and Cordes and Krajina (1968) reported finding male gametophytes on old bark or decaying wood of Sitka spruce at three sites on the west coast of Vancouver Island.

Hymenophyllaceae

9. PTERIDACEAE

Ferns delicate to coarse, deciduous, or evergreen. Fronds pinnate to decompound. Sori marginal, protected by the indusium, which opens toward the margin, or by the reflexed margins of the pinnae, or borne along the veins and lacking an indusium.
A large family of mainly terrestrial ferns, comprising over 60 genera. Some tropical and subtropical members are arborescent.

A.   Fronds covered beneath by a conspicuous white or golden yellow powder; sori borne along the veins and lacking an indusium . . . .
. . . . 7. *Pityrogramma*
A.   Fronds without a white or golden yellow powder; sori marginal.
   B.   Sori usually confluent as a marginal band.
      C.   Fronds coarse, scattered, from stout elongate and forking rhizomes . . . . 2. *Pteridium*
      C.   Fronds tufted from a very short rhizome.
         D.   Segments of frond bead-like . . . . 3. *Cheilanthes*
         D.   Segments of frond not bead-like
            E.   Pinnules and segments of frond jointed at the base . . . . 5. *Pellaea*
            E.   Pinnules and segments of frond not jointed at the base.
               F.   Stipes herbaceous, green except at the base . . . . 6. *Cryptogramma*
               F.   Stipes wiry, dark and shiny . . . .
. . . . 4. *Aspidotis*
   B.   Sori distinct, short, mostly not confluent
      G.   Stipe and fronds glabrous . . . . 8. *Adiantum*
      G.   Stipe and fronds glandular-hairy . . . .
. . . . 1. *Dennstaedtia*

1. *Dennstaedtia* Bernh.

1.   *Dennstaedtia punctilobula* (Michx.) Moore
    Dicksonia pilosiuscula Willd.
    Dicksonia punctilobula (Michx.) A. Gray
    hay-scented fern
Fig. 72 (*a*) frond and rhizome; (*b*) pinnule with sporangia. Map 70.

Fronds 30–70 cm long or longer, arising from slender, naked, freely creeping and forking rhizomes. Stipes pale brown, lustrous, chaffless. Rachis and under surface of the blades minutely glandular-pubescent. Blades lanceolate, bipinnate; pinnae lanceolate; pinnules pinnatifid with toothed lobes. Sori minute, round, situated on the upper margin of the underside of the lobes. Indusia cup-shaped.

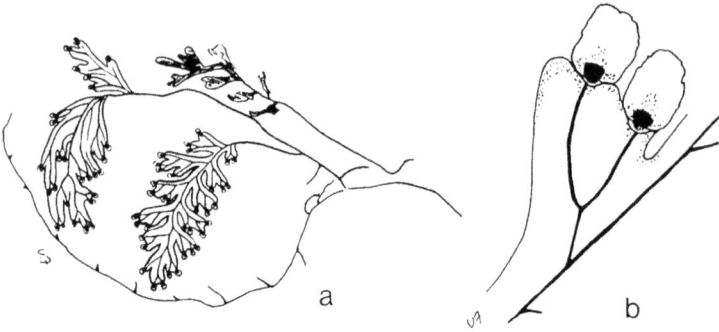

Fig. 71 *Mecodium wrightii*; (a) fronds, 1/2 × ; (b) pinnule with sporangia, 10 × .

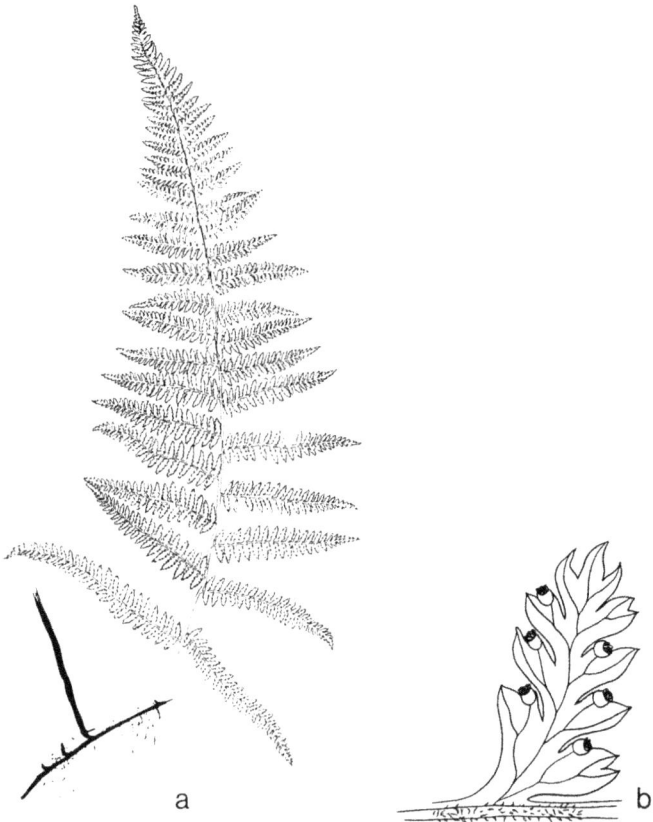

Fig. 72 *Dennstaedtia punctilobula*; (a) frond and rhizome, 1/4 × ; (b) pinnule with sporangia, 7 × .

Pteridaceae

The hay-scented fern has sometimes been confused with the lady fern and bulblet fern. It can be distinguished by the characteristic odor when crushed, the lustrous brown stipes, the glandular hairs on the rachis, which can readily be seen when held up to the light, the rather hard round sori, and its occurrence in large patches of single fronds arising from the underground rhizome.

Cytology: $n = 34$ (Britton 1964*); $n = $ ca. 33 (Cody and Mulligan 1982*).

Habitat: In gently sloping well-drained light sandy soils and around rock piles in clearings, open woods, pastures, and old fields. It is a weed of lowbush blueberry fields, rough pastures, and old fields in the eastern part of its range.

Range: In North America, from southern Newfoundland (where it is rare) and Nova Scotia to Ontario, south to Georgia and Arkansas.

Remarks: R.M. Tryon (1960) recognized 11 species of *Dennstaedtia* in the Americas, which, with the exception of *D. punctilobula*, are essentially tropical in range. A review of *D. punctilobula* in Canada was published in the series *The Biology of Canadian Weeds* (Cody et al. 1977).

2. *Pteridium* Scop.   bracken

1.   *Pteridium aquilinum* (L.) Kuhn var. *latiusculum* (Desv.) Underw.
     P. latiusculum (Desv.) Hieron.
     P. aquilinum (L.) Kuhn var. *champlainense* Boivin
     bracken
Fig. 73 (a) frond; (b) fertile pinnule; (c) venation. Map 71.

Coarse fronds 30–70 cm or more long, in extensive colonies from creeping and forking hairy rhizomes. Stipes longer or shorter than the blades. Blades triangular, usually ternate, 30–50 cm wide, bipinnate-pinnatifid to tripinnate-pinnatifid; lower pinnules more or less pinnatifid; upper pinnules entire; ultimate divisions very numerous, oblong to linear, glabrous or slightly pubescent beneath, with revolute margins. Sporangia borne in marginal sori on the undersurface of the pinnules; sporangia covered by a mostly continuous false outer indusium formed by the revolute margin and by a minute often nearly obsolete hyaline inner indusium.
Plants that grow in the shade tend to have ternate fronds, with the rachis bent so that the blade is presented to the available light, whereas plants that grow in the open tend to be upright and stiff, with shorter ascending pinnae that are twisted at right angles to the rachis.

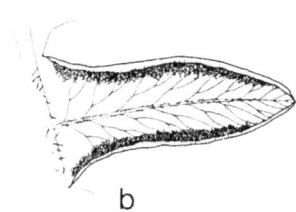

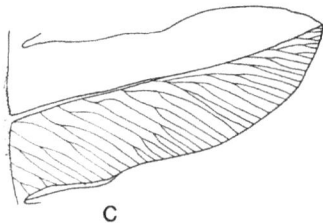

Fig. 73 *Pteridium aquilinum* var. *latiusculum*; (a) frond, 1/4 ×; (b) fertile pinnule, 3 ×; (c) venation, 3 ×.

Pteridaceae

Cytology: $n = 52$ (Britton 1953*, Cody and Mulligan 1982*).

Habitat: Occurs as a weed in pastures, on grassy slopes in abandoned fields, in burnt-over areas, in damp or more often dry, usually sterile soil, on open slopes, and in open woods and thickets. This is probably the most common Canadian fern.

Range: Worldwide distribution; var. *latiusculum* widespread in Canada from Newfoundland to eastern Manitoba and rare in eastern British Columbia and the foothills of the Rocky Mountains in Alberta. To the south, extending to North Carolina, Tennessee, Missouri, and Oklahoma.

1.1 ***Pteridium aquilinum*** (L.) Kuhn var. ***pubescens*** Underw.
   P. aquilinum (L.) Kuhn var. *lanuginosum* (Bong.) Fern.
Fig. 74, undersurface of pinnule. Map 72.

Fronds usually longer than var. *latiusculum*, 1.5 m long or longer. Blade broadly triangular, but rarely ternate. Lower surface of pinnules more or less densely villous or villous-puberulent. Inner indusium ciliate and sometimes also pubescent.

Cytology: $n = 52$ (Löve et al. 1971).

Habitat: Moist to dry woods, clearings, open slopes, and roadsides.

Range: British Columbia and southwestern Alberta, south through Washington and Oregon to California and Utah.

Remarks: R.M. Tryon (1941) treated the worldwide genus *Pteridium* as comprising a single species, *P. aquilinum*, made up of two subspecies, each with several varieties. Australian botanists, however, now treat their plant as a separate species, *P. esculentum* (Forst.) Diels, and similarly South American botanists treat their plant as *P. arachnoideum* (Kaulf.) Maxon.
   Material from the Bruce Peninsula, Ont., which R.M. Tryon (1941) called var. *pubescens*, was later included by Boivin (1952) in his var. *champlainense*. Because of the great variation in frond shape and indument in the eastern var. *latiusculum*, dependent largely on shading, we have included this material in var. *latiusculum*. A review of *Pteridium aquilinum* in Canada was published in the series *The Biology of Canadian Weeds* (Cody and Crompton 1975).
   In early spring the young fronds of *Pteridium* are highly valued as a green vegetable, particularly in Japan. However, recent studies have shown that this species is carcinogenic in rats, is responsible for enzootic bovine hematuria in cattle, and causes vitamin B_1 avitaminosis in horses and other nonruminants. It is therefore not recommended for human consumption.

3. *Cheilanthes* Sw. lip fern

Small evergreen ferns of dry rocky regions. Rhizomes short and much branched to longer and simple, bearing numerous slender, brown to blackish, hyaline scarious-margined scales. Segments small and bead-like. Sori marginal, often confluent, covered by the inrolled margin of the pinnule.

This is a genus comprising over 100 species, widely distributed around the world in mainly arid regions. The two species that occur in Alberta and British Columbia are readily distinguishable from other ferns in the region by the small bead-like segments of their pinnae.

A.   Blade tomentose below, thinly villous above; lacking scales . . . .
. . . . 1. **C. feei**
A.   Blade villous and scaly below, glabrous above . . . .
. . . . 2. **C. gracillima**

1.   **Cheilanthes feei** Moore
     slender lip fern
Fig. 75 (a) fronds; (b) upper surface of pinnule; (c) lower surface of pinnule. Map 73.

Fronds 5–20 cm long, tufted from a short and much branched rhizome. Stipes 3–10 cm long, dark purplish brown, with a few scarious-margined scales at the base and tawny multicellular hairs above. Blades 2–10 cm long, linear-oblong to ovate, tripinnate. Pinnae deltoid to ovate-oblong; rachis and lower side densely tomentose with pale brown hairs; upper surface with soft whitish hairs; ultimate segments small and rounded. Margins of segments somewhat inrolled but not covering the mature sporangia, which cover the entire lower surface.

Cheilanthes feei can be distinguished from *C. gracillima* by the absence of scales on the lower surface and thinly villous upper surface of the pinnules.

Cytology: $2n = 87$ (Knobloch 1967). An apogamous triploid species.

Habitat: Crevices of limestone or calcareous cliffs.

Range: Southern British Columbia and southwestern Alberta, south and east in the United States to California and Illinois.

Remarks: In British Columbia this species is restricted to two limestone regions, where it is quite common. It is rare in the mountains of western Alberta (Argus and White 1978).

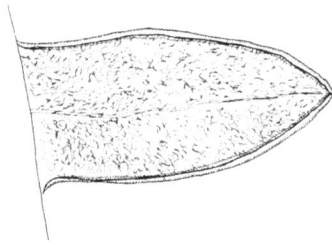

Fig. 74 *Pteridium aquilinum* var. *pubescens*; undersurface of pinnule, 3 ×.

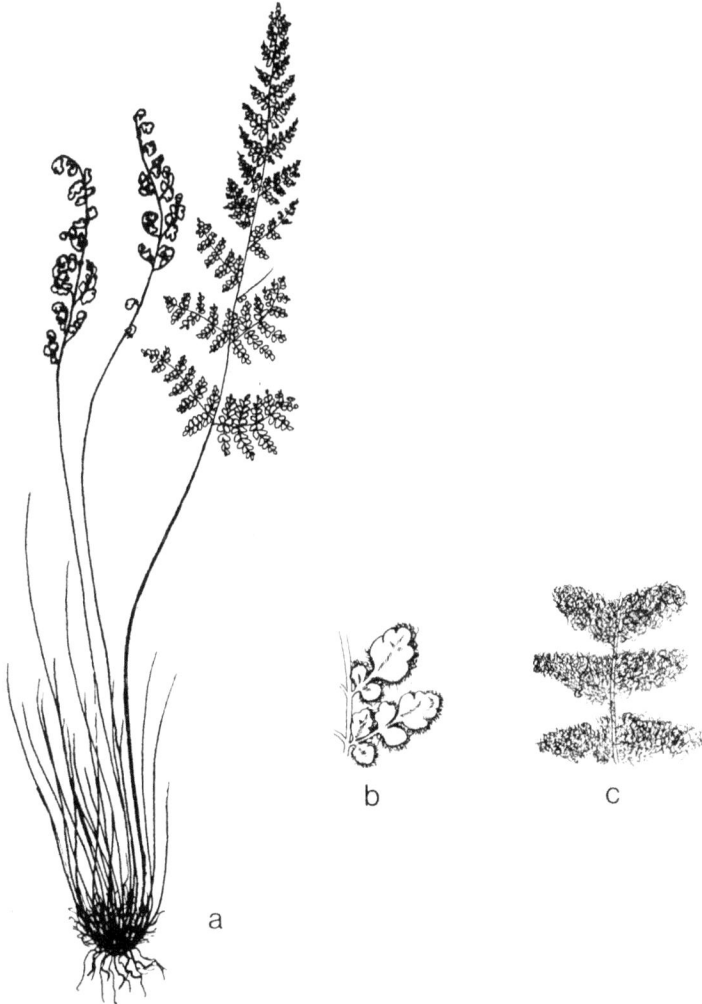

Fig. 75 *Cheilanthes feei*; (a) fronds, 4/5 ×; (b) upper surface of pinnules, 1/3 ×; (c) lower surface of pinnules, 4 ×.

Pteridaceae

2. **Cheilanthes gracillima** D.C. Eat.
lace fern
Fig. 76 (a) fronds; (b) sterile pinnule; (c) fertile pinnule. Map 74.

Fronds 5–25 cm long or longer, densely tufted from the short, much branched rhizome. Stipes dark brown, glabrous, or with scattered long hairs and narrow scales. Blades bipinnate, linear to oblong-lanceolate, 3–10 cm long; rachis and costae frequently villous-puberulent, with long narrow brown scales. Pinnae lance-oblong; pinnules oblong to oval, woolly below, green and glabrous or with a few stellate hairs above, and with broadly inrolled margins. Mature sporangia covering the entire lower surface and extruding from between the folded margins.

Cheilanthes gracillima may be distinguished from *C. feei* by the scaly undersurface and glabrous upper surface of the pinnules.

Cytology: None.

Habitat: Dry crevices of cliffs and rocky slopes of usually igneous rocks.

Range: Southern British Columbia and southwestern Alberta to California, eastward to Idaho, Montana, and Nevada.

Remarks: In British Columbia *C. gracillima* is found across the extreme southern part of the province. In Alberta it is known only in Waterton Lakes National Park, where it is very rare (Argus and White 1978).

4. **Aspidotis** (Nutt. ex Hook. & Bak.) Copel.

1. **Aspidotis densa** (Brack.) Lellinger
Cheilanthes siliquosa Maxon
C. densa (Brack.) St. John
Pellaea densa (Brack.) Hook.
Cryptogramma densa (Brack.) Diels
Indian's-dream
Fig. 77 (a) fertile and sterile fronds; (b) fertile pinnule. Map 75.

Fronds dimorphic, 30 cm long or longer, densely tufted from a short, much branched, chestnut-brown, scaly rhizome. Stipes wiry, chestnut brown, lustrous, glabrous, much longer than the blades. Fertile blades tripinnate, 2–6 cm long, 1–4 cm wide, broadly ovate; pinnules narrowly linear, mucronate. Indusium marginal, thin, continuous, erose-denticulate. Sterile fronds usually much shorter than the fertile (fewer and often lacking), with segments smaller, somewhat broader, and sharply toothed or incised.

Fig. 76 *Cheilanthes gracillima*; (a) fronds, 3/4 ×; (b) sterile pinnule, 3 ×; (c) fertile pinnule, 3 ×.

Fig. 77 *Aspidotis densa*; (a) fertile and sterile fronds, 3/4 ×; (b) fertile pinnule,
10 ×.

Pteridaceae

This small fern can usually be separated from species in the genera *Cheilanthes*, *Pellaea*, and *Cryptogramma*, in which it has been included in the past, by its tripinnate fertile fronds with narrowly linear mucronate pinnules and usually by the absence of sterile fronds.

Cytology: $n = 30$ (W.H. Wagner 1963; Smith 1975).

Habitat: Exposed cliff crevices and rocky or talus slopes.

Range: Southern British Columbia, south and east to Montana, Wyoming, Utah, and California; disjunct to Gaspé, Megantic, and Wolfe counties of Quebec, where it is rare. A specimen collected by H.M. Ami labeled "rocky hillsides of Guelph dolomite, Durham, Ont. (CAN)" may represent a mislabeling because the species is not otherwise known from Ontario, although it has been searched for by numerous fern enthusiasts in the vicinity of Durham.

Remarks: As noted above, this species has been unsatisfactorily placed at various times in the genera *Pellaea*, *Cheilanthes*, and *Cryptogramma*. A recent study by Lellinger (1968) has placed it in the genus *Aspidotis*, together with two other western American species and one from Africa.

5. **Pellaea** Link   cliff-brake

Small tufted plants from compact rootstocks. Fronds firm; stipes and rachises wiry; pinnae gray green; veins free. Sori marginal and confluent under the inrolled and altered margin of the fertile pinnules.

The genus *Pellaea* section *Pellaea* was treated in a monograph by A.F. Tryon in 1957, and cytotaxonomic studies were reported by A.F. Tryon and Britton (1958). The genus is medium-sized and comprises about 80 species, but has only 16 species in the section *Pellaea*. The basic chromosome number (x) is 29, and both sexual and apogamous species and varieties are known. The centre of diversity for the section *Pellaea* is in the southwestern United States and Mexico, and Canadian species are definitely outliers from the centres of distribution. We have only two species, *P. atropurpurea* and *P. glabella*, although the latter has two varieties (some would maintain that we have four distinct species — see the Introduction). All Canadian plants have a similar appearance, a rather gray green color that blends well with the limestone rock crevices and ledges with which they are associated. The apogamous development of some of the species is an adaptation to xerophytic habitats (A.F. Tryon 1968).

A. Fronds dimorphic; stipe and rachis scurfy, with appressed
    pubescence .... 1. **P. atropurpurea**
A. Fronds monomorphic; stipe and rachis glabrous or with a few
    spreading hairs .... 2. **P. glabella**

1. **Pellaea atropurpurea** (L.) Link
    purple cliff-brake
Fig. 78 (a) sterile and fertile fronds; (b) fertile pinnule. Map 76.

Fronds dimorphic; fertile frond 10–35 cm long, 3.5–8 cm wide,
longer than the sterile frond. Stipes and rachis dark purple brown,
dull, pubescent, with more or less appressed hairs. Pinnae rigid,
evergreen, bluish green, simple above, bipinnate below; fertile pinnae
linear to oblong or narrowly ovate, with the lower pinnules stalked;
sterile pinnules ovate-oblong. Sori situated around the margins of the
fertile pinnules. Inrolled margin of pinnule forming the indusium.
    This species looks somewhat like *P. glabella* but may be
distinguished from it by the usually taller, more upright habit, with
fertile fronds that are more divided, that are darker blue green to olive
green, and that have markedly hairy stipes and rachis.

Cytology: "*n*" = 2*n* = 87 (Rigby 1973*). This species is an
apogamous triploid.

Habitat: Dry, steep, exposed, limestone rock slopes or cliffs,
limestone paving, and tops of large talus boulders.

Range: Southern Quebec (Britton et al. 1967; Brunton 1972; Brunton
and Lafontaine 1974), southern Ontario (Britton and Rigby 1968;
Soper 1963), Lake Athabaska, southwestern Alberta (Brunton 1979),
and adjacent southeastern British Columbia, south to Florida and
Arizona. Distribution of *Pellaea* in Canada is given by Rigby and
Britton (1970). For the whole distribution of the species see A.F.
Tryon (1972).

Remarks: This species and *P. glabella* are often confused. *Pellaea
atropurpurea* is a rare plant in the Canadian flora and occurs at widely
separated locations. The Lake Athabasca station is amazingly distant
from the centre of distribution in the southwestern United States
(A.F. Tryon 1972). *Pellaea atropurpurea* is rare in Ontario (Argus and
White 1977), Saskatchewan (Maher et al. 1979), and Alberta (Argus
and White 1978).

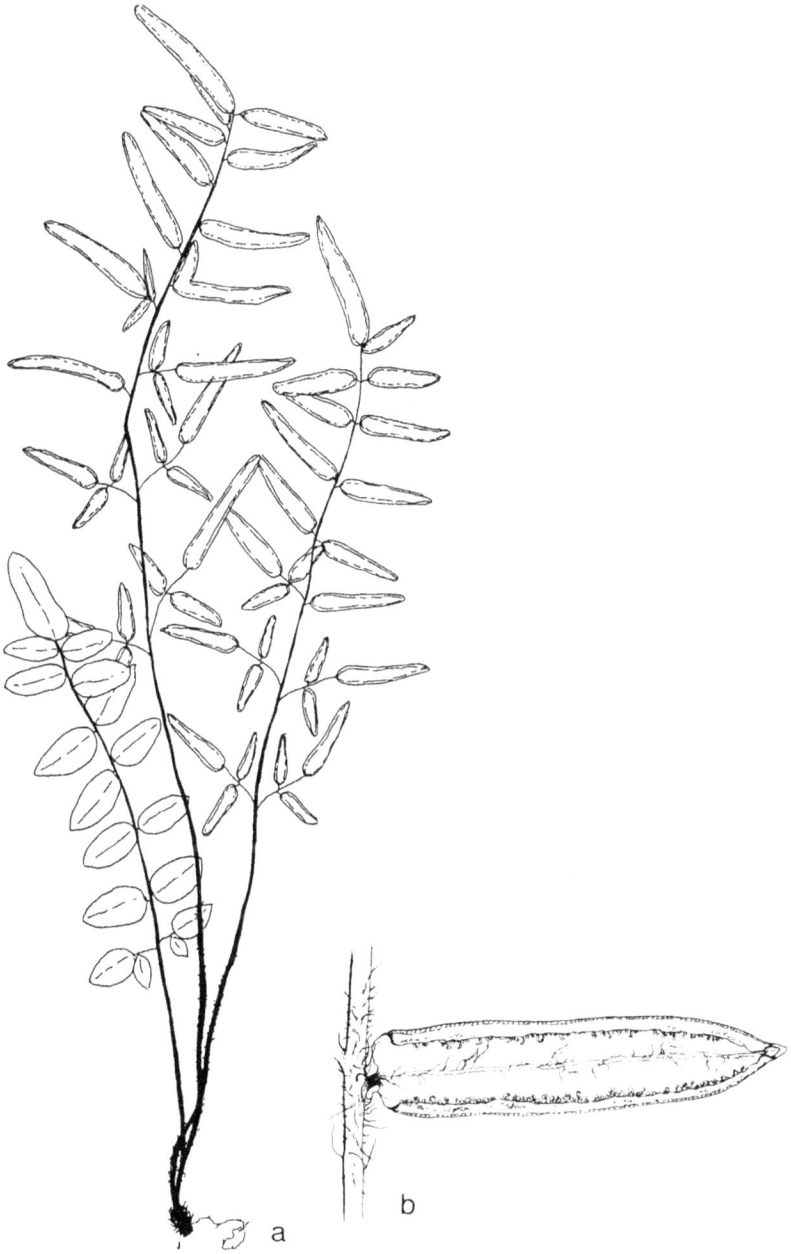

Fig. 78 *Pellaea atropurpurea*; (*a*) sterile and fertile fronds, 2/3 ×; (*b*) fertile pinna, 3 ×.

2.   *Pellaea glabella* Mett. var. *glabella*
     P. atropurpurea (L.) Link var. *bushii* Mackenzie
     smooth cliff-brake
Fig. 79 (a) fronds (b) fertile pinnule. Map 77.

Fronds similar, 10–25 cm long or longer, usually shorter than those of *P. atropurpurea*, open and spreading out beyond the rock face. Stipes and rachis dark reddish brown, smooth, and lustrous. Pinnae rigid, evergreen, bluish green, simple above, pinnate below; basal pinnae persistent, with stalk and rachis up to 5.0 cm long. Pinnules sessile or nearly so, oblong-lanceolate. Sori situated around the margins of the fertile pinnules; inrolled margin forming the indusium. Spores 32 per sporangium.

The smooth cliff-brake is a distinctive species of high, steep limestone cliffs. It appears from small, tight, crevices and blends well with the background. Quite often there are no other ferns or vegetation associated with it. Because the stipes are dark reddish brown, smooth, and shiny, the species is sometimes misidentified as purple cliff-brake.

Cytology: $"n" = 2n = 116$ (Britton 1953, Rigby 1973*). This taxon is an apogamous tetraploid, but see W.H. Wagner et al. (1965) for Missouri.

Habitat: Crevices of dry, sometimes partly shaded, limestone cliffs.

Range: Southwestern Quebec and southern Ontario (Britton and Rigby 1968; Soper 1963), south to Tennessee and Texas. (Rigby and Britton 1970; Brunton and Lafontaine 1974).

Remarks: Some researchers consider the plant quite "unfernlike." The tough, evergreen leaves with few divisions and distinctive coloration are reminiscent of some of the Old World pteridophytes, which also can invade the masonry of forts, castles, and stone walls.

2.1   *Pellaea glabella* Mett. ex Kuhn var. *nana* (Richards.) Cody
      P. glabella Mett. ex Kuhn var. *occidentalis* (E. Nels.) Butters
      P. occidentalis (E. Nels.) Rydb.
      P. pumila Rydb.
Fig. 80 (a) fronds; (b) fertile pinnae. Map 78.

Differs from var. *glabella* by its thin, brittle, and golden brown stipes, forming dense thickly clumped "pincushions" flush with the rock face, and by its mitten-shaped sessile pinnae perpendicular to the rachis. Spores 64 per sporangium.

The smaller size of the plants and their distribution are aids for rapid identification.

Pteridaceae

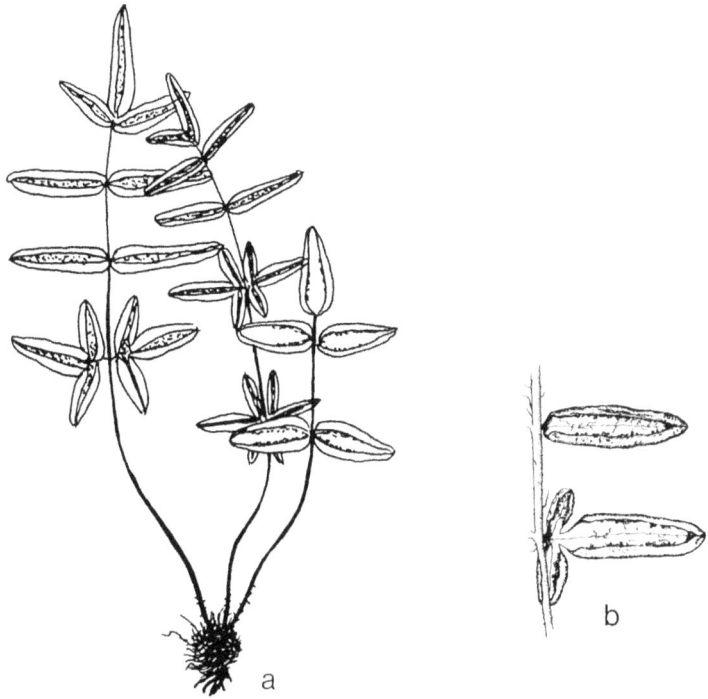

Fig. 79 *Pellaea glabella* var. *glabella*; (a) fronds, 1 × ; (b) fertile pinna, 1 1/2 × .

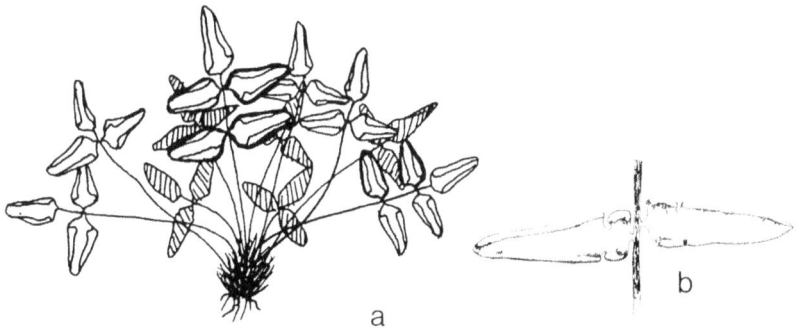

Fig. 80 *Pellaea glabella* var. *nana*; (a) fronds, 1 × ; (b) fertile pinnae, 1 1/2 × .

Pteridaceae

Cytology: $n = 29$ (A.F. Tryon and Britton 1958). This is a basic sexual diploid taxon.

Habitat: Dry, exposed crevices of limestone cliffs.

Range: Manitoba to southwestern District of Mackenzie and southwestern Alberta (Brunton 1979), south to Wyoming and South Dakota.

Remarks: Some researchers maintain that this basic diploid taxon should be given specific rank. That view stresses the differences between var. *nana* and var. *glabella*. At the same time, var. *nana* has many similarities to var. *glabella*, and they therefore must be closely allied genetically. We have followed A.F. Tryon (1957) and accepted the latter position. Variety *nana* is rare in the District of Mackenzie (Cody 1979), Manitoba (White and Johnson 1980), and Saskatchewan (Maher et al. 1979).

2.2 ***Pellaea glabella*** Mett. ex Kuhn var. ***simplex*** (E. Nels.) Butters
    P. atropurpurea (L.) Link var. *simplex* (Butters) Morton
    P. suksdorfiana Butters
Fig. 81 (*a*) fronds; (*b*) fertile pinnules. Map 79.

Differs from var. *glabella* by its usually withered basal pinnae with stalk and rachis up to 1.0 cm long. Stipes sturdy, reddish brown to brown purple, rarely sparsely pubescent; old stipes conspicuous and often silver in color. Pinnae oblong-lanceolate, petioled, acute to the rachis. Spores 32 per sporangium.

Cytology: $n = 2n = 116$ (A.F. Tryon and Britton 1958). An apogamous tetraploid taxon.

Habitat: Crevices of shaded cool, east- or north-facing calcareous cliffs, often overlooking water.

Range: Southwestern Alberta (Brunton 1979) and southern interior British Columbia, south to Washington, Utah, Arizona, New Mexico, and Colorado.

Remarks: Some researchers recognize var. *simplex* as *P. suksdorfiana*. The remarks under *P. glabella* var. *nana* are pertinent here. For var. *simplex*, however, the usual definition of a biological species does not apply because it is apogamous.

Fig. 81 *Pellaea glabella* var. *simplex*; (a) fronds, 1 × ; (b) fertile pinna, 2 × .

6. *Cryptogramma* R. Br.

Small rock ferns with dimorphic fronds, from short much branched or slender elongate rhizomes. Blades glabrous, evergreen, or deciduous; veins free. Sori marginal, covered by a continuous indusium formed by the reflexed margin.

This is a small genus of widespread distribution in boreal and alpine situations. The South American and Himalayan representatives are closely related to *C. crispa*. They are found in rocky habitats.

A. Fronds scattered on an elongate slender rhizome, deciduous . . . .
. . . . 1. *C. stelleri*
A. Fronds densely tufted from a short much branched rhizome, evergreen . . . . 2. *C. crispa*

1. ***Cryptogramma stelleri*** (Gmel.) Prantl
   slender cliff-brake
Fig. 82 (*a*) sterile and fertile fronds; (*b*) fertile pinnules. Map 80.

Fronds dimorphic, scattered along the horizontal rhizome. Sterile fronds almost flaccid, 3–10 cm long; stipes pale to purplish; blades ovate to ovate-deltoid, bipinnate; pinnules oblong, ovate, or obovate flabelliform. Fertile fronds stiffer than sterile fronds, 9–21 cm long; pinnules lanceolate to oblong. Sori situated around margins of fertile pinnules; inrolled margin forming a false indusium.

The slender cliff-brake may be recognized by its delicate fronds, which are scattered along an elongate slender rhizome. The fronds may easily be overlooked, particularly as they turn brown later in the season.

Cytology: $n = 30$ (Britton 1964*; Cody and Mulligan 1982*). This is a basic diploid species.

Habitat: Moist, shaded, usually calcareous crevices and cliffs.

Range: Circumpolar; in North America from Alaska to Newfoundland and Labrador, south into the northern United States, but interrupted in distribution because of habitat limitations.

Remarks: This species is rare in the District of Mackenzie (Cody 1979) and Nova Scotia (Maher et al. 1978).

2. ***Cryptogramma crispa*** (L.) R. Br. var. ***acrostichoides*** (R. Br.)
   C.B. Clarke
   C. acrostichoides R. Br.

Fig. 82 *Cryptogramma stelleri*; (a) sterile and fertile fronds, 1×; (b) fertile pinnules, 3×.

mountain-parsley; parsley fern
Fig. 83 (a) sterile and fertile fronds; (b) fertile pinnule. Map 81.

Fronds dimorphic, winter green, densely clustered from short-creeping mostly ascending rhizomes. Sterile fronds up to 15 cm long; stipes straw-colored; blades ovate to ovate-lanceolate, bipinnate-pinnatifid; pinnae short-petioled; ultimate segments thick, ovate, oblong, or obovate, obtuse, with crenate or toothed margins. Fertile fronds standing stiffly above the sterile, with fewer linear-oblong entire segments; margins of segments broadly reflexed often to the midrib, but opening as the sporangia mature. Sori eventually covering the surface of the fertile pinnule.

Mountain-parsley may be recognized by its densely bunched, crisp, glabrous, winter green fronds, which usually stand out from the dull, rocky substrate.

Cytology: $n = 30$ (R.L. Taylor and Mulligan 1968*; Cody and Mulligan 1982*).

Habitat: Crevices, ledges, and talus slopes and in pockets of organic soil in the Precambrian region.

Range: In western North America from southern Alaska, through the mountains of British Columbia and western Alberta, south to California and New Mexico, across the Precambrian Shield to northwestern Ontario, northeastern Minnesota, and Isle Royale, Mich. (Marquis and Voss 1981); also in Kamchatka.

2.1 **Cryptogramma crispa** (L.) R. Br. var. **sitchensis** (Rupr.) C. Chr. Fig. 84 (a) sterile and fertile fronds; (b) sterile pinnules. Map 82.

Similar to var. *acrostichoides*, but with the sterile fronds broadly triangular, finely dissected; ultimate segments obovate.

Habitat: In the Mackenzie Mountains, N.W.T., on limestone talus and moraines.

Range: Southern Alaska to the Mackenzie Mountains, south into northern British Columbia.

Remarks: The Eurasian var. *crispa* is tetraploid. It tends to have softer sterile leaves that are often more dissected, and the rhizome scales are uniformly brown. Variety *acrostichoides* is rare in Ontario (Argus and White 1977). A report by Macoun (1890) of its occurrence at McLeod's Harbour on Manitoulin Island was refuted by Soper (1963).

Pteridaceae

Fig. 83 *Cryptogramma crispa* var. *acrostichoides*; (a) sterile and fertile fronds,
2/3 × ; (b) fertile pinnule, 5 × .

Fig. 84 *Cryptogramma crispa* var. *sitchensis*; (*a*) sterile and fertile fronds, 1/2 ×; (*b*) sterile pinnules, 5 ×.

Pteridaceae

Calder and Taylor (1968) consider that variety *sitchensis* is a "weak segregate hardly worthy of recognition," whereas others, e.g. Löve (in Löve et al. 1977) treat var. *acrostichoides* as a species, because its morphology and chromosome number are distinct from *C. crispa* of Europe. It is true that var. *acrostichoides* is a basic diploid entity, but we would like to see a thorough biosystematic study that compares all three varieties before definite taxonomic conclusions are reached.

Variety *sitchensis* is rare in the District of Mackenzie (Cody 1979).

7. *Pityrogramma* Link

1. **Pityrogramma triangularis** (Kaulf.) Maxon
   Gymnogramma triangularis Kaulf.
   goldback fern
   Fig. 85 (*a*) fronds; (*b*) portion of undersurface of pinna. Map 83.

Fronds to 30 cm long or longer, tufted from the thickish somewhat ascending rhizome. Stipes much longer than the blade, stiff and wiry, lustrous dark brown, glabrous except at the base. Blade deltoid, pinnate; lowest pinnae pinnate, with the two lower first pinnules longer than the rest; the remaining pinnae pinnatifid; segments blunt, coriaceous; margins narrowly revolute; upper surface glabrous; lower surface with a white or yellowish waxy powder. Sporangia borne along the veins, confluent in age. Indusium lacking.

The white or yellowish waxy powder on the undersurface of the deltoid blades is characteristic.

Cytology: $n = 30, 60$ (Alt and Grant 1960).

Habitat: Open to partly shaded rocky slopes and crevices.

Range: Southwestern coastal British Columbia, south to California, and inland to Arizona, Nevada, and Utah. This very distinctive species seems to be restricted to the floristic region in Canada where madrona (*Arbutus*), Canada's only broad-leaved evergreen tree, grows (see also *Isoetes nuttallii*).

Remarks: *Pityrogramma triangularis* belongs to a genus of about 15 species of small, mainly tropical ferns. Alt and Grant (1960) have shown that there are both diploids and tetraploids of *P. triangularis* from north to south in California, with the tetraploid occurring near the coast. No discernible characters with which the two races can be separated have as yet been detected. Chromosome counts have apparently not been made on plants from north of California, and so we have no knowledge of which race occurs in British Columbia.

Fig. 85 *Pityrogramma triangularis*; (a) fronds, 2/3 × ; (b) portion of undersurface of pinna, 5 × .

According to T.M.C. Taylor (1970), the fronds curl up in dry weather, showing their characteristic light-colored undersurface, which has a whitish or yellowish powder on it.

8. *Adiantum* L. maidenhair fern

Delicate ferns. Fronds produced in rows from slender creeping rhizomes. Veins free-forking. Sori oblong, borne along the upper margin of the pinnules; each sorus covered by an indusium that arises from the inrolled margin.

The genus *Adiantum* is worldwide in distribution and numbers over 200 species. Both our species have been used in horticulture, and in Canada, *A. pedatum* does well in shaded spots in the fern lover's garden.

A. Frond with a simple main zigzag rachis continuing the arching to pendulous stipe .... 1. **A. capillus-veneris**

A. Frond palmately forking at the summit of the upright stipe .....
.... 2. **A. pedatum**

1. **Adiantum capillus-veneris** L.
   Venus'-hair fern
Fig. 86 (a) fronds; (b) pinnules with sporangia. Map 84.

Fronds to 40 cm long or longer, often pendulous from a slender elongate rhizome. Stipe lustrous blackish brown, continuing into a zigzag rachis. Blade ovate-lanceolate, 2-3 pinnate at the base, to simply pinnate above. Pinnules rhombic-ovate with irregularly jagged lobes; veins flabellate, forking from the base. Sori oblong to lunate on the outer margins of the pinnules, which are inrolled to form an indusium.

The zigzag rachis, which is simply pinnate above, readily separates this species from the maidenhair fern, *A. pedatum*.

Cytology: $n = 30$ (Britton 1953).

Habitat: Runnels of hot springs.

Range: Circumpolar: Primarily in warm temperate regions on wet cliffs and seeps, extending northward to Virginia, Missouri, Colorado, Utah, and California, and introduced farther north in sewers and as a weed in greenhouses. In Canada known only from Fairmont Hot Springs in British Columbia, where it is in danger of being extirpated.

Remarks: This species is rare and endangered in British Columbia.

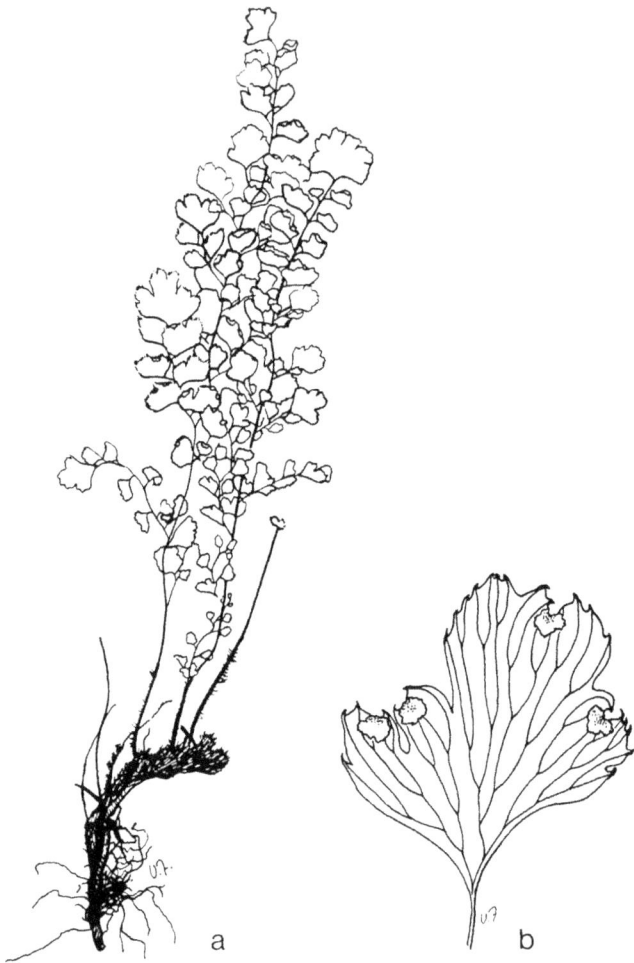

Fig. 86 *Adiantum capillus-veneris*; (a) fronds, 2/3 × ; (b) pinnules with sporangia 5 × .

Pteridaceae

2. **Adiantum pedatum** L. ssp. **pedatum**
   maidenhair fern
Fig. 87 (a) frond; (b) pinnule with sporangia. Map 85.

Fronds 30–55 cm long, in colonies arising from horizontal rhizomes. Stipes lustrous purple brown, forking at the summit into two arching rachises, each of which is divided several times into spreading divisions, thus forming a semicircular blade 15–35 cm wide or wider. Pinnules short-stalked, obliquely triangular oblong; terminal pinnule fan-shaped; main vein along the lower margin; upper margin cleft, with lobes thus formed blunt. Sori elongate, borne on the upper margins of the lobes of the pinnules. Indusium formed by the inrolled margin.

The usually arching and palmately divided lustrous purple brown rachises and the fan-shaped pinnules with the main vein along the lower margin set this fern apart from all others in Canada.

Cytology: $n = 29$ (Britton 1953*, Cody and Mulligan 1982*).

Habitat: Wooded, sometimes rocky slopes in humus-rich soil.

Range: Nova Scotia to Ontario, south in the United States to Georgia, Alabama, Mississippi, Louisiana, and Oklahoma.

2.1 **Adiantum pedatum** L. ssp. **aleuticum** (Rupr.) Calder & Taylor
     A. pedatum L. var. rangiferinum Burgess
Fig. 88 (a) frond; (b) pinnule with sporangia. Map 86.

Differs from ssp. pedatum in the branches strongly ascending rather than widely divergent. Compared with ssp. pedatum, pinnae usually fewer and pinnules (10) 12–20 (23) mm long, more deeply lobed, with their tips acute rather than rounded and with the sinuses between the lobes usually broader.

Cytology: $n = 29$ (R.L. Taylor and Mulligan 1968*).

Habitat: In usually shaded humus-rich soil on ledges and in rocky woods from sea level to the treeline.

Range: Western Alberta, British Columbia, and Alaska, south in the United States to California.

2.2 **Adiantum pedatum** L. var. **subpumilum** W.H. Wagner
Fig. 89, frond. Map 87.

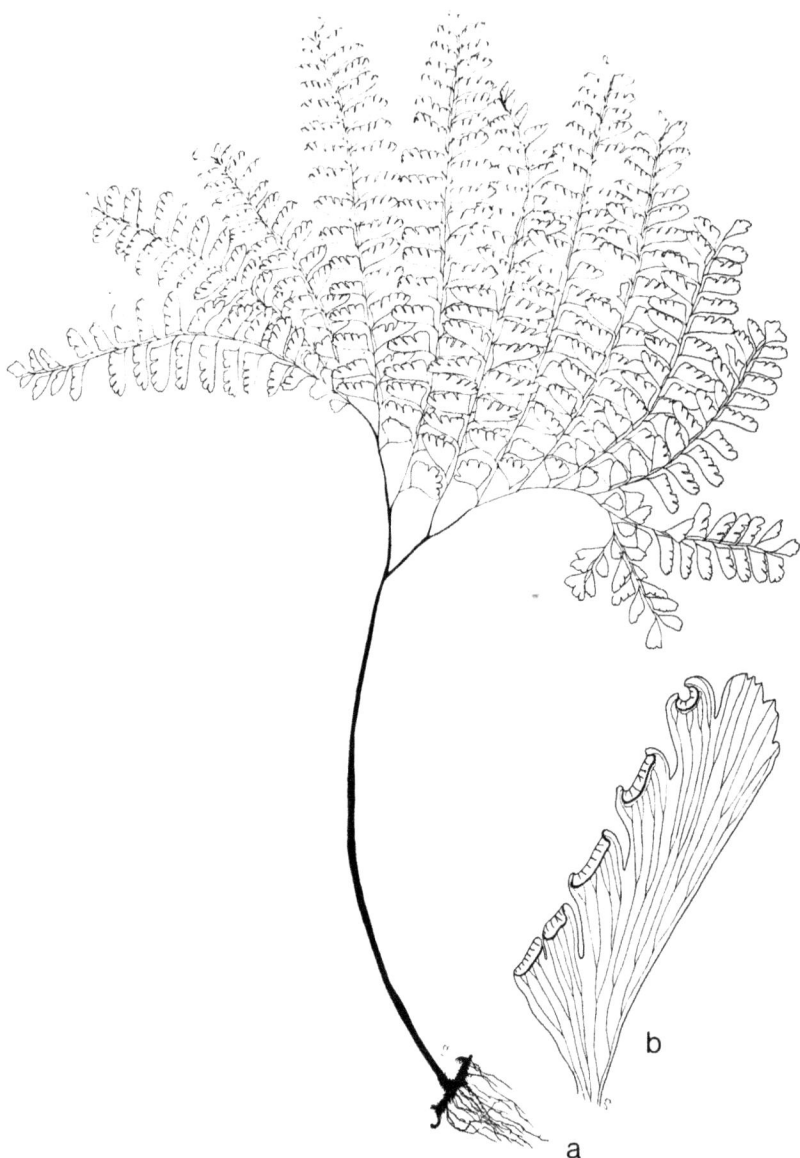

Fig. 87 *Adiantum pedatum* ssp. *pedatum*; (a) frond, 1/3×; (b) pinnule with sporangia, 3×.

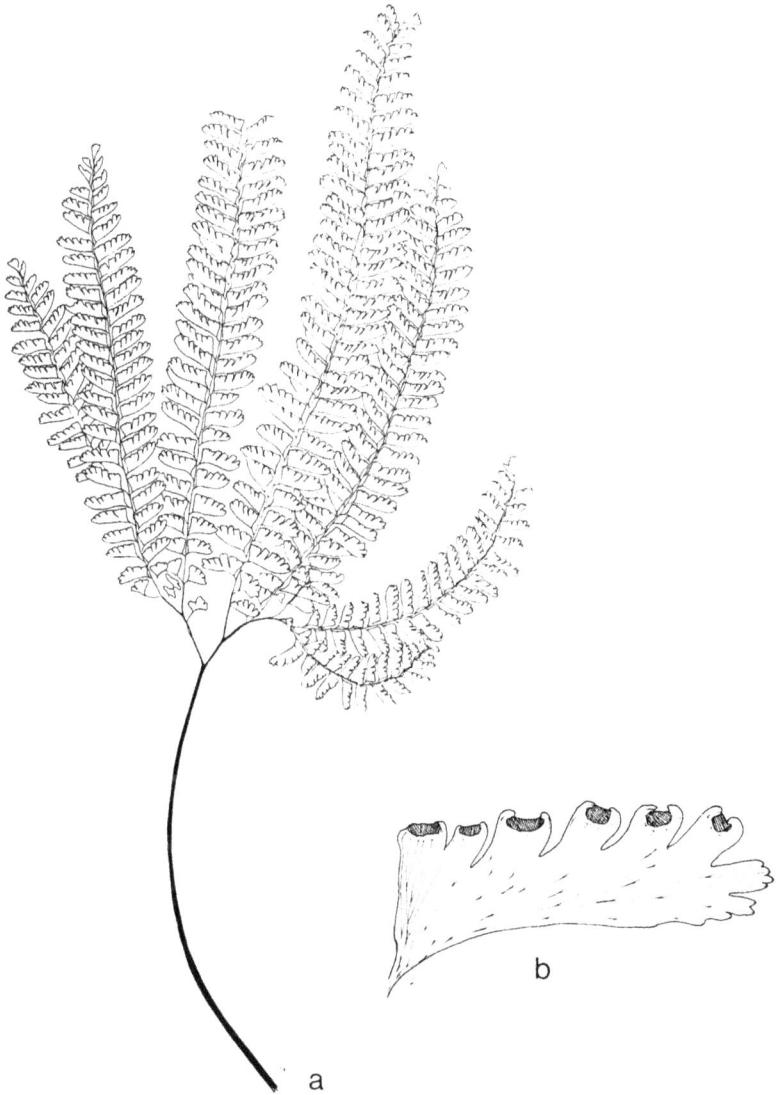

Fig. 88 *Adiantum pedatum* ssp. *aleuticum*; (a) frond, 1/3 ×; (b) pinnule with sporangia, 5 ×.

A dwarf variety that differs in its smaller stature and smaller imbricate pinnules. Pinnules with fewer vein forkings; vein forkings occurring in the distal parts of the pinnules.

Cytology: $n = 29$ (Wagner and Boydston 1978).

Habitat: Wet, exposed cliffs on exposed metamorphic coastal rocks.

Range: In nature known only in the Brooks Peninsula, northwest Vancouver Island, B.C. (type locality); also known in cultivation but from unknown sources (W.H. Wagner and Boydston 1978).

2.3   *Adiantum pedatum* L. ssp. *calderi* Cody
Fig. 90, frond. Map 88.

Differs from ssp. *aleuticum*, with which it has been associated, by its generally shorter stature, stiffly crowded stipes, bluish green glaucus fronds, consistently smaller pinnules (middle pinnules 7–12 (17) mm long), and conspicuous indusia.

Cytology: $n = 29$ (Cody and Mulligan 1982*).

Habitat: Serpentine and dolomite talus slopes, tablelands, and rocky woods.

Range: Western Newfoundland, Gaspé Peninsula, Eastern Townships of Quebec, and adjacent northern Vermont, and disjunct to serpentines in northern California and Washington (Cody 1982).

Remarks: A large form with deeply lacerate pinnules up to 2.5 cm long was described from British Columbia as var. *rangiferinum* by Burgess (1886).   This has been included in the synonymy of ssp. *aleuticum*.

Pteridaceae

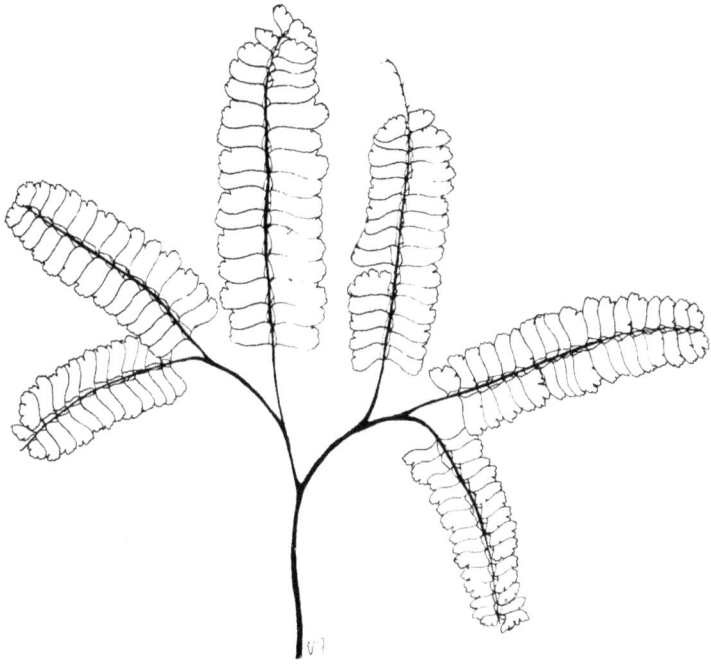

Fig. 89   *Adiantum pedatum* var. *subpumilum*; frond, 2/3 × .

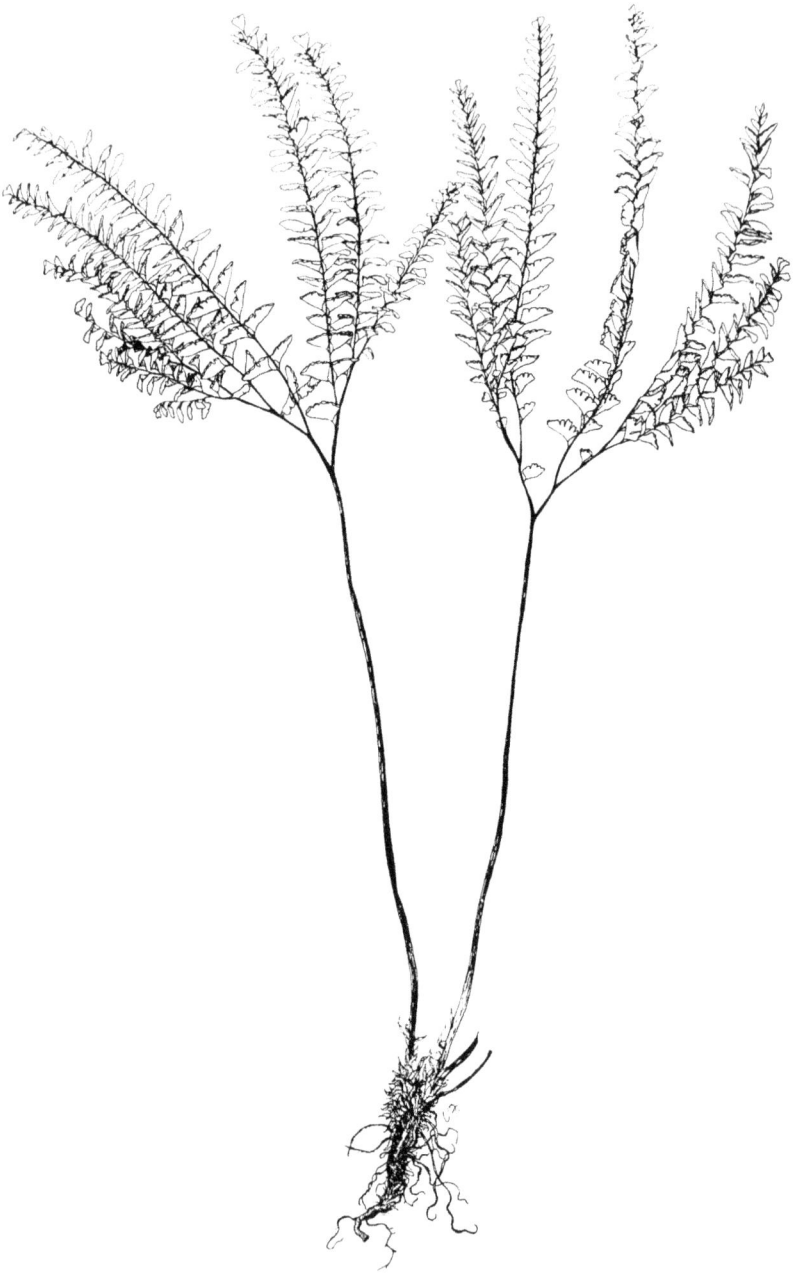

Fig. 90 *Adiantum pedatum* ssp. *calderi*; fronds, 1/2 ×.

Pteridaceae

10. ASPIDIACEAE

Small to large ferns. Fronds pinnate to decompound, forming a crown at the top of a stout rhizome or singly along a fine rhizome. Sori dorsal, usually roundish, but sometimes somewhat elongate. Indusium opening on one side, peltate or absent.

Aspidiaceae is a large family of mainly tropical and subtropical terrestrial ferns comprising over 60 genera.

A. Sporangia partly or wholly covered by the rolled-up pinnules; pinnules forming globular berry-like divisions of the stiff fertile frond.
    B. Fronds in vase-like clumps; simple pinnate fertile fronds surrounded by tall regularly pinnate sterile ones .... 1. *Matteuccia*
    B. Fronds solitary or scattered along the rhizome; sterile fronds coarsely pinnatifid; fertile fronds bipinnate .... 2. *Onoclea*
A. Sporangia not in hard rolled-up berry-like divisions.
    C. Sori round or nearly so.
        D. Indusia present.
            E. Indusia segmented .... 3. *Woodsia*
            E. Indusia not segmented.
                F. Indusium hood-shaped, attached by its base on the side toward the midrib .... 9. *Cystopteris*
                F. Indusium round, reniform, or elongate.
                    G. Fronds scattered along a thin cord-like rhizome (or tufted from a stout rhizome in *T. limbosperma*) .... 7. *Thelypteris*
                    G. Fronds tufted or forming a crown at the end of a stout rhizome.
                        H. Sori elongate, often curved over the ends of the veins; indusia attached on one side .... 10. *Athyrium*
                        H. Sori round.
                          I. Indusia reniform or with a deep sinus.
                            J. Veins reaching the margin .... 7. *Thelypteris limbosperma*
                            J. Veins not reaching the margin .... 5. *Dryopteris*

                              I.   Indusia round, without a
                                   deep sinus . . . .
                                   . . . . 4. **Polystichum**
          D. Indusia absent.
              K.   Fronds in a crown at the end of a stout rhizome . .
                   . . . . 10. **Athyrium**
              K.   Fronds singly along a cord-like rhizome.
                   L.   Fronds more or less ternate . . . .
                        . . . . 6. **Gymnocarpium**
                   L.   Fronds pinnate-pinnatifid . . . .
                        . . . . 8. **Phegopteris**
          C. Sori elongate or horseshoe-shaped . . . . 10. **Athyrium**

                              1. *Matteuccia* Todaro

1.   *Matteuccia struthiopteris* (L.) Todaro var. *pensylvanica* (Willd.)
        Morton
     Pteretis pensylvanica (Willd.) Fern.
     P. nodulosa (Michx.) Nieuwl.
     Onoclea struthiopteris (L.) Hoffm. var. *pensylvanica* (Willd.)
        Boivin
     ostrich fern
Fig. 91 (*a*) upper portion of sterile frond; (*b*) fertile frond; (*c*) portion of
fertile pinna. Map 89.

     Fronds dimorphic, forming a crown at the end of the stout widely
creeping and forking rhizome. Sterile fronds up to 1.2 m long or longer
and 12–24 cm wide, pinnate-pinnatifid, abruptly narrowed to the base;
pinnae broadly linear, acuminate; pinnules oblong, bluntish. Fertile
fronds much shorter than the sterile, persistent over winter; pinnae
greenish, becoming dark brown at maturity. Veins free, not forked, on
both sterile and fertile pinnae. Sori borne on the margins of the
shallowly lobed, tightly inrolled, and pod-like pinnae.
     A full crown of fronds is somewhat reminiscent of a large
headdress. Sterile fronds might be mistaken for *Osmunda
cinnamomea*, but can readily be distinguished from it by the shape of
the blade, which tapers to the base and is sharply cut off at the tip.

Cytology: $n = 40$ (Britton 1953*).

Habitat: Damp shady places, roadside ditches, and floodplains of
streams and rivers.

Range: *Matteuccia struthiopteris* s.l. circumpolar; var. *pensylvanica*
from Newfoundland to British Columbia, southwestern District of
Mackenzie, southeastern Yukon Territory and Alaska, south to
Virginia, Ohio, Indiana, Illinois, Missouri, and South Dakota.

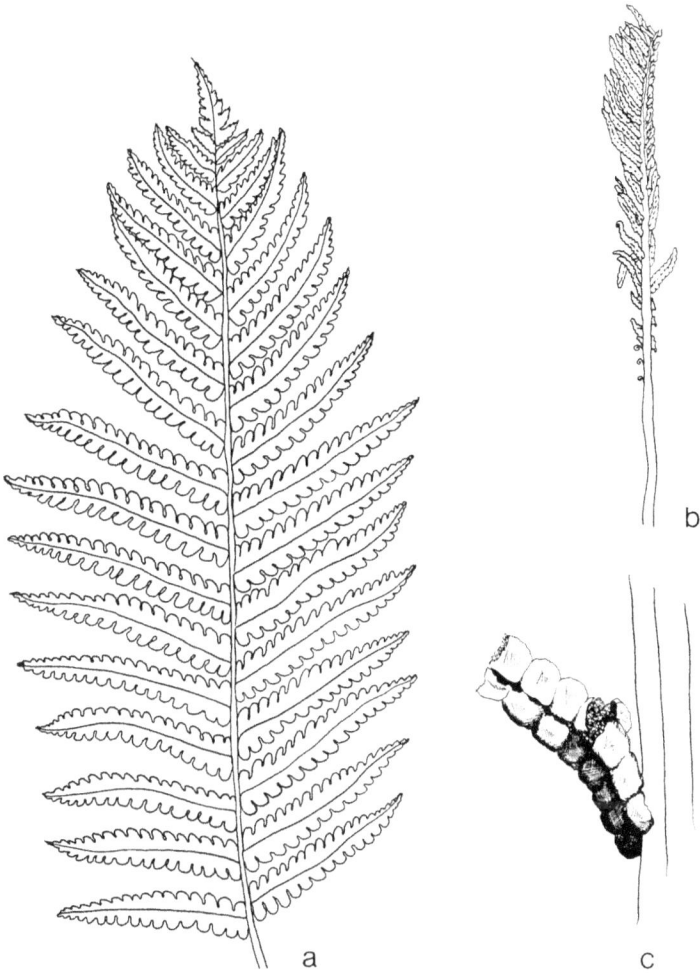

Fig. 91 *Matteuccia struthiopteris* var. *pensylvanica*; (a) upper portion of sterile frond, 1/3 ×; (b) fertile frond, 1/3 ×; (c) portion of fertile pinna, 2 ×.

Remarks: Morton (1950) quotes E.T. Wherry as calling the ostrich fern "our most renamed fern." It has been variously placed in *Onoclea*, *Struthiopteris*, *Pteretis*, and *Matteuccia*. Fernald (1945) maintained that the North American fern was specifically distinct from the European plant. The differences, as pointed out by Morton (1950) are slight, and the varietal level is more satisfactory. The young fiddleheads of this fern are used as a vegetable, particularly in the Maritime Provinces. They may be prepared fresh, or preserved by either freezing or canning. Cruise (1972) reported that the removal of young fronds of the ostrich fern seemed to cause spore-bearing fronds to appear earlier and in greater numbers. The species is common in eastern Canada but rare west of Manitoba. It is rare in the Yukon (Douglas et al. 1981).

2. *Onoclea* L.

1. ***Onoclea sensibilis*** L.
   sensitive fern
Fig. 92 (a) sterile and fertile fronds; (b) portion of fertile pinna. Map 90.

Fronds dimorphic, borne several together on slender creeping rhizomes. Sterile fronds up to 80 cm long; blades 12–30 cm long, 15–30 cm wide, broadest at the base, pinnate at the base, pinnatifid above; rachis winged, with the wing becoming broader toward the tip; pinnae wavy-margined or coarsely toothed. Fertile fronds persistent over the winter, shorter than the sterile; pinnules greenish, becoming blackish at maturity, modified and inrolled to form berry-like structures. Veins free on the fertile fronds and netted on the sterile fronds. Sori borne within the tightly inrolled, berry-like pinnules.
   Forma *obtusilobata* (Schkukr) Gilbert has intermediate fronds between the normal fertile phase and the normal sterile phase; it occurs with the typical form and may be the result of damage to young fronds.
   Where *Woodwardia areolata* occurs in Nova Scotia, it might possibly be confused with the sensitive fern. The latter can, however, be readily distinguished by its entire rather than minutely serrate pinnae margins and by its basal pinnae, which are subopposite rather than alternate.

Cytology: $n = 37$ (Cody and Mulligan 1982*).

Habitat: Forms large patches in low places in woodlands, wet meadows, and roadside ditches. In ditches and meadows it often reaches proportions that make it an undesirable weed.

Fig. 92 *Onoclea sensibilis*; (a) sterile and fertile fronds, 1/3 × ; (b) portion of fertile pinna, 2 × .

Range: Eastern North America, southern Labrador and Newfoundland to southeastern Manitoba, south to Florida, Louisiana, and Texas; also in eastern Asia.

Remarks: The common name, sensitive fern, is derived from the fact that the fronds, although coarse, are sensitive and blacken when touched by the first frost.

3. *Woodsia* R. Br.   woodsia

Small tufted ferns with free veins arising from compact rootstocks. Indusium of thread-like or plate-like segments, attached below, and more or less arched over the round sori.

The genus *Woodsia* is a medium-sized genus of perhaps 40 species in the world and was treated in a monograph by Brown (1964). In Canada there are only five, six, or seven species, depending on which authority you consult. Amateurs find the species difficult to recognize because some plants superficially resemble the abundant *Cystopteris fragilis* s.l. (fragile fern), and many keys stress soral characters such as the indusium, which in mature specimens may be lost or damaged. To distinguish *Woodsia* from *Cystopteris* one should check to see if the indusium is attached below the sorus (*Woodsia*) or is hooded (*Cystopteris*), i.e., attached at one side and arching over the sorus. Failing this, one can compare the stipes by holding them up to the light in the field. They are opaque in *Woodsia* and translucent in *Cystopteris*. The venation is useful also. In the former the veins are less distinct and appear to stop short of the margin, whereas in the latter, they are distinct right to the margin. Also, in *Woodsia*, one expects to find either even or uneven stubble from the remains of old stipe bases.

The species in the genus fall naturally into two groups (see key under A). In the first group, there is an articulation point towards the base of the stipe, and when old fronds drop off, an even stubble is left. The other group lacks these joints and is the uneven stubble group.

We have three species in the first group: two basic diploids, *W. glabella* and *W. ilvensis*, and a derived tetraploid, *W. alpina*.

The second group consists of two basic ancestral diploids that are abundant in Western Canada, *W. scopulina* and *W. oregana*, and two tetraploids, *W. oregana* var. *cathcartiana* and *W. obtusa*. The last occurs fairly frequently southeast of Canada but has very few stations in Canada. In this respect, it is analogous to *Asplenium platyneuron*.

There are also a number of hybrids known, both within each group and between the groups, but on the whole they have not been studied extensively with the use of modern experimental methods.

A.   Stipes jointed at the base, with persistent bases appearing about the same length.

Aspidiaceae

B. Fronds delicate, glabrous; stipes and rachises green or stramineous; rachises chaffless .... 1. *W. glabella*
B. Fronds more or less firm; stipes and rachises brown; rachises chaffy, at least towards the base.
    C. Fronds hairy and usually chaffy below; stipes usually very chaffy .... 2. *W. ilvensis*
    C. Fronds glabrous or glabrate, chaffless; stipes chaffless or with a few deciduous scales .... 3. *W. alpina*
A. Stipes not jointed at the base, with persistent broken bases of various lengths.
    D. Pinnae and rachis bearing glands (which often stain drying papers yellow) and white articulate hairs ....
    .... 5. *W. scopulina*
    D. Pinnae and rachis with or without a fine glandular pubescence and lacking white articulate hairs.
        E. Pinnules broadly rounded; indusia of a few broad segments .... 6. *W. obtusa*
        E. Pinnules slightly lobed or finely toothed; indusia of narrow and thread-like segments ..... 4. *W. oregana*

1. **Woodsia glabella** R. Br.
smooth woodsia
Fig. 93 (a) fronds; (b) fertile pinnae. Map 91.

Fronds to 16 cm long or longer, 1.5 cm wide, linear to linear-lanceolate; pinnae thin-membranous, suborbicular to ovate, toothed or lobed, glabrous. Stipes jointed near the base, usually with chaff only below the joint. Sori distinct or confluent. Indusia composed of 5–8 ciliate-like segments.

This small and attractive species, with its green rachis and stipe, is not likely to be confused with the other species of *Woodsia*. It is, however, sometimes confused with *Asplenium viride* (see under that species).

Cytology: $n = 39$ (Britton 1964*; Cody and Mulligan 1982*); $n = 38$ (Löve and Löve 1976*).

Habitat: In moss or humus among rocks or in protected, cool, moist calcareous crevices.

Range: Circumpolar; in North America from Greenland and Newfoundland to Alaska south to the Gaspé Peninsula, Que., New York, Ontario, Minnesota, and northern British Columbia.

Remarks: This basic diploid ancestral species would seem to have the same chromosome number here as in Europe. It is rare in Manitoba (White and Johnson 1980), Nova Scotia (Maher et al. 1978), Ontario (Argus and White 1977), Saskatchewan (Maher et al. 1979), and Alberta (Argus and White 1978).

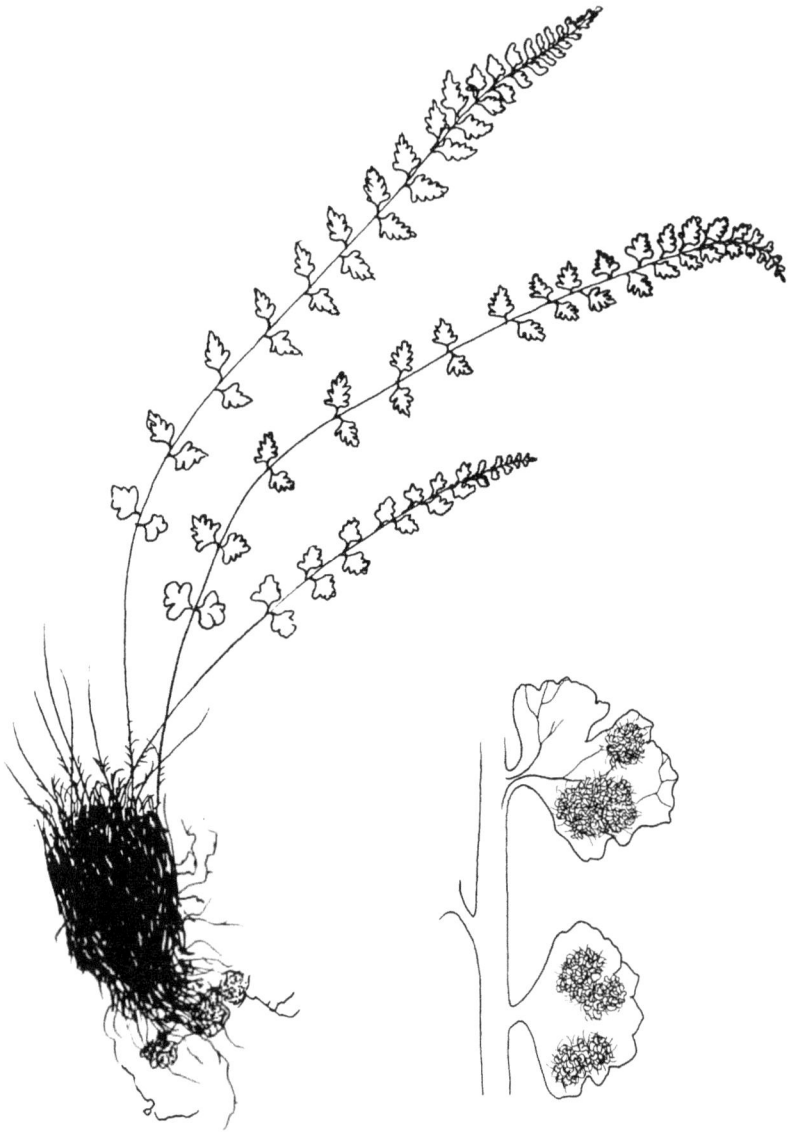

Fig. 93 *Woodsia glabella*; (a) fronds, 1 × ; (b) fertile pinnae, 5 × .

2. **Woodsia ilvensis** (L.) R. Br.
rusty woodsia
Fig. 94 (a) fronds; (b) fertile pinna; (c) sorus. Map 92.

Fronds 5–25 cm long or longer, 2–3 cm wide, oblong-lanceolate, pinnate-pinnatifid to bipinnate; pinnae oblong-lanceolate; margins of the segments crenate and usually somewhat inrolled. Stipes jointed, with the old stipe-bases persistent. Rachis and undersurface of the blade usually brown-chaffy. Sori round, numerous, and close together on the undersurface. Indusia of up to 20 long ciliate-like segments.

The rusty woodsia is noted for being both scaly and glandular. It is one of the most abundant ferns on the cliffs and talus slopes north of Lake Superior. In eastern Canada, it is certainly the species that the amateur is likely to see first in large numbers. The species is so rare in Great Britain that almost every plant known has been tabulated.

Cytology: $n = 41$ (Cody and Mulligan 1982*).

Habitat: Dry, often exposed, usually acid rocks and crevices of cliff faces.

Range: Circumpolar; in North America from Greenland to Alaska, south to North Carolina, Michigan, Illinois, Banff, Alta., and central British Columbia.

Remarks: The species is quite variable in size, form, and degree of chaffiness. At times, plants that grow in the shade look quite unlike those from exposed sites. It is rare in the Yukon (Douglas et al. 1981) and Alberta (Argus and White 1978).

3. **Woodsia alpina** (Bolton) S.F. Gray
W. bellii (Lawson) A.E. Porsild
northern woodsia
Fig. 95 (a) fronds; (b) fertile pinna; (c) sorus. Map 93.

Fronds up to 15 cm long, 0.5–2.5 cm broad, linear to oblong-lanceolate; pinnae suborbicular to oblong or lanceolate, crenate to pinnatifid, flat, glabrous, and with no chaff. Stipes without chaff or somewhat scaly, jointed near the base. Sori separate or confluent and occurring near the margins. Indusia of ciliated plate-like lobes.

Woodsia alpina is considered to be the derived allotetraploid from a cross of *W. glabella* × *ilvensis*. It is a variable species that looks like a more robust *W. glabella* with shining brown to purple-colored stipes and thicker blades. The middle pinnae are 2- to 3-lobed, whereas in *W. ilvensis* they are 3- to 6-lobed.

Cytology: $n = 82$ (Löve and Löve 1976*).

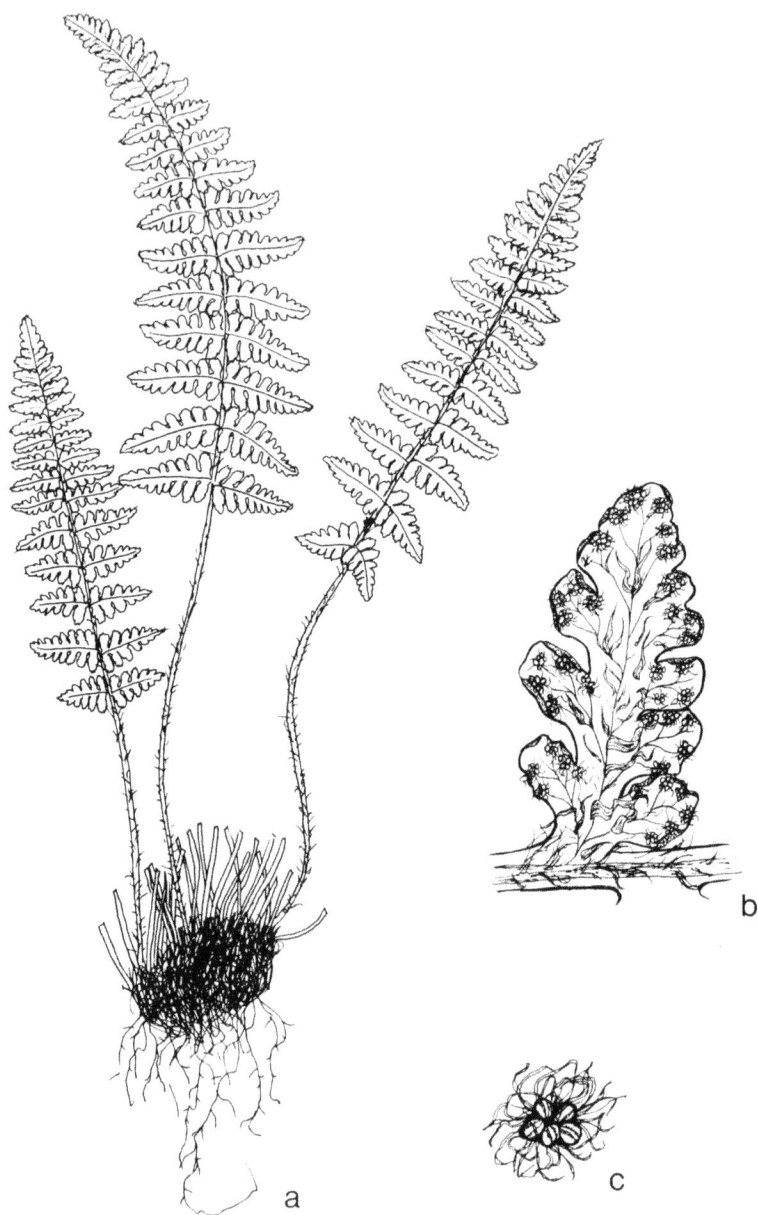

Fig. 94 *Woodsia ilvensis*; (a) fronds, 2/3 × ; (b) fertile pinna, 4 × ; (c) sorus, 10 × .

Aspidiaceae

Fig. 95 *Woodsia alpina*; (a) fronds, 1 ×; (b) fertile pinna, 1 ×; (c) sorus, 10 ×.

Aspidiaceae

Habitat: Rock crevices and rock screes, usually on calcareous or nonacid rocks.

Range: Circumpolar; in North America from Greenland to Alaska, south to New York, Ontario, Michigan, and Minnesota. The known distribution was extended to southern Ontario by Catling (1975).

Remarks: Lawson (1864) described some material from the Gaspé as varietally distinct from *W. alpina* in Scotland. Porsild (1945) raised this variety to specific rank as *W. bellii*. We are following Brown (1964), who followed R.M. Tryon (1948), in not recognizing these plants (which have less chaffy, more delicate, larger fronds) other than to indicate that they seem to be expressions of a less rigorous climate in the south. The situation is analogous to that of var. *remotiuscula* of *Dryopteris fragrans*. *Woodsia alpina* is rare in Manitoba (White and Johnson 1980), Nova Scotia (Maher et al. 1978), and Ontario (Argus and White 1977).

4. **Woodsia oregana** D.C. Eat.
   Oregon woodsia
Fig. 96 (a) fronds; (b) fertile pinna; (c) sorus. Map 94.

Fronds 10–30 cm long or longer, 1.0–3.5 cm wide. Blades linear-lanceolate. Pinnae opposite, remote, triangular-oblong. Pinnules oblong, blunt, with marginal crenulate-serrate teeth often inrolled. Stipes not jointed. Rachis dark brown at the base, becoming straw-colored above, glabrous or somewhat finely glandular, usually without scales. Sori round, medial. Indusia of narrow and threadlike segments.

This western species looks somewhat like *W. ilvensis* but is usually without scales and is a characteristic plant of calcareous rather than more acid substrates. Unlike *W. ilvensis* it belongs to the group with uneven stubble.

Cytology: $n = 38$ (Brown 1964; Cody and Mulligan 1982*). Basic diploid species.

Habitat: More or less protected crevices of calcareous ledges and cliffs.

Range: Gaspé, Que., Ottawa District, Algonquin Park, upper Great Lakes, Alberta and British Columbia, south to Oklahoma and New Mexico.

Aspidiaceae

Fig. 96 *Woodsia oregana*; (a) fronds, 1 × ; (b) fertile pinna, 5 × ; (c) sorus, 10 × .

Remarks: Brown (1964) believed that the tetraploid taxon *cathcartiana* had a very small distribution in only two counties along the St. Croix River in Minnesota and Wisconsin, so that *W. oregana* var. *oregana* extended from British Columbia to the Gaspé. Our interpretation is quite different. We know the material in Canada from Manitoulin Island belongs to the taxon *cathcartiana*. Also, Ontario material is highly glandular and was at one time referred to *W. pusilla* var. *cathcartiana* (T.M.C. Taylor 1947). The few collections from the Prairie Provinces (Cody and Lafontaine 1975) and the material from eastern Canada are in need of further study. We prefer at this time to restrict the name *W. oregana* var. *oregana* to the western diploid, and var. *cathcartiana* (Robins.) Morton to the tetraploid, which is highly glandular. Plants in the Great Lakes region, and presumably those farther east, belong to var. *cathcartiana*, but this should be investigated. *Woodsia oregana* is rare in Manitoba (White and Johnson 1980), Ontario (Argus and White 1977), and Saskatchewan (Maher et al. 1979).

5. **Woodsia scopulina** D.C. Eat.
    W. oregana D.C. Eat. var. *lyallii* (Hook.) Boivin
    W. appalachiana T.M.C. Taylor
    Rocky Mountain woodsia
Fig. 97 (*a*) fronds; (*b*) fertile pinna. Map 95.

Fronds to 40 cm long or longer, 1.5–8 cm wide. Pinnae oblong-lanceolate to ovate; pinnules oblong, denticulate; rachis and blade with scattered white multicellular hairs mixed with the glandular pubescence (which often stains drying papers yellow). Sori round, near the margins. Indusia composed of flat plate-like segments mostly hidden under the sori.
    When it is in suitable sites, the Rocky Mountain woodsia is usually a larger plant than the Oregon woodsia. The nonjointed, shiny, chestnut-colored stipes are characteristic. On the rachis and blade are prominent, white, articulated scales, which are diagnostic.

Cytology: $n = 38$ (R.L. Taylor and Brockman 1966*; Cody and Mulligan 1982*).

Habitat: Among rocks and in crevices of cliffs (usually calcareous).

Range: Gaspé County, Que., Algonquin Park and Thunder Bay District, Ont., western Alberta, British Columbia, southern Yukon and Alaska, south to Arkansas, Tennessee, California, and New Mexico.

Aspidiaceae

Fig. 97 *Woodsia scopulina*; (a) fronds, 1/2 ×; (b) fertile pinnule, 3 ×.

Remarks: T.M.C. Taylor (1947) described the plants in the Appalachians as *W. appalachiana* and noted that the indusial segments were broader and the rhizome scales narrower than those in typical *W. scopulina*. We are following Brown (1964) in not recognizing this species, although we know of no recent comprehensive study that compares the eastern Canadian plants with those in the Appalachians, and in turn with the western plants. Most authorities consider *W. scopulina* to be a very distinctive species (Brown 1964). Boivin (1966) referred it to a variety of *W. oregana*. *Woodsia scopulina* is rare in the Yukon (Douglas et al. 1981), Manitoba (White and Johnson 1980), Ontario (Argus and White 1977), and Saskatchewan (Maher et al. 1979).

6. **Woodsia obtusa** (Spreng.) Torr.
    blunt-lobed woodsia
Fig. 98 (a) fronds; (b) fertile pinna. Map 96.

Fronds 10–30 cm long, 2–10 cm wide. Blades broadly lanceolate, pinnate; pinnae mostly remote; lower pinnae triangular; median and upper pinnae ovate-lanceolate to oblong, pinnatifid, or pinnate at the base. Pinnules oblong, obtuse. Stipes not jointed. Rachis straw-colored, glandular-pubescent. Sori round, near the margins. Indusia covering the sori, later splitting into several jagged lobes.

Woodsia obtusa is an erect rather robust species. In Canada it is highly restricted in distribution (Britton 1977; Lafontaine 1973) and is a talus species rather than a cliff species. In aspect it looks somewhat like *Cystopteris fragilis*, with which it often grows. The stiffer aspect and the glands and scales on the axes and veins are good field characters (see also *Woodsia* compared with *Cystopteris* under comments on the genus *Woodsia*).

Cytology: $n = 76$ (Brown 1964). A tetraploid species. Brown (1964) is uncertain as to its origin, but considers it to be clearly related to *W. oregana*.

Habitat: Shaded ledges and rocky slopes.

Range: Southwestern Quebec and southern Ontario, south to Georgia, Alabama, and Texas.

Remarks: The Ontario habitats seem disturbed and might indicate that the species is a recent arrival. It is rare in Ontario (Argus and White 1977) and certainly rare in Quebec.

Fig. 98 *Woodsia obtusa*; (a) fronds, 1/2 × ; (b) fertile pinna, 4 ×.

Within the group, three species have an even stubble. One might expect three hybrids. The basic cross of the two diploids would be *W. glabella* × *ilvensis* or *W.* × *tryonis* Boivin; it was collected at Silver Islet, Thunder Bay District, Ont. (*Tryon and Faber 4962*) (see R.M. Tryon 1948). This plant was sterile and had the morphology of *W. alpina*, as expected. Unfortunately, no cytology is known for the hybrid combination (one would expect all unpaired chromosomes), and Brown (1964) cited *Tryon and Faber 4962* under *W.* × *gracilis*.

There should also be two backcrosses of *W. alpina* to its two parents. *Woodsia alpina* × *ilvensis* is *W.* × *gracilis* (Lawson) Butters and is reported from Thunder Cape, Thunder Bay District, Ont., and Rivière du Loup, Que., by Brown (1964). *Woodsia alpina* × *glabella* was reported by Soper and Maycock (1963) from Algoma District, Ont. Their cytology is unknown, but they should be sterile triploids.

Hybrids between the even and uneven stubble groups are known. The most often cited hybrid is *W.* × *abbeae* Butters, which was considered to be one of the following: *W. ilvensis* × *scopulina* by R.M. Tryon (1948); possibly *W. ilvensis* × *oregana* by Hagenah (1963); and more specifically, *W. ilvensis* × *oregana* var. *cathcartiana* by W.H. Wagner and F.S. Wagner (unpublished). It was studied cytologically and was said to be triploid. We consider *W. confusa* Taylor and *W. oregana* var. *squammosa* Boivin to be synonyms of *W.* × *abbeae*.

The type of *W.* × *maxonii* Tryon (*W. oregana* × *scopulina*) Tryon (1948) was collected on Sleeping Giant, Thunder Bay District, Ont.

Woodsia hybrids are in need of further study with the use of modern experimental methods.

4. *Polystichum* Roth

Ferns rather large, tufted, evergreen, and leather-textured, with usually scaly stipes, arising from short stout chaffy rhizomes. Sori round. Indusia round, attached at the center.

The four species and all their hybrids known in Europe have been extensively analyzed cytogenetically (Lovis 1977). The chromosome numbers of over 75 species in the world are known, which is perhaps a little over half the total number of species. The genus is a large one, almost as large as *Dryopteris*, and is quite complex in western North America. In fact, one can make an analogy between the complexities of *Dryopteris* in eastern North America and the relative dearth of species in the west, with the reverse situation in *Polystichum*. In each case, derived tetraploids are known, and in each case, there is even a hexaploid. Indeed, in *Polystichum* outside Canada, octaploids are known (Lovis 1977). Another parallel for the amateur is the recognition here of a large number of biological species that have been delineated from the cytogenetic analyses. Our present understanding of the western *Polystichum* species is presented by

D.H. Wagner (1979) and will be briefly outlined. After extensive studies, Wagner recognized *P. imbricans* as an important basic species rather than as a variety of *P. munitum*. Accordingly, for the smaller talus and cliff species we have the phylogenetic schemes shown in Diagram 3.

There are still some questions regarding these schemes. *Polystichum braunii* and *P. andersonii* are often in the same ecological niche in British Columbia, and although D.H. Wagner (1979) is convinced that they are quite distinct in morphology and have entirely separate origins, others find them quite difficult to identify easily. D.H. Wagner (1979) also believes that he has a good lead from an old herbarium specimen to find the ancestral diploid species that we have called species W. Yet another problem is the relationship of *P. lemmonii* to *P. mohrioides*, although again D.H. Wagner (1979) is quite convinced that they are separate entities. The latter species has an amazing range, right down the Andes chain to southern South America.

Accordingly, there are 10 species in western Canada of which three, *P. californicum*, *P. lemmonii*, and *P. scopulinum*, are of rare to very rare status. *Polystichum kruckebergii* and *P. setigerum* are of local occurrence in British Columbia, so that it was quite possible for the junior author to spend 6 months in British Columbia traveling around Vancouver and Victoria and see only *P. munitum* and *P. imbricans* of the 10 species recorded for the province.

In eastern Canada there are only four species: *P. braunii*, *P. lonchitis*, *P. scopulinum*, and *P. acrostichoides*. The most common and familiar of these is *P. acrostichoides*, a characteristic plant of maple and beech woods. It is a basic diploid species that as yet has not been implicated as an ancestral diploid species in the evolution of the polyploids.

The genus *Polystichum* is noted for hybridization, and so if one is in a region where there are a number of species growing together, there is a good opportunity for interspecific hybrids to be present. W.H. Wagner (1973*b*) studied a number of hybrids at one locality in Washington.

A. Sori borne on reduced upper pinnae . . . . 1. **P. acrostichoides**
A. Sori borne on the backs of unmodified pinnae.
    B. Fronds pinnate; pinnae entire, denticulate, or serrate, spinulose.
        C. Pinnae mostly oblong-lanceolate, progressively reduced towards the base, with the lowest pinnae subtriangular to broadly trowel-shaped symmetrical; spinulose tips of teeth of pinnae spreading . . . .
. . . . 2. **P. lonchitis**
        C. Pinnae linear-attenuate, not much reduced below, with the lowest pinnae ovate to lanceolate-falcate, auriculate, asymmetrical; spinulose tips of teeth of pinnae incurved.

D. Stipe and rachis persistently chaffy; pinnae acuminate, cuneate at the base; indusium ciliate .... 7. *P. munitum*
D. Stipe and rachis often naked; pinnae cuspidate or apiculate, oblique at the base; indusium entire to sharply toothed .... 4. *P. imbricans*
B. Fronds bipinnatifid or bipinnate; pinnae spinulose or not.
E. Pinnae not at all spinulose .... 3. *P. lemmonii*
E. Pinnae apiculate to spinulose.
F. Fronds bipinnate; pinnules distinct, sessile or petiolate .... 10. *P. braunii*
F. Fronds bipinnatifid; pinnules adnate to the costa.
G. Pinnae with conspicuous filiform scales on the lower surface.
H. Fronds with a proliferous bud on the rachis about one-third of the way down from the tip .... 8. *P. andersonii*
H. Fronds lacking a proliferous bud.
I. Pinnae incised to the costa; pinnules slightly toothed ....
.... 11. *P. setigerum*
I. Pinnae not incised to the costa ..
.... 9. *P. californicum*
G. Pinnae lacking filiform scales.
J. Pinnae acute at the apex; teeth coarse, spreading .... 5. *P. kruckebergii*
J. Pinnae obtuse at the apex, occasionally cuspidate; teeth fine, incurved .... 6. *P. scopulinum*

1. *Polystichum acrostichoides* (Michx.) Schott
Christmas fern
Fig. 99 (*a*) sterile and fertile fronds; (*b*) portion of fertile pinna with immature sori; (*c*) portion of fertile pinna with mature sori. Map 97.

Fronds 35–65 cm long. Stipes and rachis chaffy. Blades lanceolate, 7–12 cm wide or wider, simply pinnate; pinnae oblong to lanceolate, acute or sometimes bluntish at the tip, auricled at the base on the upper side; margins serrulate-bristly. Sori borne on reduced upper pinnae distinct or more often confluent.
Forma *incisum* (Gray) Gilbert has the pinnae coarsely toothed, the fertile pinnae usually less reduced, and the sori usually less confluent.
This medium-sized species with dark green, subevergreen fronds is sufficiently striking to be soon familiar to all amateurs walking in maple-beech woods in southeastern Canada. The first impression is a

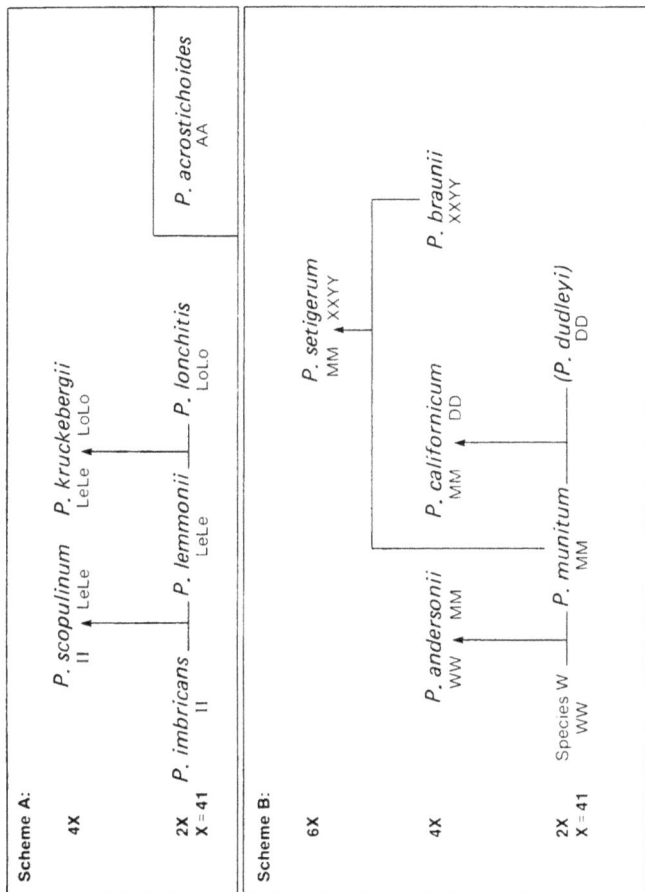

Scheme A:

4X

P. scopulinum P. kruckebergii
II   LeLe LoLo

2X
X = 41

P. imbricans ____ P. lemmonii ____ P. lonchitis
II   LeLe   LoLo

P. acrostichoides
AA

Scheme B:

6X

P. setigerum
MM XXYY

P. braunii
XXYY

4X

P. andersonii   P. californicum
WW MM   MM DD

2X
X = 41

Species W ____ P. munitum ____ (P. dudleyi)
WW   MM   DD

Note: Although genomes are shown in this diagram for the derived alloploids. Lovis (1977) believes that basic *Polystichum* species exhibit some pairing of chromosomes that indicate segmental allopolyploidy rather than genomic allopolyploidy. The letters used above are therefore a simplification of the system

Diagram 3 Evolutionary schemes for *Polystichum*, after W.H. Wagner (1973b) and D.H. Wagner (1979).

Aspidiaceae 185

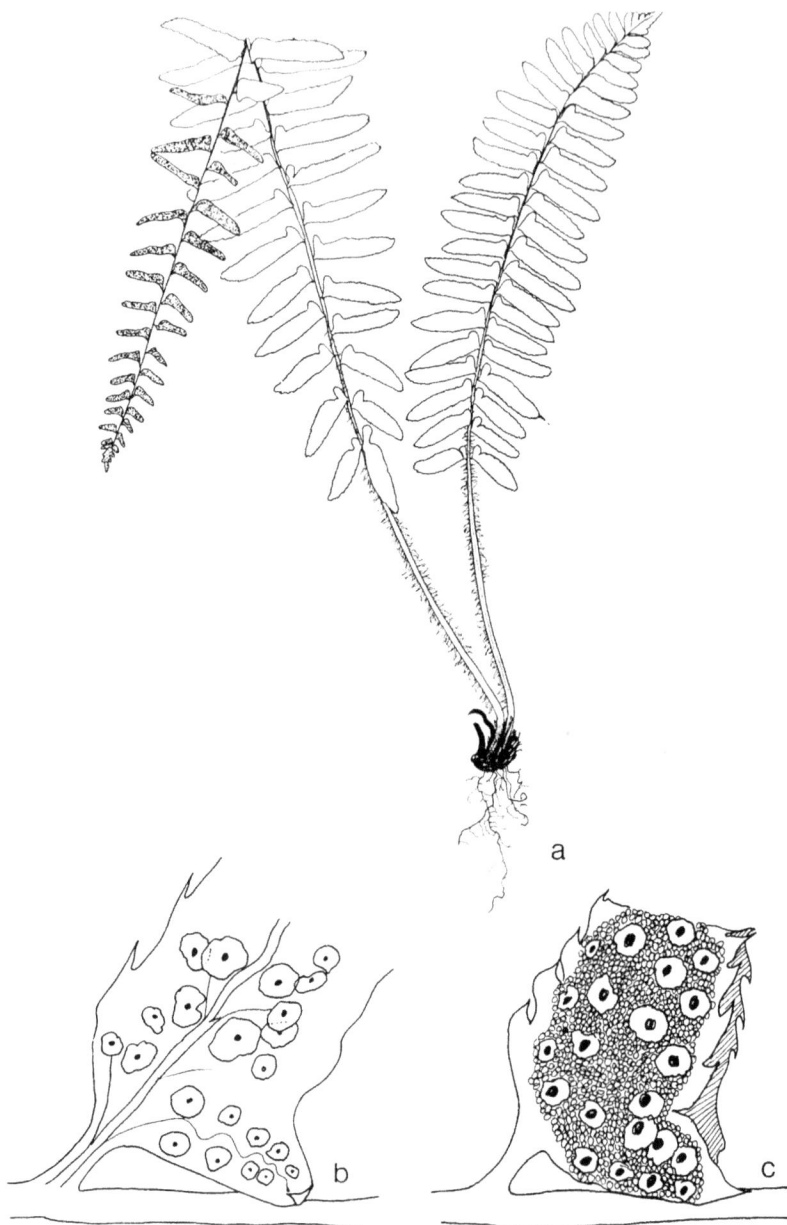

Fig. 99 *Polystichum acrostichoides*; (a) sterile and fertile fronds, 1/3 ×; (b) portion of fertile pinna with immature sori, 5 ×; (c) portion of fertile pinna with mature sori, 5 ×.

Aspidiaceae

darker green, tougher Boston fern that grows scattered about in open woods. The Boston fern is a tropical genus, *Nephrolepis*, but there is a superficial similarity in aspect.

Cytology: $n = 41$ (Britton 1953*; Cody and Mulligan 1982*).

Habitat: Rich woods and humus-rich rocky slopes.

Range: Nova Scotia to southern Ontario, south to northern Florida and eastern Texas. The forma *incisum* may be found through the range of the species.

Remarks: There are many described forms (Weatherby 1936 describes seven of them) that are considered to be ecological variants or, in some cases, mutations. The observation of variation in the leaf form of the species is an interesting hobby for the amateur if a limited number of fern species are available near at hand.

2.   **Polystichum lonchitis** (L.) Roth
     Aspidium lonchitis (L.) Sw.
     Dryopteris lonchitis (L.) O. Kuntze
     Holly fern
Fig. 100 (*a*) fertile frond; (*b*) pinna with mature sori; (*c*) sorus; (*d*) undersurface of sterile pinna. Map 98.

   Fronds 10–60 cm long. Stipes very short, chaffy. Blades linear to narrowly linear-oblanceolate, acuminate, tapering to the base. Middle and upper pinnae oblong-lanceolate, falcate; bases of pinnae auriculate above and cuneate below; basal pinnae deltoid, often very small, equilateral, serrate-dentate; teeth spreading-spinulose. Sori round, in two rows, occurring midway between the midvein and the margin. Indusium entire.
   Field characters are the medium to small size of the plants, the short pinnae, the extremely short stipe, and the pinnae towards the base reduced to small triangular auricles. The species is aptly called the holly fern because of its shiny, lustrous green, tough blades.

Cytology: $n = 41$ (Britton 1964; Cody and Mulligan 1982*, eastern Canada; R.L. Taylor and Mulligan 1968*, western Canada).

Habitat: Limestone cliffs, moist rocky slopes, talus slopes, and occasionally coniferous woods.

Range: Circumpolar; in North America in Greenland, western Newfoundland, Cape Breton Island, Gaspé Peninsula, central Quebec–Labrador (Waterway and Lei 1982), Bruce Peninsula and

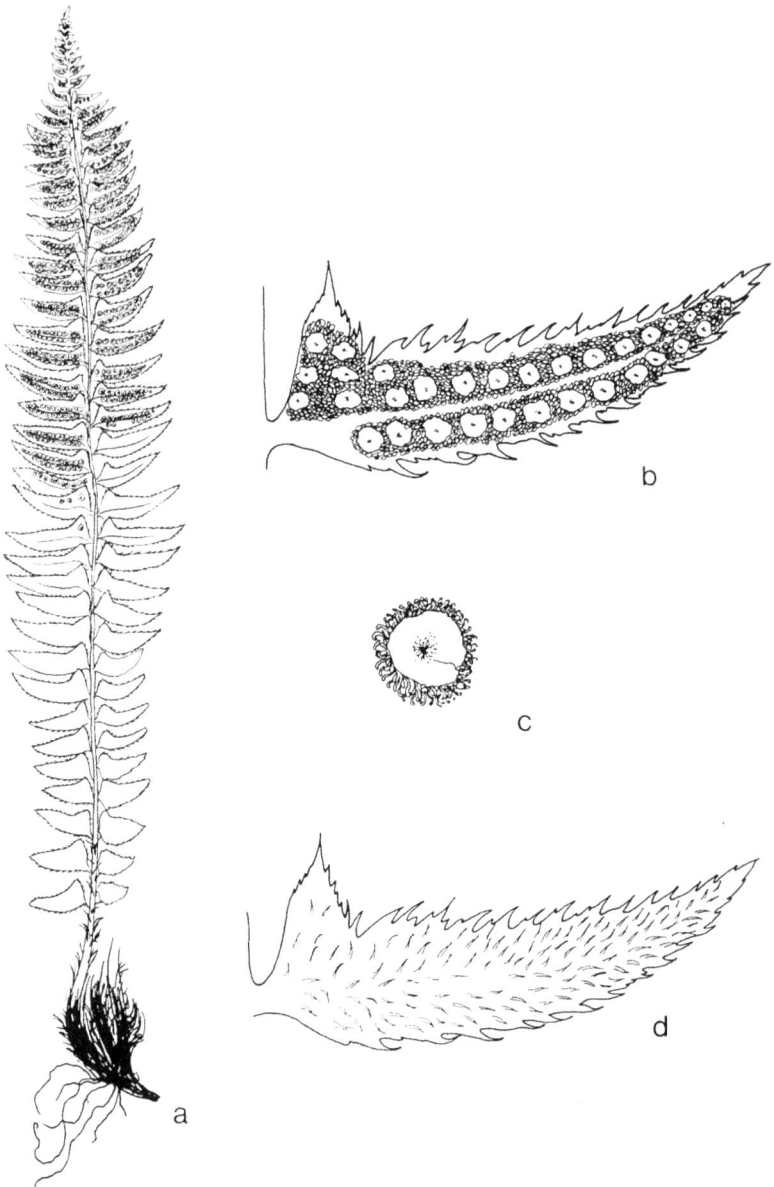

Fig. 100 *Polystichum lonchitis*; (*a*) fertile frond, 1/2 ×; (*b*) pinna with mature sori, 4 ×; (*c*) sorus, 15 ×; (*d*) lower surface of sterile pinna, 4 ×.

Algoma District, Ont., Keweenaw County, Mich., southwestern Alberta (Brunton 1978), British Columbia, Yukon Territory to Kenai Peninsula, Alaska, south in the western United States to Colorado, Utah, and southern California.

Remarks: This characteristic fern of cool, northern, limestone habitats has a broad distribution. In Ontario it is often a companion plant for Hart's-tongue (Soper 1954). The species is rare in the Yukon (Douglas et al. 1981), Nova Scotia (Maher et al. 1978), and Ontario (Argus and White 1977).

3. **Polystichum lemmonii** Underw.
   P. mohrioides (Bory) Presl var. *lemmonii* (Underw.) Fern.
   P. mohrioides auth. non (Bory) Presl
   Fig. 101 (a) frond; (b) fertile pinna. Map 99.

Fronds 15–35 cm long or longer, densely clustered from a short ascending rhizome. Stipes glandular, puberulent, and very chaffy at the base. Blades linear to narrowly lance-oblong; pinnae deeply pinnatifid or the lower pinnae pinnate; ultimate segments oval, obtuse, crenate, or crenately lobed; lobes lacking mucronate or spinulose tips. Sori on the middle and upper pinnae, towards the base of the pinnules. Indusia large, entire, or obscurely erose-toothed.

This species is confined to ultramafic rocks (Kruckeberg 1964). In comparison with *P. imbricans*, it is a small species, with fronds less than 30 cm long. The pinnules are rounded and overlapping and lack spines. Both the upper and lower epidermis have unicellular glands.

Cytology: $n = 41$ (W.H. Wagner 1973b). This is a basic ancestral species with genomes LeLe.

Habitat: Open serpentine and asbestos subalpine slopes.

Range: In Canada known only in the Okanagan Divide in southern British Columbia; in the United States, from northern Washington to northern California.

Remarks: This is a rare species in all of western North America and the distribution barely extends into Canada. We are following D.H. Wagner (1979) in considering this species distinct from *P. mohrioides*. The latter is known from the Andes south to Chile.

4. **Polystichum imbricans** (D.C. Eat.) D.H. Wagner
   P. munitum (Kaulf.) Presl var. *imbricans* (D.C. Eat.) Maxon
   Fig. 102 (a) frond; (b) fertile pinna. Map 100.

Fig. 101 *Polystichum lemmonii*; (*a*) frond, 1/2 × ; (*b*) fertile pinna, 2 × .

Aspidiaceae

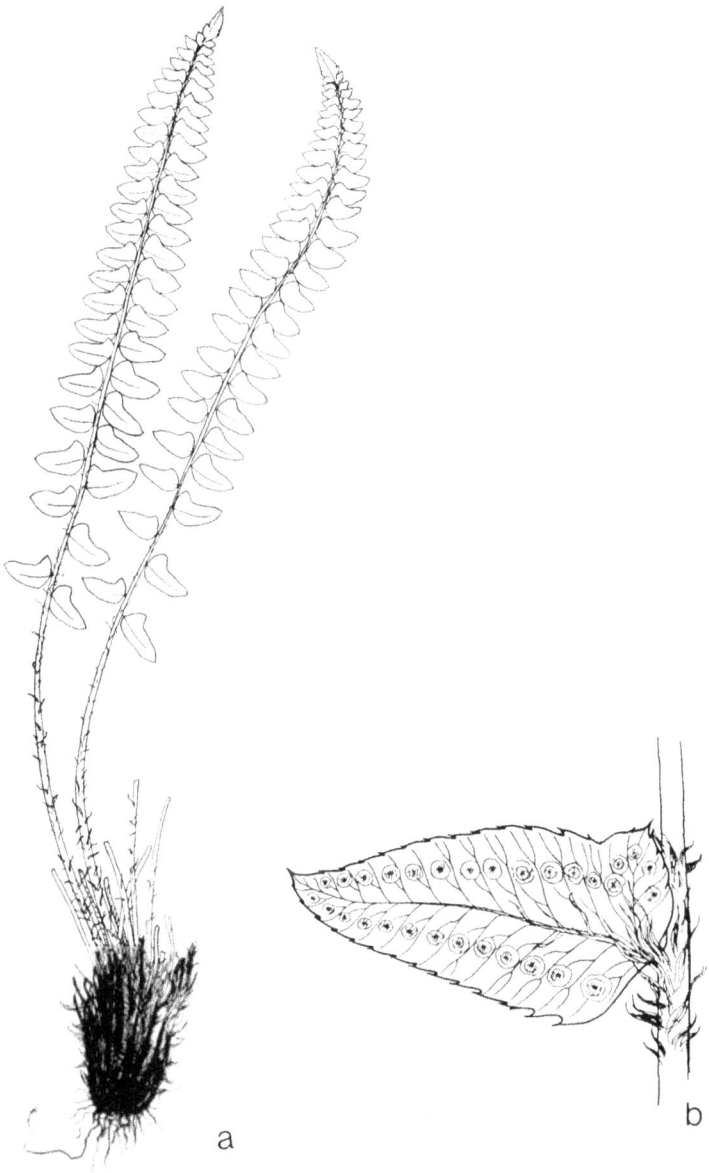

Fig. 102 *Polystichum imbricans*; (a) frond, 1/3 × ; (b) fertile pinna, 3 × .

Fronds similar to *P. munitum*, linear-lanceolate, but shorter, up to 60 cm long. Stipes less chaffy than *P. munitum*, with the upper part and rachis smooth or nearly so. Pinnae usually overlapping and folded inward, lanceolate, 2–4 cm long or longer, auriculate above, abruptly tapering to the spinulose tip; teeth incurved, spinulose. Sori midway between the margin and midvein. Indusia entire.

Plants are generally smaller than those of *P. munitum* and have crowded ascending pinnae, stiffly erect habit, flat pinnae or pinnae slightly cupped on the upper side; the upper surface of the pinna is perpendicular to the rachis. D.H. Wagner (1979) has studied *P. imbricans* extensively and recommends using a combination of characters for absolute identification. He lists eight characters in order of reliability for exact determination.

Cytology: $n = 41$ (Taylor and Lang 1963*). This is a basic ancestral species with genomes II.

Habitat: Usually in the open in rock crevices, clearings, and dry rocky coniferous woods.

Range: Southern coastal British Columbia, south to southern California.

Remarks: In making determinations, make sure you use technical characters, because plants growing in shady, moist situations superficially mimic *P. munitum*. Conversely, *P. munitum* in drier, more exposed locations tends to look like *P. imbricans*.

5. **Polystichum kruckebergii** W.H. Wagner
   Kruckeberg's holly fern
Fig. 103(*a*) fronds; (*b*) fertile pinna. Map 101.

Fronds up to 30 cm long, few together, tufted from a small stout erect rhizome. Stipe short, scaly. Blade linear-lanceolate; pinnae overlapping, ovate-triangular, conspicuously spreading, toothed; teeth tips cartilaginous; larger pinnae frequently with one or more pairs of basal pinnules. Sori borne in two rows on the backs of the pinnae on the upper half of the frond, becoming confluent. Indusia with entire wavy margins.

Polystichum kruckebergii was described by W.H. Wagner (1966*a*), and the type chosen was a collection from near Lillooet in southwestern British Columbia. W.H. Wagner (1966*a*) carefully delineates it from *P. scopulinum*, noting that the former is usually smaller, with shorter and less oblong pinnae; in a median pinna the number of teeth per side is approximately 6 rather than 12 (8–25), as in *P. scopulinum*. The margins of *P. kruckebergii* are markedly bristly, and the pinna tips are more pointed.

Fig. 103 *Polystichum kruckebergii*; (a) fronds, 1/2 ×; (b) fertile pinna, 10 ×.

Cytology: $n = 82$ (W.H. Wagner 1973*b*). Derived tetraploid LoLoLeLe from *P. lonchitis* and *P. lemmonii* (W.H. Wagner 1973*b*).

Habitat: Subalpine cliffs and talus slopes.

Range: Central British Columbia (A.L. Kruckeberg 1982) to northern California, east to Idaho and Utah.

Remarks: This is a rare species that should be looked for on ultramafic rocks. Its known distribution in Canada is very limited.

6. **Polystichum scopulinum** (D.C. Eat.) Maxon
   P. mohrioides (Bory) Presl var. *scopulinum* (D.C. Eat.) Fern.
   Aspidium aculeatum (L.) Roth var. *scopulinum* D.C. Eat.
   crag holly fern
Fig. 104 (*a*) fronds; (*b*) fertile pinna. Map 102.

Fronds 15–40 cm long, densely tufted from short erect or decumbent scaly rhizomes. Stipes densely chaffy at the base; scales sparse and deciduous above. Blades narrowly lanceolate, slightly tapered to the base and tip; pinnae usually folded inwards and upwards, deltoid-ovate to deltoid-oblong, pinnately lobed, especially towards the base; teeth with a cartilaginous tip. Sori borne on the middle and upper pinnae, in two median rows. Indusia thin, erose-dentate.

The species is intermediate between *P. imbricans* and *P. lemmonii*. Characteristics from the former include the folding inward and upward of the pinnae and the leathery to fleshy texture of the blade. The pinnae are oblong and have about 12 short but distinct cartilaginous teeth per side.

Cytology: $n = 82$ (W.H. Wagner 1973*b*). A derived tetraploid of constitution II LeLe from *P. imbricans* and *P. lemmonii* (D.H. Wagner 1979), not *P. munitum* and *P. mohrioides* (W.H. Wagner 1973*b*).

Habitat: Crevices of cliffs and rocky slopes, often of ultramafic, or at least basic, rocks.

Range: In western North America from southernmost British Columbia to southern California, Idaho, and Utah; disjunct in the Gaspé Peninsula, Que., and western Newfoundland.

Remarks: *Polystichum scopulinum* has a much wider distribution than its presumed parents. The disjunct station on Mont-Albert, Que., has been a noted topic through the years. The species should be looked for when on serpentine rocks.

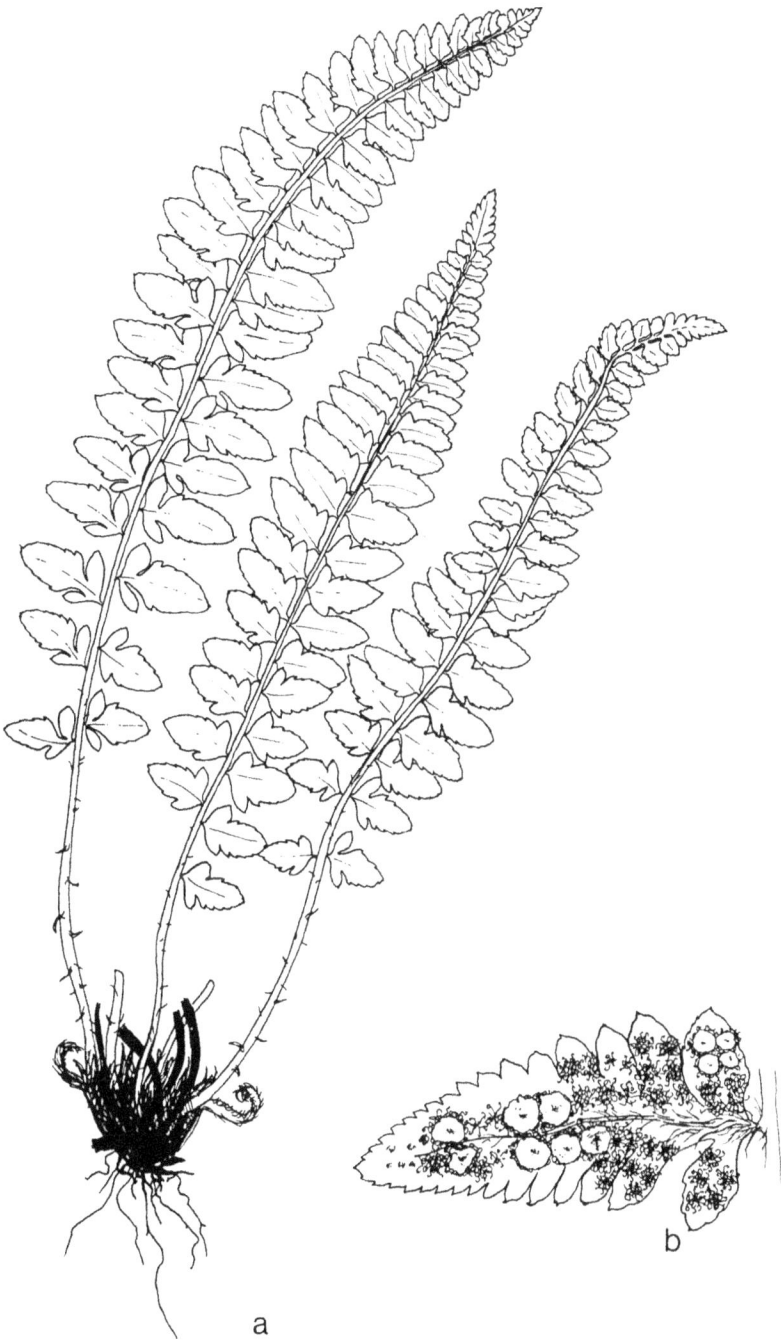

Fig. 104 *Polystichum scopulinum*; (a) fronds, 2/3 × ; (b) fertile pinna, 2 1/2 ×.

Aspidiaceae

7.   *Polystichum munitum* (Kaulf.) Presl
     sword fern
Fig. 105 (a) frond; (b) fertile pinna. Map 103.

Fronds 20–150 cm long, forming a stiffly erect crown at the stout woody scaly rhizome. Stipes densely chaffy. Blade linear-lanceolate, short acuminate, pinnate; pinnae linear-attenuate, auriculate at the base above, cuneate below, sharply serrate. Sori large, situated midway between the margin and the midvein. Indusium fimbriate-margined.

Large plants (over 1 m) present no identification problems. The fronds are once pinnate and the pinnae are sharply serrate. The sharply serrate pinnae are a variable feature, and some plants have deeply serrate or even incised pinnae (Calder and Taylor 1968). There is no difficulty in seeing impressive colonies of this species in easily accessible localities in the coastal forests of British Columbia, e.g. Stanley Park, Vancouver, and Pacific Rim National Park. Subalpine plants superficially resemble *P. lonchitis*.

Cytology: $n = 41$ (W.H. Wagner 1973b; Cody and Mulligan 1982*). This is an important basic ancestral diploid species MM.

Habitat: Moist coniferous woods and shaded slopes; particularly common along roadside clearings in southern coastal British Columbia.

Range: Alaskan Panhandle, south near the coast to Baja California, Mexico; inland in southern British Columbia and to northern Idaho and northwestern Montana.

Remarks: This species is a striking component of the western coastal forests.

8.   *Polystichum andersonii* Hopkins
     P. braunii (Spenner) Fée ssp. *andersonii* (Hopkins) Calder
          & Taylor
     Anderson's holly fern
Fig. 106 (a) frond; (b) fertile pinna. Map 104.

Fronds to 1 m long or longer, usually with a proliferous scaly bud near the apex. Stipes about one-fifth the length of the frond, persistently chaffy. Blades lanceolate, narrowed towards the base; pinnae oblong-lanceolate; lowermost pinnae subtriangular, deeply cut to the costa, but with segments rarely undercut; segments with spinulose teeth; both surfaces with filiform scales; rachis and costa with broader scales. Sori 1–8 on the lateral segments of the middle and upper pinnae. Indusia erose-dentate; teeth gland-tipped.

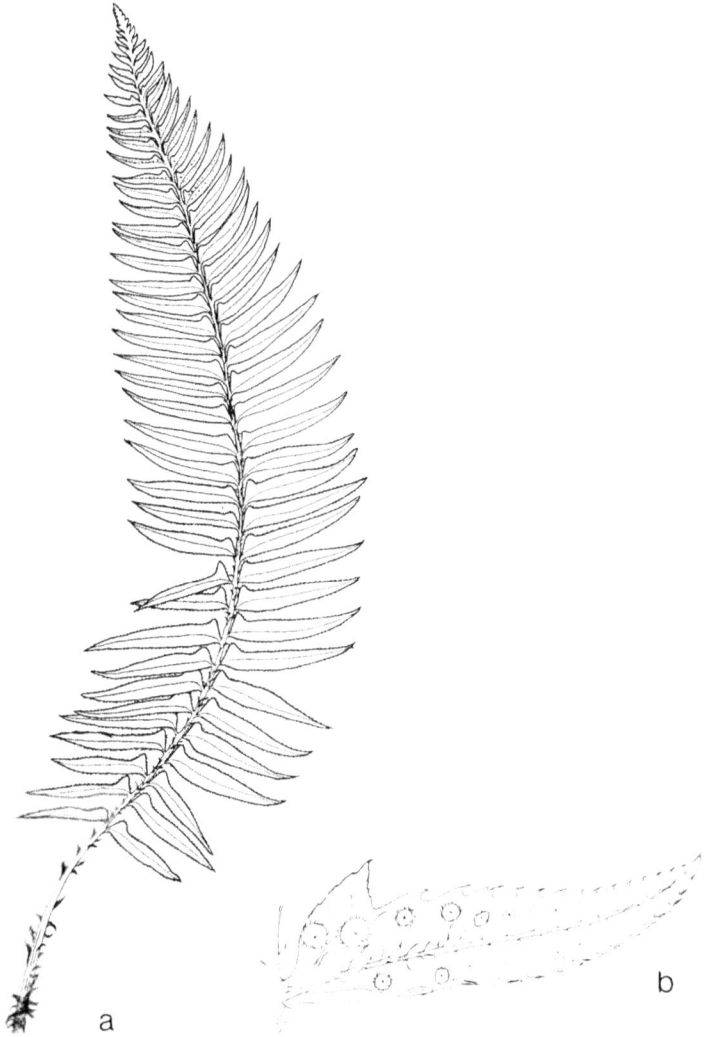

Fig. 105   *Polystichum munitum*; (a) frond, 1/2 ×; (b) fertile pinna, 3 × .

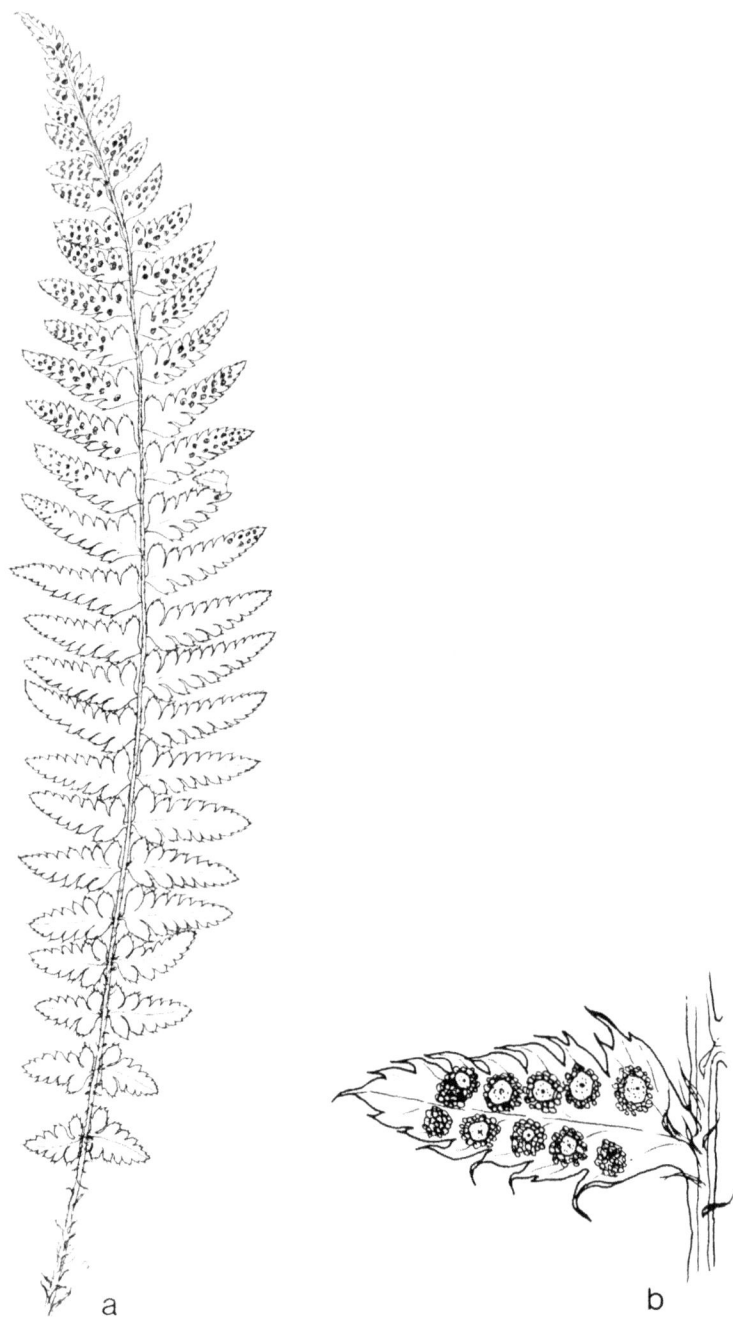

Fig. 106 *Polystichum andersonii;* (a) frond, 1/2 × ; (b) fertile pinna, 3 × .

Aspidiaceae

This species has often been referred to subspecific or varietal status under *P. braunii*. Taylor (1970) suggests that it may prove to be "only a geographical variant."

Polystichum andersonii is identified by the presence of one or more proliferous buds on the rachis, and the basal distal or upper pinnules on the pinnae are longer than the adjacent ones. D.H. Wagner (1979) considers the species to be quite distinct from *P. braunii* and postulates that it is an allotetraploid, with one parent being *P. munitum* and the other an undescribed species to which we have referred as Species W in Scheme B. W.H. Wagner (1973*b*) postulated quite a different origin for *P. andersonii*, saying it seemed to be too far removed from *P. munitum* to have that species as an ancestor.

Cytology: $n = 82$ (Taylor and Lang 1963*). Postulated genomes MMWW.

Habitat: Moist woods and shaded rocky slopes in the mountains.

Range: Alaskan Panhandle, south to Oregon, Idaho, and Montana.

Remarks: An analysis that compares hybrids of *P. andersonii* with those of *P. braunii*, as well as with those of other species, would clarify the origin of *P. andersonii*. The origin of the species, as presented by D.H. Wagner (1979), is still hypothetical.

9. **Polystichum californicum** (D.C. Eat.) Diels
   Aspidium californicum D.C. Eat.
   California holly fern
Fig. 107 (*a*) frond; (*b*) fertile pinna. Map 105.

Fronds 40–75 cm long from the erect rhizome. Stipes about one-third the length of the frond and chaffy, especially towards the base; upper part of stipe becoming naked. Blades linear-lanceolate to lanceolate, little narrowed towards the base; pinnae deeply cut, with segments often slightly undercut, overlapping, and toothed in the upper part; teeth with short ascending or incurved spinulose tips and with filiform hairs below and along the costa above. Basal pinnule on the upper side of the pinnae usually somewhat enlarged. Sori in two rows on the segments of the middle and upper pinnae. Indusia large, ciliate.

Scheme B of Diagram 3 suggests that *P. munitum* crossed with *P. dudleyi* (not in Canada), and the hybrid gave rise to *P. californicum* (W.H. Wagner 1973*b*). D.H. Wagner (1979) postulated that *P. dudleyi* might have crossed with *P. imbricans* to give rise to another allotetraploid now included in *P. californicum*. He rejected the second hypothesis at that time because all the *P. californicum* plants that he studied were uniform with the chemical analysis that he used.

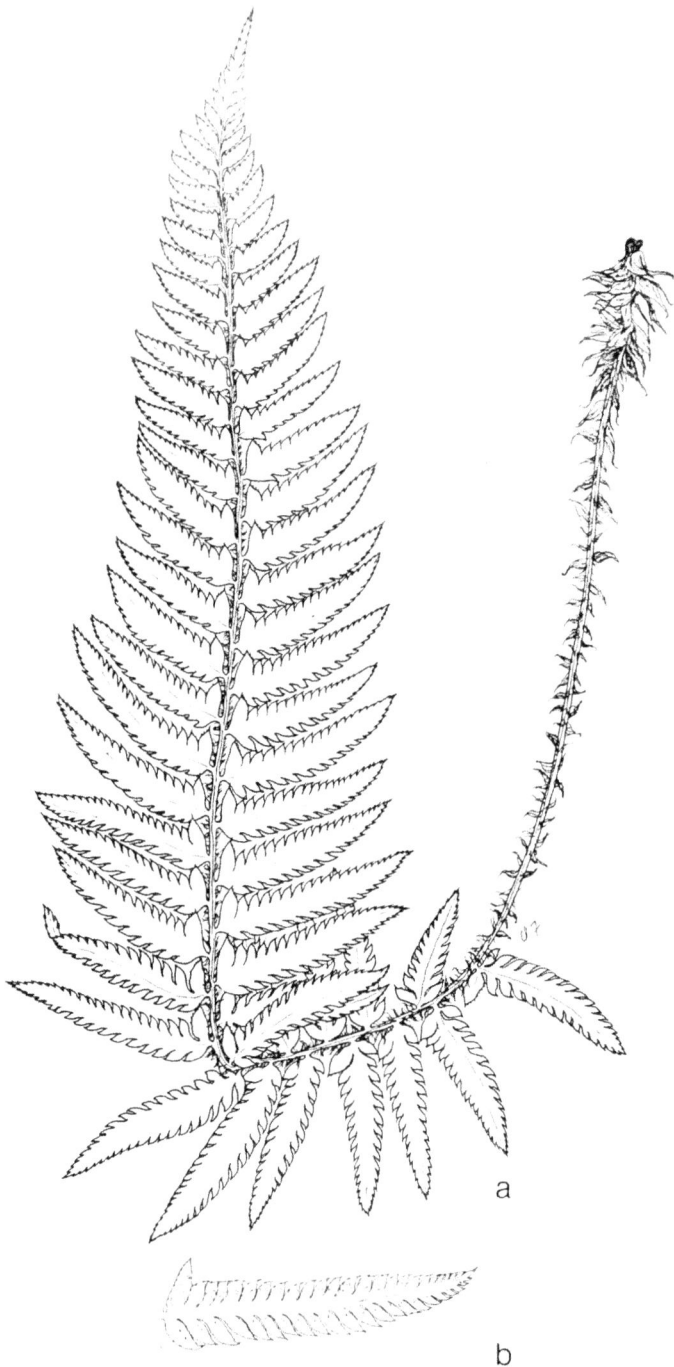

Fig. 107   *Polystichum californicum*; (a) frond, 1/2 × ; (b) fertile pinna, 1 1/2 × .

Aspidiaceae

Recognition of the smaller, northern forms of this species presents great difficulty. If a colony of plants can be found that key out to this species, it will be necessary to use all the methods we now have available to analyze the plants (see remarks that follow).

Cytology: $n = 82$ (W.H. Wagner 1973*b*). Derived allotetraploid MMDD.

Habitat: Lowland coastal forests of Canada.

Range: Known in Canada only in Texada Island, B.C.; in the United States south in the mountains to central California.

Remarks: This species was not included in Canada's flora by T.M.C. Taylor (1970). Identification problems are certainly apparent when D.H. Wagner (1979) refers to "recent misidentifications of northern populations from B.C., Washington and California of this species as *P. scopulinum*." Further study is required. Inclusion of this species in the Canadian flora rests on one old specimen from Texada Island — *Anderson 666* (V).

10. **Polystichum braunii** (Spenner) Fée
    P. braunii (Spenner) Fée var. *purshii* Fern.
    Braun's holly fern
    Fig. 108 (*a*) frond; (*b*) portion of fertile pinnule; (*c*) sorus. Map 106.

Fronds to 1 m long forming a crown at the end of the stout ascending rhizome. Stipe about one-sixth the length of the frond, persistently chaffy. Blades broadly lanceolate, narrowed at the base; rachis with persistent dense chaff; pinnae slenderly lanceolate; middle and upper pinnae gradually tapering, with the lower straight-sided and abruptly tapering to the apex; pinnae generally once pinnate; pinnules petiolate or rarely slightly decurrent, narrowly ovate to trapezoid-oblong, obtuse, nearly rectangular at the base and slightly auricled on the upper side, sharply serrate with incurved bristle-tipped teeth. Sori in two rows near the midrib. Indusia often erose.

This large and handsome species, with fully bipinnate blades and bristle-tipped teeth, is readily identified, provided it can be distinguished from *P. andersonii* and *P. setigerum*. The lack of proliferous buds and of enlarged proximal pinnules on the basal pinnae is useful in this regard.

Cytology: $n = 82$ (Taylor and Lang 1963; R.L. Taylor and Mulligan 1968*, western Canada; Cody and Mulligan 1982,* eastern Canada). The ancestral genomes of the European plants have not been identified. Lovis (1977) considers it to be a segmental allotetraploid

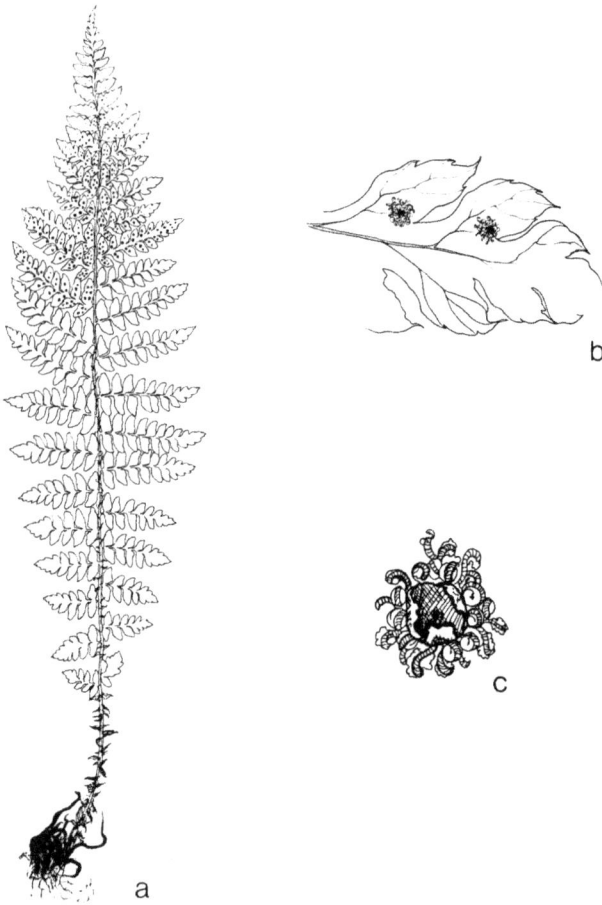

Fig. 108 *Polystichum braunii*; (a) frond, 1/3 × ; (b) portion of fertile pinnule, 4 × ; (c) sorus, 20 × .

Aspidiaceae

and gives it the formula BBBB (the origin of B is obscure, but it is presumably an ancestral diploid, *P. braunii*). We have tentatively designated it as XXYY, although the origins of X and Y are obscure.

Habitat: Rich woods and shaded talus slopes.

Range: Circumpolar; in North America from Labrador and Newfoundland to Thunder Bay District, Ont., south to Pennsylvania and Michigan; western British Columbia north to the Kenai Peninsula and Kodiak Island, Alaska.

Remarks: We are in agreement with D.H. Wagner (1979) that recognition of a var. or ssp. *purshii*, based on plants with an increased proportion of broad to filiform laminar scales (Calder and Taylor 1968a), serves no useful purpose and should be dropped. It is not possible to designate the plants in western Canada as either all var. *braunii* or as all var. *purshii*. *Polystichum braunii* is rare in Ontario. (Argus and White 1977).

11. **Polystichum setigerum** (Presl) Presl
   P. alaskense Maxon
   P. braunii (Spenner) Fée ssp. *alaskense* (Maxon) Calder & Taylor
   Alaskan holly fern
Fig. 109 (*a*) frond; (*b*) portion of fertile pinna. Map 107.

Fronds 1 m long or longer. Stipes about one-fifth the length of the frond, persistently chaffy. Blades lanceolate, narrowed at the base; lower pinnae often deflexed; pinnae oblong-lanceolate, pinnatifid, with segments undercut; teeth bristle-tipped; pinnae scaly on both surfaces, with filiform scales and broader scales on the rachis and the costa. Sori in two rows on the pinnules of the upper half of the frond. Indusia ciliate.
   Polystichum setigerum is still another entity in Scheme B that looks very like *P. braunii* and *P. andersonii* – so much so, that Calder and Taylor (1968a) treated it as ssp. *alaskense* (Maxon) Calder & Taylor. Because it is a derived hexaploid, it is now treated as a full species, *P. setigerum*, based on a plant collected by Thaddaeus Haenke in 1791 at Nootka Sound, B.C. Maxon in 1918 named this species *P. alaskense*. D.H. Wagner (1979) states that the best features for determining this species are the lack of a proliferous bud (a *P. andersonii* feature), a degree of cutting or incision of the pinnae (similar to that of *P. braunii*), and the enlarged proximal pinnules on the basal pinnae (as in *P. andersonii*).

Cytology: $n = 123$ (D.H. Wagner 1979*). Perhaps derived from *P. munitum* and *P. braunii*.

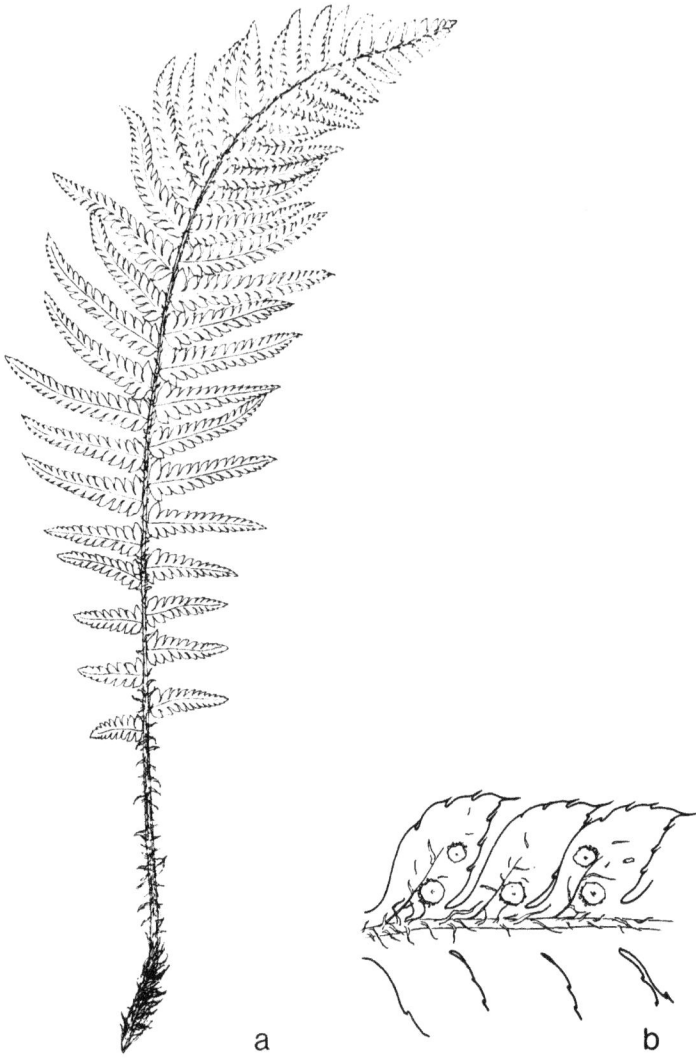

Fig. 109 *Polystichum setigerum*; (a) frond, 1/3 × ; (b) portion of fertile pinna, 3 × .

Aspidiaceae

Habitat: Lowland coastal forests in dense woods and on shaded rocky slopes; at times growing with *P. andersonii* and *P. braunii*.

Range: Alaskan Panhandle, south to southern British Columbia; disjunct to Attu Island in the Aleutian Islands.

Remarks: It is unlikely that the last word has been written on all the confusing species in Scheme B of Diagram 3. D.H. Wagner (1979) admits that *P. lonchitis* rather than *P. munitum* might be a possible parent and that *P. andersonii* is also a possibility rather than *P. braunii*. Only by an extensive cytogenetic analysis of hybrids, with the use of the modern arsenal of SEM and chemistry, will we be able to solve this problem.

Hybrids of *Polystichum*

Polystichum × *hagenahii* Cody (*P. acrostichoides* × *lonchitis*) is known only from the type locality, Cape Crocker Indian Reserve, Bruce County, Ont.

Specimens of *P. braunii* × *acrostichoides* have been seen from Inverness County, N.S., and Waterloo, Que.

Polystichum munitum × *imbricans* is known from Mount Newton, Vancouver Island, B.C. (D.H. Wagner 1979).

W.H. Wagner (1973*b*) discusses a large number of sterile hybrids that he has studied in the western United States, many of which could occur in Canada. Still other hybrids have been reported in the literature (Knobloch 1976).

5. *Dryopteris* Adans.   wood fern

Usually large (one species is small) ferns with fronds arising in clusters from stout creeping or erect rootstocks. Stipes continuous with the rootstock, not jointed. Blades bipinnatifid or pinnate to nearly bipinnate, glabrous, or somewhat pubescent. Indusium roundish reniform, attached in the centre, covering the rounded sori. Veins usually free, simple, or forked.

The genus has been extensively studied cytologically in Europe starting with Manton (1950), followed by her graduate student S. Walker (1961) and Walker's graduate student, M. Gibby (Gibby and Walker 1977). In North America, Britton (1953) and Wagner (1970) have studied *Dryopteris* in some detail. Our current understanding of the evolution of the species in eastern North America is summarized in Wagner (1970) and Lovis (1977). Widén in Finland has studied the chemistry of the phloroglucinols of the various species of the world, which has been useful in our understanding of the relationships within the genus (see review by von Euw et al. 1980). The external morphology of the spores has been examined by SEM (Britton 1972*a*, 1972*b*; Britton and Jermy 1974).

In the phylogenetic scheme (see Diagram 4) there is still one ancestral genome (B) missing (S of W.H. Wagner 1970); otherwise the species seem quite well analyzed from a few artificial hybrids and from many natural ones.

Of the seven extant diploids, only three are known that have not participated further in evolution. These are *Dryopteris arguta*, *D. fragrans*, and *D. marginalis*. The others are ancestral to the derived alloploids. Although *D. ludoviciana* is confined to southeastern United States and so is not part of our flora, its influence is considered to be present in the origin of *D. cristata* and from this, in turn, of *D. clintoniana*. The evolution of the species can be shown schematically as in Diagram 5.

Readers familiar with previous treatments of this group can appreciate that problems in identification have arisen regarding the so-called "*D. spinulosa* complex", i.e., those species to the left of species B. Other problems have arisen in the past between *D. cristata* and *D. clintoniana*, two species to the right of species B. At this time, there are still problems in distinguishing between *D. expansa* and *D. campyloptera*, and some researchers are impressed by the differences between *D. expansa* in eastern Canada compared with *D. expansa* in western Canada. Carlson and W.H. Wagner (1982) have recently compared the distributions of the North American members of this genus.

A. Blades usually small, copiously scaly on the under surface; old fronds or their bases forming a conspicuously persistent curled tuft at the base of the plant; indusia large, glandular . . . .
. . . . 2. **D. fragrans**
A. Blades usually large, scales few or absent.
    B. Sori marginal or nearly so; blade leathery, grayish green, paler beneath; ultimate segments of pinnae round-lobed . . .
    . . . . 8. **D. marginalis**
    B. Sori medial to submedial; pinnae with sharp-toothed segments.
        C. Basal pinnules on basal pinnae sessile or adnate.
            D. Fronds dimorphic; the sterile fronds shorter and more lax; pinnae of fertile fronds often in a nearly horizontal position . . . . 10. **D. cristata**
            D. Sterile and fertile fronds similar; pinnae of the fertile fronds in the same plane as the blade or nearly so.
                E. Blade broadest near the middle; stipe much shorter than the blade . . . . . 7. **D. filix-mas**
                E. Blade broadest or nearly so at the base; stipe longer than the blade.

F.   Stipe up to half the length of the blade; teeth of pinnules spine-like . . . .
. . . . 1. *D. arguta*
F.   Stipe about as long as the blade; teeth of pinnules not spine-like.
    G.   Blade reduced rather gradually to the apex; pinnae broadly triangular to long-triangular, broadest at the base . . . .
. . . . 1. *D. clintoniana*
    G.   Blade abruptly reduced to an acuminate apex; pinnae narrowly lanceolate to narrowly oblong-lanceolate, broadest at the middle . . . . 9. *D. goldiana*
C.   Basal pinnules on basal pinnae stalked.
    H.   Indusia and blade (especially at the base of the pinnae) definitely to densely glandular . . . .
. . . . 3. *D. intermedia*
    H.   Indusia glabrous; blade usually glabrous, occasionally slightly glandular.
        I.   Lower basal pinnule on each basal pinna closer to the second upper pinnule than to the basal upper one
        J.   Blades ovate-triangular, arching, short-stiped . . . . . 5. *D. campyloptera*
        J.   Blades broadly triangular to broadly oblong, nearly upright, long-stiped . . .
. . . . 4. *D. expansa*
        I.   Lower basal pinnule on each basal pinna closer to the upper basal pinnule than to the second upper one . . . . 6. *D. carthusiana*

1.   **Dryopteris arguta** (Kaulf.) Maxon
Aspidium rigidum Am. auth.
D. rigida (Sw.) A. Gray var. *arguta* (Kaulf.) Underw.
coastal shield fern
Fig. 110 (*a*) frond; (*b*) fertile pinnule. Map 108.

Fronds up to 70 cm long, evergreen, tufted from the short-creeping stout rhizome. Stipe stout, up to half the length of the blade, scaly. Blade widest towards the base, twice pinnate; pinnae oblong-lanceolate, long-acuminate; pinnules oblong, mostly rounded-obtuse, serrate to pinnately incised; veinlets spreading, all ending in salient often cartilaginous spine-like teeth. Sori large, medial. Indusia pale greenish yellow, glabrous, but with somewhat glandular margins.

Ancestral diploid (2X) species are:
$2n = 82$; $n = 41 = X$ (basic chromosome number)

Species	Genomes
D. arguta	AA
D. expansa	EE
D. fragrans	FF
D. goldiana	GG
D. intermedia	II
D. ludoviciana	LL (not in Canada)
D. marginalis	MM
Species B	BB extinct?

Derived allotetraploid (4X) species are:
$2n = 164$, $n = 82$

Species	Genomes
D. campyloptera	IIEE
D. carthusiana	IIBB
D. celsa	LLGG (not in Canada)
D. cristata	LLBB
D. filix-mas	OOCC (D. oreades X caucasica, neither in North America)

Derived allohexaploid (6X) species is:
$2n = 246$, $n = 123$

Species	Genomes
D. clintoniana	LLGGBB

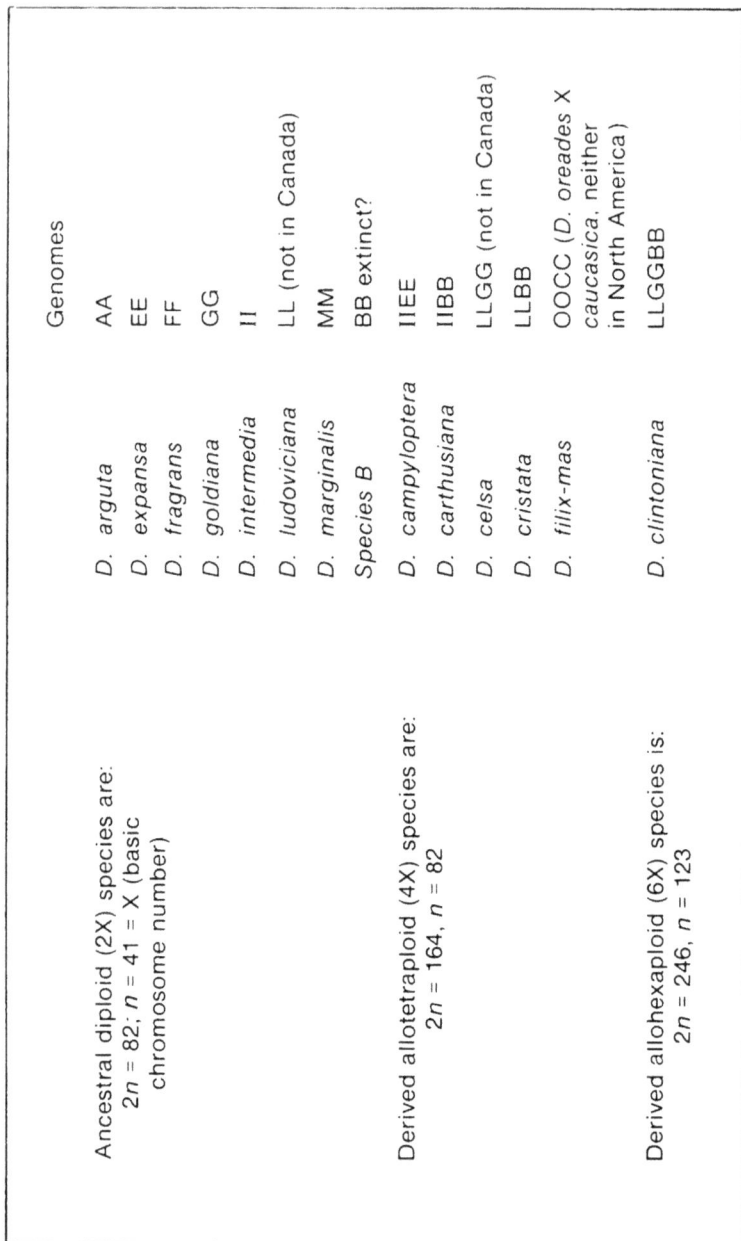

Diagram 4 Genomes of species of _Dryopteris_.

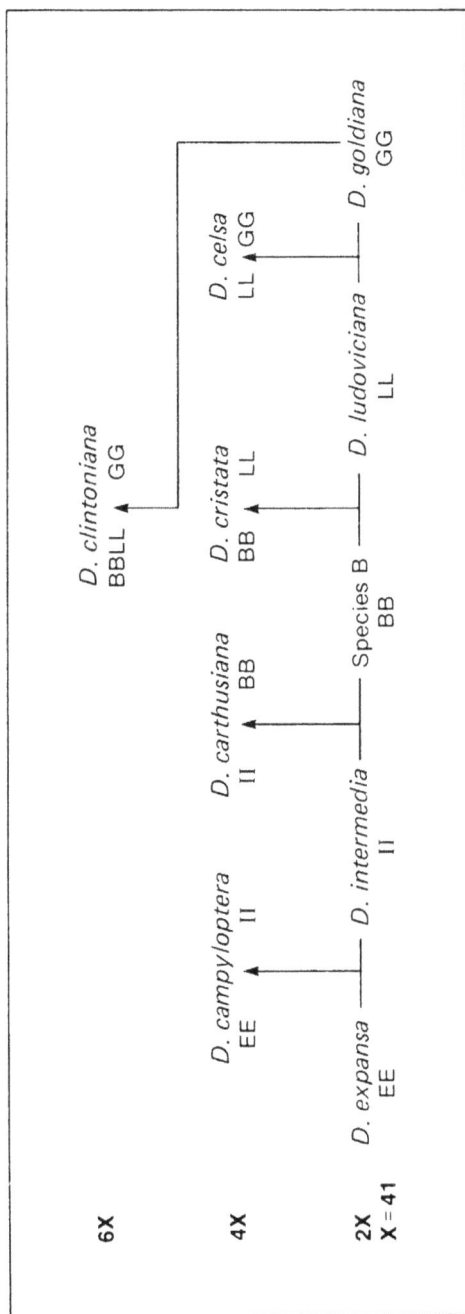

Diagram 5 Evolutionary scheme for *Dryopteris*, after Walker (1961); W.H. Wagner (1970); Gibby (1977); Gibby and Walker (1977).

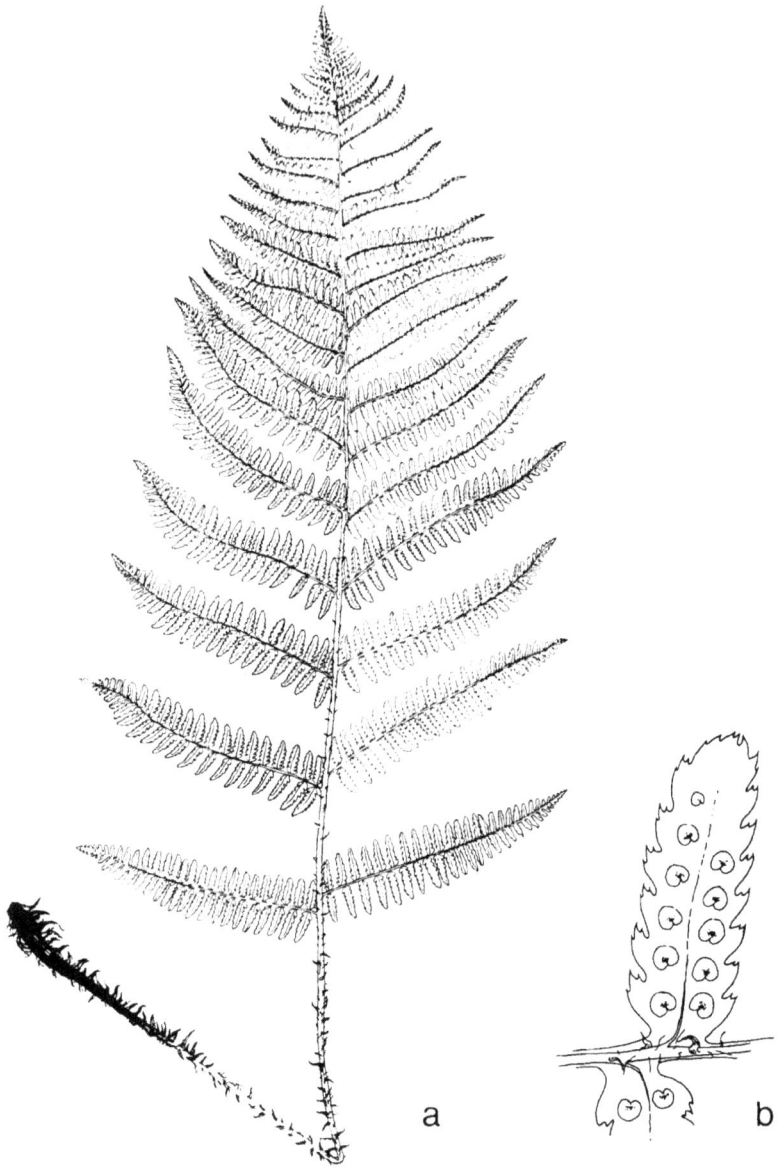

Fig. 110 *Dryopteris arguta*; (a) fronds, 1/4 × ; (b) fertile pinnule, 3 × .

Aspidiaceae

This fern might be confused with *D. filix-mas*; however, it differs from that species by having the blade widest towards the base and the presence of spine-like teeth on the pinnules. Also, on the West Coast it is more strongly evergreen than *D. filix-mas*.

Cytology: *n* = 41 (W.H. Wagner and Chen 1964).

Habitat: Deep humus among broken rocks and in rocky woods along the coast.

Range: Apparently limited in Canada to the southeast coast of Vancouver Island and the islands of the Gulf of Georgia, B.C.; south in the United States to California and inland to Arizona.

Remarks: This is a rare fern in Canada. We have seen a large colony near Nanaimo, B.C., which must have been established for a long time.   It seemed surprising, when we searched the area, that superficially similar habitats did not harbor this species. This lack of aggression is of course one reason for its rarity.

2.   ***Dryopteris fragrans*** (L.) Schott
      D. fragrans (L.) Schott var. *remotiuscula* Komarov
      Aspidium fragrans (L.) Sw.
      Thelypteris fragrans (L.) Nieuwl.
      fragrant cliff fern
Fig. 111 (*a*) frond; (*b*) portion of fertile pinnule. Map 109.

Fronds up to 30 cm long or longer, forming a spreading or ascending crown from a stout rhizome; old fronds curled, shriveled, and persistent. Stipes 1-15 cm long, glandular, and chaffy. Blades coriaceous, tapering from the middle to the base and apex; pinnae overlapping and often inrolled, densely chaffy with brown to reddish scales; pinnae oblong-lanceolate, pinnately incised or crenate; rachises and pinnae glandular. Indusia large and often overlapping, whitish, becoming brown, with their margins often ragged.

Dryopteris fragrans is always a pleasant surprise to anyone visiting a rocky environment. When climbing talus slopes or skirting cliffs, one is rewarded by seeing this fern emerging from a crevice or from under a talus boulder.   At times, it might be mistaken for *Woodsia ilvensis*, but the tell-tale curled dead fronds hanging below the plant make for an easy field check, even from some distance.

Cytology: *n* = 41 (Britton and Soper 1966*; T.M.C. Taylor and Lang 1963*). This species is not ancestral to any of our other species.

Habitat: Cliffs and talus slopes (often somewhat calcareous).

Fig. 111 *Dryopteris fragrans;* (a) frond, 2/3 × ; (b) portion of fertile pinnule, 3 × .

Aspidiaceae

Range: Circumpolar; in North America from Greenland to Alaska, south to Newfoundland, New York, Wisconsin, Minnesota, and northern British Columbia. Extremely abundant near Lake Superior.

Remarks: We consider the southern var. *remotiuscula* Komarov, which is larger, more lax, and has more distant pinnae, to be a response to the longer growing season in the southern part of its range. This expression is clinal, with no clear demarcation geographically. We prefer to ignore the variety. *Dryopteris fragrans* is rare in Nova Scotia (Maher et al. 1978) and Alberta (Argus and White 1978).

3. **Dryopteris intermedia** (Muhl.) A. Gray
   D. spinulosa (O.F. Muell.) Watt var. *intermedia* (Muhl.) Underw.
   Aspidium spinulosum (O.F. Muell.) Sw. var. *intermedium*
      (Muhl.) D.C. Eat.
   evergreen wood fern
Fig. 112 (*a*) frond; (*b*) fertile pinnule. Map 110.

   Fronds up to 70 cm long or longer, winter green, forming a crown at the end of the stout rhizome. Stipe scaly, particularly towards the base, and one-quarter to one-third the length of the frond. Blade oblong-ovate to lanceolate, more or less acuminate, twice pinnate-pinnatifid, usually glandular, particularly near the bases of the pinnae; pinnae at right angles to the rachis, lanceolate to triangular-ovate; inner lower pinnules on the basal pinnae usually shorter than the others. Indusium glandular.
   Good field characters are the extremely lacy appearance, the deep bluish green of the subevergreen fronds and, more particularly, the short inner lower pinnules and the glandularity.

Cytology: $n = 41$ (Britton and Soper 1966*). This is an important ancestral diploid species that has contributed genomes to *D. campyloptera* and *D. carthusiana*. *Dryopteris intermedia* has the same genomes and the same phloroglucinol chemistry as two Old World species, *D. azorica* (only in the Azores) and *D. maderensis* (Gibby and Walker 1977).

Habitat: Moist woods, swamps, and bogs.

Range: Newfoundland to Ontario, west to Minnesota, south in the United States to North Carolina, Tennessee, and Alabama.

Remarks: This extremely attractive plant is a characteristic species of eastern North America. It is an important species for interspecific hybrids, giving rise to two particularly abundant hybrids, *D.* × *triploidea* and *D.* × *boottii*.

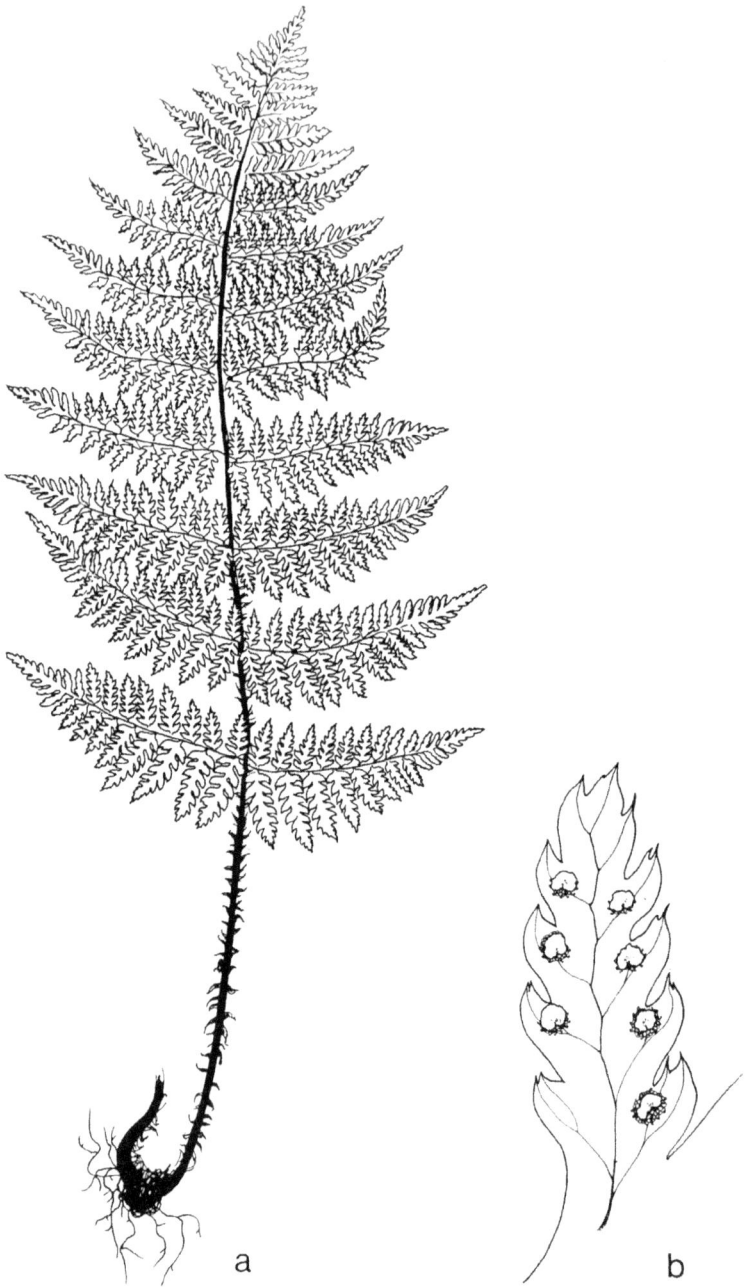

Fig. 112   *Dryopteris intermedia*; (a) frond, 1/2 × ; (b) fertile pinnule, 3 × .

Aspidiaceae

4. **Dryopteris expansa** (Presl) Fraser-Jenkins & Jermy
D. assimilis S. Walker
D. dilatata Am. auth. pro parte
D. austriaca Am. auth. pro parte
northern wood fern
Fig. 113 (*a*) basal portion of frond; (*b*) portion of fertile pinnule.
Map 111.

Fronds to 1 m long, winter green in the west, forming a large more or less upright crown at the end of the stout erect or ascending chaffy rhizome. Stipes usually shorter than the blade, with brownish often dark-centred ovate-lanceolate scales. Blades broadly triangular to ovate or broadly oblong, abruptly acuminate, twice pinnate-pinnatifid to tripinnate; pinnae short-stalked, acuminate; basal pinnae broadly ovate or triangular, inequilateral; lower basal pinnule on each basal pinna closer to the second upper pinnule than to the basal upper one. Ultimate segments of pinnae serrate; teeth mucronate. Sori medial. Indusia glabrous, with some populations finely glandular.

Field recognition in western Canada is simplified because this is a common large plant with a lacy distinct aspect, and there are few places where any other species could be confused with it. In eastern Canada, it is another matter. There the plant is quite variable in aspect and often it looks quite like *D. campyloptera*. Typically, *D. expansa* is more erect, and the superior pinnules next to the rachis do not overlap the rachis as much as in *D. campyloptera*. Also, the petiole is often longer and the blade is usually more elongated, i.e., less triangular than in *D. campyloptera*.

Cytology: $n = 41$ (Britton and Widén 1974*; Mulligan and Cody 1968*). Ancestral diploid, part parent to *D. campyloptera*.

Habitat: Cool moist woods and thickets.

Range: Circumpolar with gaps; southern Greenland, Labrador, and northern Newfoundland to Algoma, Thunder Bay, and Rainy River districts, Ont., western Alberta, British Columbia, Yukon, and Alaska.

Remarks: This is a difficult plant to identify when it is found where the distributions of *D. expansa* and *D. campyloptera* overlap. Amateurs are likely to be unhappy with decisions that lean so heavily on cytology. They could make a contribution here by comparing the two species, one in the Lake Superior basin and the other on Prince Edward Island for example, and by pointing out useful field characters to make the separation simpler. The species is rare in the District of Mackenzie (Cody 1979).

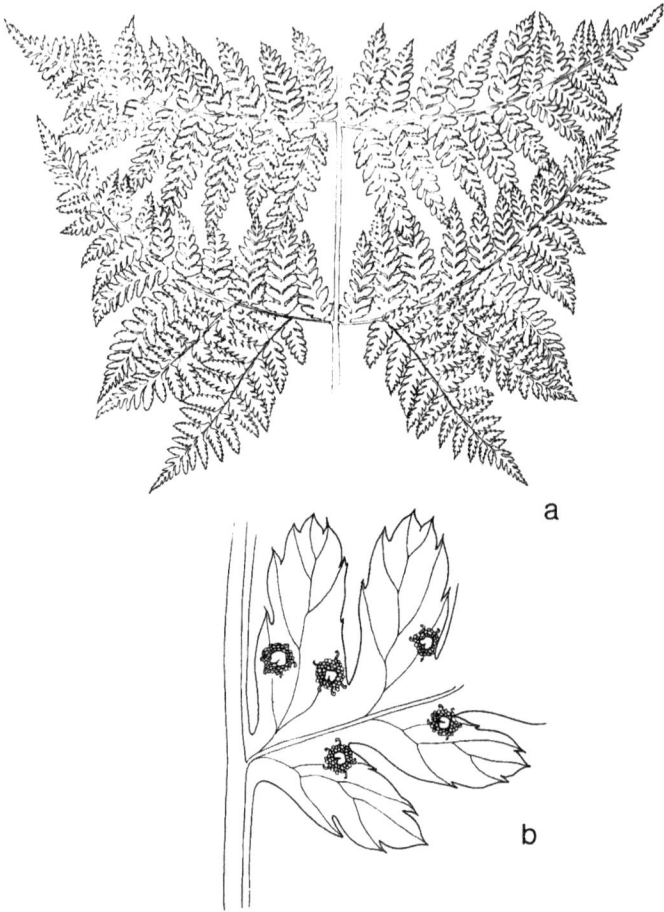

Fig. 113 *Dryopteris expansa*; (*a*) basal portion of frond, 1/3 ×; (*b*) portion of fertile pinnule, 5 ×.

5. **Dryopteris campyloptera** Clarkson
    D. spinulosa (O.F. Muell.) Watt. var. *americana* (Fisch.) Fern.
    D. austriaca Am. auth. pro parte
    Appalachian mountain wood-fern or eastern spreading
       wood-fern
Fig. 114, basal pinna. Map 112.

Fronds to 65 cm long, deciduous, forming an arching crown at the end of the stout chaffy rhizome. Stipes shorter than the blade; scales light brown, attenuate. Blade not glandular, ovate to ovate-triangular, tripinnatifid, or with the basal pinnae sometimes tripinnate; pinnae short-stalked, broadly lanceolate, attenuate; basal pinnae triangular, with the basal upper and lower pinnules remote and the inferior 2–4 times as long as the superior; ultimate segments oblong, obtuse, sharply toothed or cleft; teeth spinulose-tipped. Sori medial. Indusia glabrous or rarely with a few glands.

This fern has the laciness of *D. intermedia* and the width and stature of *D. expansa*. The species is most easily identified by the long basal pinnules next to the stipe on the lowermost pinnae, which are often remote from the superior pinnules opposite, and by its somewhat triangular blade, short petiole, and spreading habit (less erect). Unfortunately, some individuals intergrade annoyingly with the ancestral parent, *D. expansa* (especially in the northern part of the range of *D. campyloptera*), and with luxuriant plants of *D. intermedia*.

Cytology: $n = 82$ (Britton and Widén 1974*). Genomes EEII from *D. expansa* and *D. intermedia* (Gibby and Walker 1977).

Habitat: Cool, rocky woodlands at sea level in the north, but restricted to higher elevations in the south of its range.

Range: Eastern North America; southern Labrador, Newfoundland, Nova Scotia, Prince Edward Island, New Brunswick, and southern Quebec (not positively identified in Ontario as yet), south in the United States in the Appalachian region, to Tennessee and North Carolina.

Remarks: This large and graceful species is most abundant in Canada in the cool maple and yellow birch woods of the Laurentian Mountains north of Montreal, Gaspé Peninsula (base of Mont-Albert), Que., Cape Breton, N.S., and sheltered valleys of Newfoundland.

6. **Dryopteris carthusiana** (Vill.) H.P. Fuchs
    D. spinulosa (O.F. Muell.) Watt
    Thelypteris spinulosa (O.F. Muell.) Nieuwl.
    Aspidium spinulosum (O.F. Muell.) Sw.
    spinulose wood fern
Fig. 115 (*a*) frond; (*b*) portion of fertile pinnule. Map 113.

U.7.

Fig. 114 *Dryopteris campyloptera*; basal pinna, 4/5 × .

Aspidiaceae

Fig. 115 *Dryopteris carthusiana*; (a) frond, 1/3 ×; (b) portion of fertile pinnule, 3 ×.

Fronds 30–80 cm long, forming a crown at the top of a stout ascending rhizome. Stipes with ovate brown scales, particularly near the base. Blades lanceolate, 10–20 cm wide, bipinnate or bipinnate-pinnatifid; pinnules oblong, with spine-tipped teeth; blade and rachis not glandular; inner lower pinnule of basal pinnae usually longer than the next one to it. Sori round. Indusia not glandular.

The well-known species *Dryopteris spinulosa*, now with a changed name (*D. carthusiana*) because of the rules of priority, has a much reduced variation when such taxa as *D. intermedia, D.* × *triploidea, D.* × *uliginosa*, and *D. campyloptera* are removed from consideration. When it is compared with *D. intermedia, D. carthusiana* is less lacy and the blade is paler, more yellow green, and has less divergent teeth. The fronds and indusia should be almost completely devoid of glands.

Cytology: $n = 82$ (Britton and Soper 1966*; Cody and Mulligan 1982*). Genomes IIBB.

Habitat: Moist to wet woodlands, thickets, and streambanks.

Range: Circumpolar; in North America from Labrador to locally in British Columbia, south in the United States to northern South Carolina, Kentucky, Arkansas, and Missouri, west to Montana, Idaho, and Washington.

Remarks: The populations in western Canada are rather small and widely separated. If the broad distribution of this species in the world is taken into consideration, the species must be an ancient allotetraploid, and so its place of origin is obscure. We cannot state whether it is of New World or Old World origin. It is rare in the District of Mackenzie (Cody 1979).

7. **Dryopteris filix-mas** (L.) Schott
   Aspidium filix-mas (L.) Sw.
   Thelypteris filix-mas (L.) Nieuwl.
   male fern
Fig. 116 (a) frond; (b) portion of fertile pinna. Map 114.

Fronds up to 1 m long or longer, forming a crown from a stout ascending scaly rhizome. Stipe usually short, thickly covered with long-attenuate pale brown scales and shorter setiform scales. Blades lanceolate to lance-oblong, narrowed towards the base, acuminate, dark green above; pinnae lance-linear; lower pinnae short and more ovate-lanceolate; pinnules oblong, obtuse, crenate, or serrate. Sori medial, usually only on the lower three-quarters of the pinnules on the upper half of the frond. Indusia glabrous.

Aspidiaceae

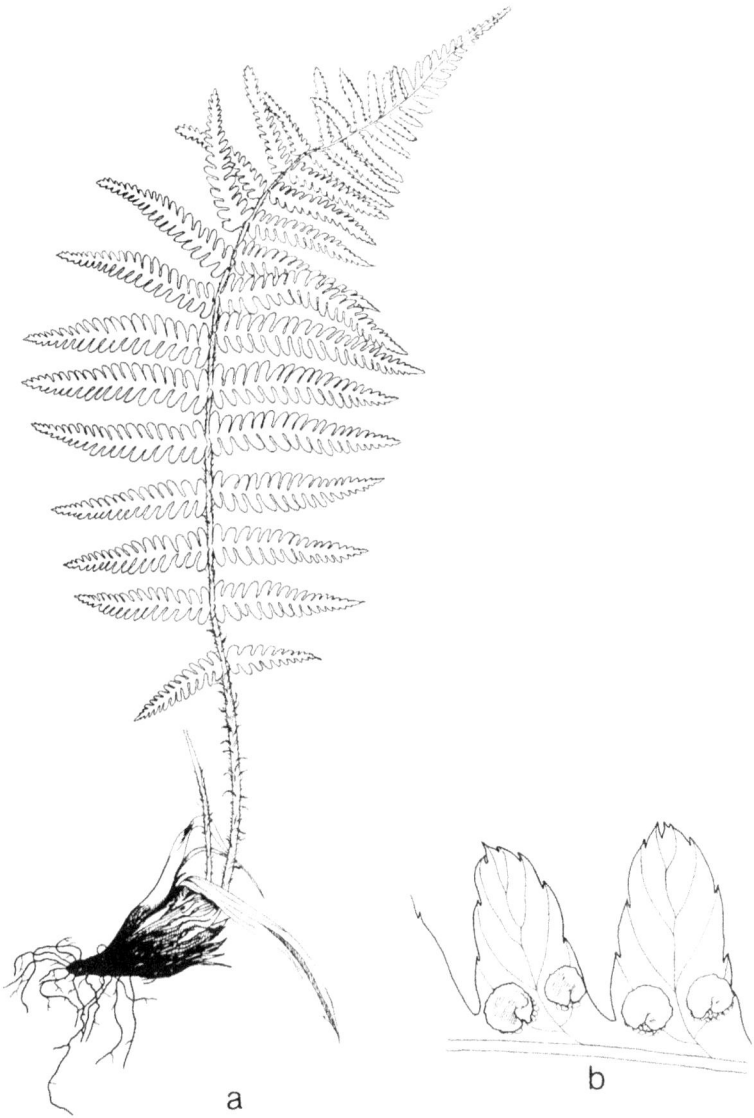

Fig. 116 *Dryopteris filix-mas*; (a) frond, 1/3 × ; (b) fertile pinnules, 7 × .

The double taper to the fronds and the vegetative growth of the plant, which produces a confused crown or patch, are characteristics of this species. Some researchers note a superficial resemblance to *D. marginalis*, but the sori are not submarginal and the plant is much less leathery.

Cytology: $n = 82$ (Britton and Soper 1966*). The European plants have an ancestry from two diploids, *D. oreades* (*abbreviata*) and *D. caucasica*.

Habitat: Rich woods and rocky slopes of valleys (chiefly on limestone in eastern Canada).

Range: Circumpolar; in North America from southern Greenland, western Newfoundland, Cape Breton Island, N.S., Gaspé Peninsula, Que., Bruce, Grey, and Simcoe counties and Michipicoten Island in Lake Superior, Ont., northern Saskatchewan, Waterton Lakes National Park, Alta., and southern British Columbia, south in the United States to Maine, Vermont, Michigan (rare in northeastern United States), California, Arizona, and Texas, and more widespread in the western mountains.

Remarks: The chemistry of the phloroglucinols, spores, and cytology are all reasonably similar to those for the species in Europe. The North American species would seem to be part of the broader distribution of the Eurasian one. The fact that *D. filix-mas* crosses so readily with *D. marginalis* suggests that these two species share a very ancient relationship, i.e., perhaps there was a common ancestor of both *D. oreades* and *D. marginalis*. *Dryopteris filix-mas* is rare in Ontario (Argus and White 1977) and Alberta (Argus and White 1978).

8. **Dryopteris marginalis** (L.) Gray
    Thelypteris marginalis (L.) Nieuwl.
    marginal shield fern
Fig. 117 (a) frond; (b) fertile pinnule. Map 115.

Fronds 25–60 cm long or longer, crowded to form a crown on the stout ascending rhizome; lower part of the stipe covered with thin, light brown lance-linear scales. Blades 9–20 cm wide or wider, dark green above, gray green below, leathery, lanceolate to oblong-ovate, bipinnate; pinnae lanceolate; pinnules oblong, entire to deeply lobed. Sori situated near the margin. Indusia smooth, whitish, becoming light brown at maturity.
    The marginal or leathery wood fern is extremely familiar to amateurs in eastern Canada. The leathery or spongy character of the subevergreen fronds and the submarginal sori are easy and reliable field characters.

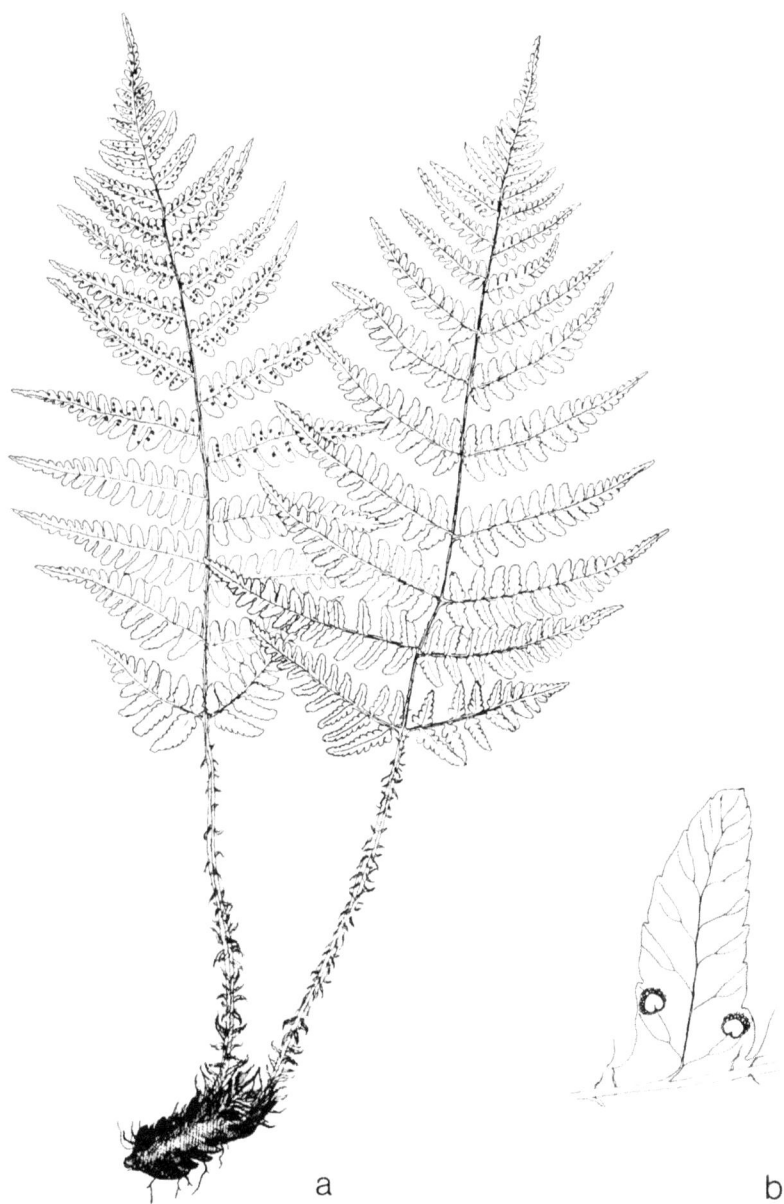

Fig. 117 *Dryopteris marginalis*; (a) frond, 1/3 × ; (b) fertile pinnule, 4 × .

Cytology: $n = 41$ (Britton and Soper 1966*; Cody and Mulligan 1982*). This distinctive diploid species is not an ancestor of any of our other species.

Habitat: Rocky woods and shaded ledges and occasionally in swamps.

Range: Newfoundland, Gaspé, Que., Nova Scotia to Ontario, west to Wisconsin, south in the United States to Georgia, Alabama, Arkansas, Oklahoma, and Kansas.

Remarks: A characteristic species of open woods in eastern Canada, it can flourish in somewhat drier locations than its relatives, although it often grows intermixed with another frequent easterner, *D. intermedia*. There are some named forms that are more dissected than the typical plant.

9.   *Dryopteris goldiana* (Hook.) Gray
      Thelypteris goldiana (Hook.) Nieuwl.
      Goldie's fern
Fig. 118 (a) frond; (b) fertile pinnule. Map 116.

      Fronds up to 1 m long, crowded at the top of the stout ascending rhizome; lower part of the stipe covered with dark brown to blackish lance-acuminate scales. Blades ovate-lanceolate, 20–40 cm wide, pinnate-pinnatifid; pinnae broadly oblong-lanceolate; pinnules linear-oblong, usually crenulate or serrated on the margins. Sori round, situated near the midrib.
      Goldie's fern has long been considered "one of the very finest and largest of the species in the Eastern States, being surpassed in these respects only by the osmundas and the ostrich fern" (Eaton 1879). The sides of the blade are parallel, and the blade narrows to an apex rather abruptly, so that amateurs refer to it as "being choked in the head." Some collectors have commented on the play of dark to bright green when a clump is viewed from a short distance. Mature specimens are easy to identify.

Cytology: $n = 41$ (Britton and Soper 1966*; Cody and Mulligan 1982*). Ancestral diploid, considered a part parent to both *D. celsa* and *D. clintoniana*.

Habitat: Ravines in rich, moist woods and bordering swampy woods.

Range: New Brunswick, southwestern Quebec and southern Ontario, south in the United States to North Carolina, Kentucky, Missouri, and Minnesota.

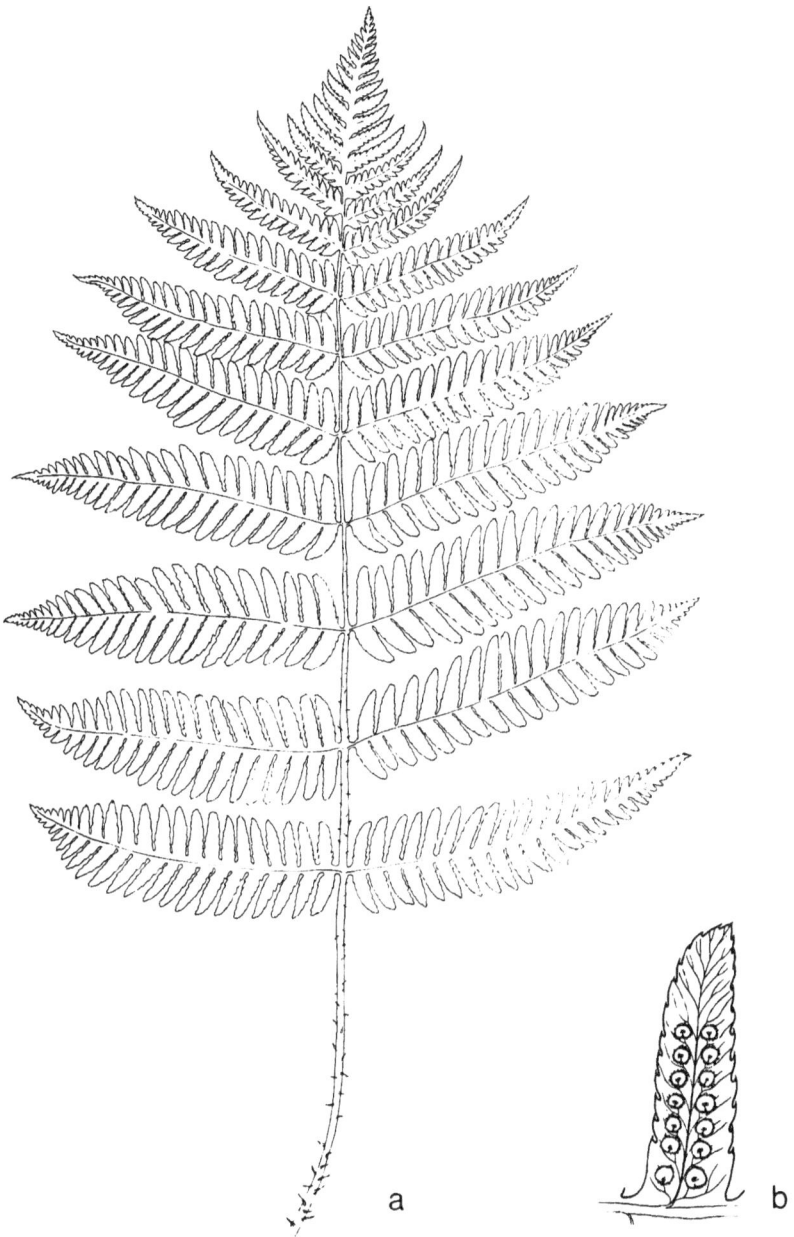

Fig. 118   *Dryopteris goldiana*; (a) frond, 1/3 × ; (b) fertile pinnule, 1 1/2 × .

Remarks: This fern was found by John Goldie near Montreal in 1818 and was described and named by Hooker. Goldie was on a field trip that included walking from Montreal to Niagara Falls, then to Pittsburg, and back to Montreal. *Dryopteris goldiana* is rare in Ontario (Argus and White 1977).

10. **Dryopteris cristata** (L.) Gray
Thelypteris cristata (L.) Nieuwl.
crested wood fern
Fig. 119 (a) fronds; (b) portion of fertile pinna. Map 117.

Fronds 25–70 cm long, forming a crown at the top of the stout ascending rhizome; fertile frond longer than the sterile frond. Stipes with pale brown ovate-lanceolate scales. Blades linear-oblong to narrowly lance-oblong, 6–15 cm wide, pinnate-pinnatifid; basal pinnae short, triangular. Pinnae of fertile fronds turned at right angles to the rachis. Pinnules oblong, obtuse, serrate. Sori round, situated midway between the margin and midvein. Indusia glabrous.

Typical characteristics are the extremely narrow upright fertile fronds, with pinnae that can be perpendicular to the ground, giving a venetian blind effect. Intergradations with *D. clintoniana* (in southern Ontario and southwestern Quebec) are usually hybrids or are poorly developed plants of that species. The narrow, glossy, sterile leaves with much reduced basal pinnae are features to note and they distinguish it from *D. clintoniana*.

Cytology: $n = 82$ (Britton and Soper 1966*; Cody and Mulligan 1982*). Genomes LLBB.

Habitat: Thickets and wet woods to boggy or swampy open ground.

Range: Newfoundland to southeastern British Columbia, south in the United States to North Carolina, Tennessee, disjunct in Nebraska, and local in Idaho, Montana, and southeastern British Columbia; Europe.

Remarks: A characteristic species, often occurring in small numbers in *Alnus* thickets and sphagnum edges of lakes. It rarely makes solid patches as does *D. intermedia* or some of the other species of the genus. *Dryopteris cristata* is rare in Alberta (Argus and White 1978).

Aspidiaceae

Fig. 119 *Dryopteris cristata*; (a) fronds, 1/3 × ; (b) portion of fertile pinna, 1 1/2 × .

11. **Dryopteris clintoniana** (D.C. Eat.) Dowell
Dryopteris cristata (L.) Gray var. *clintoniana* (D.C. Eat.)
   Underw.
Thelypteris cristata (L.) Nieuwl. var. *clintoniana* (D.C. Eat.)
   Weath.
Clinton's wood fern
Fig. 120 (a) frond; (b) fertile pinnules. Map 118.

Fronds 30–80 cm long or longer, forming a crown at the top of the stout ascending rhizome; fertile and sterile fronds similar. Stipes scaly at the base; scales darker and shining at the middle. Blades lanceolate, up to 20 cm wide, pinnate-pinnatifid; basal pinnae little reduced, gradually acuminate at the apex; pinnae oblong-lanceolate, acuminate; segments united by a narrow wing, oblong, obtuse, incurved serrate or biserrate, with subspinulose teeth. Sori medial. Indusia glabrous.

The long and relatively broad fronds without dimorphism are characteristic of this species. Difficulties in identification arise mainly from poorly developed plants. Individual fronds on a plant vary a great deal, and often there will be only one or two large fronds present.

Cytology: $n = 123$ (Britton and Soper 1966*). Genomes LLBBGG. Considered to have arisen from a cross of *D. cristata* × *goldiana*.

Habitat: Swamps and rich wet woods.

Range: New Brunswick, southern Quebec, and southern Ontario, south in the United States to Maine, Pennsylvania, New Jersey, Ohio, and northwest Indiana.

Remarks: The species is common only in a limited area of central southeastern Canada—in Ontario, south of the Precambrian Shield. Canadian plants, which are on the northern edge of the distribution of the species, are smaller and have fewer segments per pinna than those from farther south.

Hybrids of *Dryopteris*

There are some well-known interspecific hybrids in *Dryopteris* (Montgomery 1982), and as many as 31 different hybrid combinations have been mentioned in the literature for temperate North America. In Canada, we know of no hybrids of the western *D. arguta*, but even so, we are left with 10 species that in theory could give rise to 9-8-7-6-5-4-3-2-1 = 45 separate hybrids. The number known in Canada is much less than that, approximately 16 or 17, and commonly occurring hybrids are very few.

Aspidiaceae

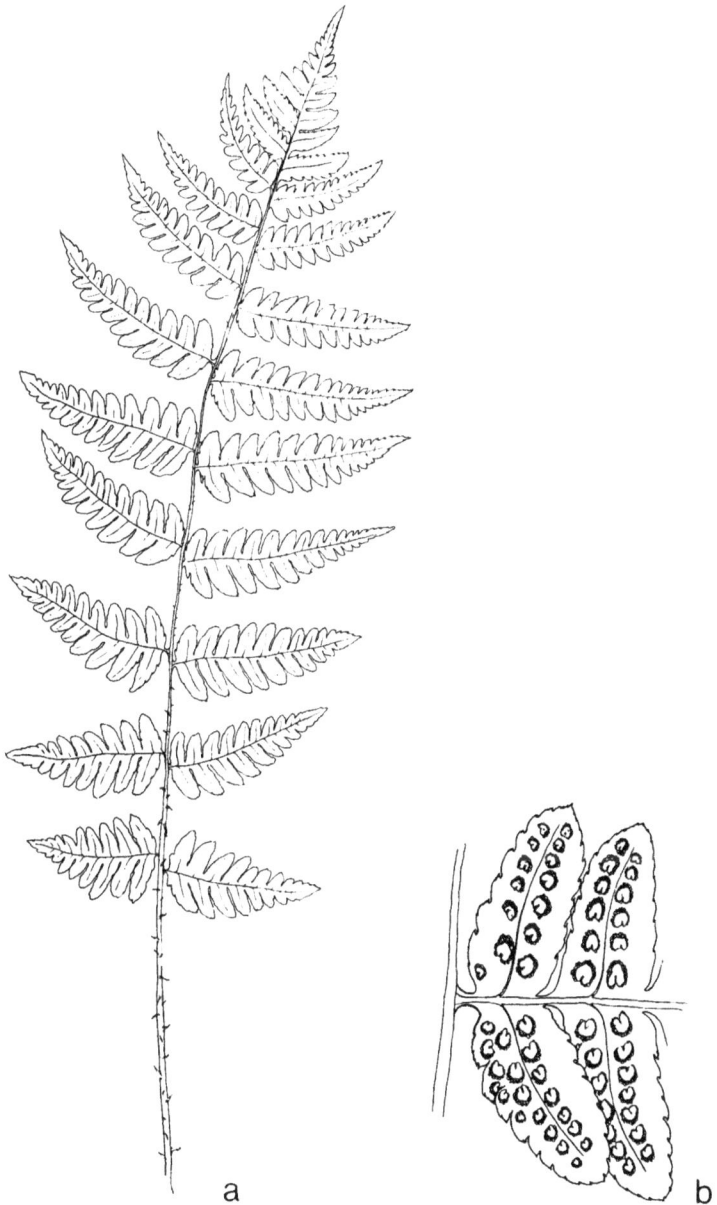

Fig. 120 *Dryopteris clintoniana*; (a) frond, 1/3 × ; (b) fertile pinnules, 1 1/3 × .

In morphology, hybrids usually have aborted spores and possess characteristics from each parent.

Since *D. marginalis* is such a distinctive species, with its leathery fronds, submarginal sori, and deep bluish green color, hybrids with this species are most easily recognized and will be considered first.

D. campyloptera × *marginalis* occurs very rarely in Virginia and Pennsylvania.

D. carthusiana × *marginalis* (*D.* × *pittsfordensis* Slosson) occurs rarely in Ontario, New England, south to West Virginia, Michigan, and Wisconsin.

D. clintoniana × *marginalis* (*D.* × *burgessii* Boivin) occurs infrequently in Quebec, Ontario, New Hampshire to Michigan, south to New Jersey and Pennsylvania.

D. cristata × *marginalis* (*D.* × *slossonae* Wherry) occurs rarely in New Brunswick, Ontario, west to Wisconsin, south to Virginia and Ohio.

D. expansa × *marginalis* occurs rarely in Michigan.

D. filix-mas × *marginalis* is abundant within the range of *D. filix-mas* in Ontario, Vermont, New York, and Michigan.

D. fragrans × *marginalis* (*D.* × *algonquinensis* Britton) is known only from the type locality in Algonquin Park, Ont.

D. goldiana × *marginalis* (*D.* × *neo-wherryi* Wagner) occurs rarely in Ontario, New England, south to West Virginia and North Carolina, west to Illinois and Arizona.

D. intermedia × *marginalis* occurs rarely in Ontario, Vermont, south to Virginia, Indiana, and Michigan.

Another good hybridizer is *D. intermedia*, which is noted for its subevergreen fronds, lacy texture, and very particularly the glandular indusia. Hybrids with this species are glandular and show some influence of the mentioned characteristics of finely dissected blades and subevergreenness.

D. campyloptera × *intermedia* occurs very rarely in Virginia, Pennsylvania, and North Carolina.

D. carthusiana × *intermedia* (*D.* × *triploidea* Wherry) is Canada's most common hybrid, occurring from Nova Scotia to northwestern Ontario (R.M. Tryon and Britton 1966). It is also found in New England, south to North Carolina, west to Kentucky and Minnesota.

D. clintoniana × *intermedia* (*D.* × *dowellii* Wherry) occurs frequently in Ontario, New Hampshire to Michigan, south to New Jersey and Pennsylvania.

D. cristata × *intermedia* (*D.* × *boottii* (Tuckerm.) Underw.) is probably Canada's second most frequent hybrid, occurring from Newfoundland to Ontario. It is also found in Wisconsin, south to Virginia, West Virginia, and Tennessee.

D. expansa × *intermedia* occurs rarely in Michigan.

D. fragrans × *intermedia* is cited by R.M. Tryon (1942). It was rejected by W.H. Wagner and Chen (1965) and is reported only from Sibley Peninsula, Ont.

D. filix-mas × *intermedia* is unknown but should be searched for where the ranges coincide.

D. goldiana × *intermedia* extends from New England to Michigan and Ohio.

Since 11 of the 16 crosses in Canada have been considered, the other five will be shown without all the other possibilities, as follows.

D. clintoniana crosses:

D. carthusiana × *clintoniana* (*D.* × *benedictii* (Farw.) Wherry). Occurs rarely in Ontario, New Hampshire, and Vermont, south to Virginia and west to Michigan.

D. clintoniana × *cristata*. Occurs frequently in Ontario, New England to Michigan, south to Pennsylvania.

D. clintoniana × *goldiana*. Occurs rarely in Ontario, New Jersey, New York, Pennsylvania, Michigan, and doubtfully in Tennessee.

Miscellaneous:

D. carthusiana × *cristata* (*D.* × *uliginosa* (A. Br.) Druce). Occurs rarely in Ontario, Maine to Virginia and West Virginia, west to Minnesota and North Dakota.

D. campyloptera × *expansa*. Rare; known only from Gaspé West County, Que.

D. carthusiana × *goldiana*. Very rare; Vermont(?).

D. filix-mas × *goldiana*. Rare; reported from Vermont.

Note: In the United States there are also the *D. celsa* hybrids, so that the total is approximately 30–31 of known or reported hybrids.

6. **Gymnocarpium** Newm.   oak fern

Small ferns with fronds delicate, glabrous, or glandular, arising singly from slender rootstocks. Sori round. Indusium absent. Veins free, simple, or forking.

This is a small genus of perhaps fewer than 10 species, with its greatest diversity in Asia. Sarvela (1978) lists 17 species names, and in his synopsis and key arrives at six species for the genus. These have been considered previously under a very large number of generic names, e.g., *Polypodium*, *Dryopteris*, *Lastrea*, *Phegopteris*, *Thelypteris*, *Currania*, and *Carpogymnia*, to mention some. This is a clear indication of the uncertain affinity of the genus. There seems general acceptance now of the genus name *Gymnocarpium*.

In the Canadian flora there is one common species, *G. dryopteris*, the oak fern, and two others that are much less frequent, *G. robertianum*, the limestone oak fern, and *G. jessoense* ssp. *parvulum*, the Nahanni oak fern.

A. Blades membranous, with the two lower divisions nearly as long as the terminal one; rachis essentially glabrous . . . .
. . . . 1. *G. dryopteris*

A. Blades firm and somewhat stiff, with the two lower divisions about half as long as the terminal one; rachis at least at the junction of the second and third pinna pair densely glandular.

B. Proximal basal pinnules of the lowermost pair of pinnae usually much longer than the corresponding upper pinnule; lobes or pinnules in central part of the basal pinnae at right angles to the rachis; upper surface of blade moderately glandular; rachis and lower surface of blade densely glandular . . . . 3. *G. robertianum*

B. Proximal basal pinnules of the lowermost pair of pinnae usually only slightly longer than the corresponding upper pinnules; lobes or pinnules in central part of the basal pinnae oblique to the rachis, or curved; upper surface of blade glabrous; rachis and lower surface of blade moderately densely glandular . . . .
. . . . 2. *G. jessoense* ssp. *parvulum*

1. *Gymnocarpium dryopteris* (L.) Newm. ssp. *dryopteris*
Dryopteris disjuncta Am. auth.
D. linnaeana C. Chr.
Thelypteris dryopteris (L.) Slosson
Carpogymnia dryopteris (L.) Löve & Löve
oak fern
Fig. 121 (a) frond; (b) fertile pinnule. Map 119.

Fronds up to 30 cm long or longer, arising singly from a slender forking blackish rhizome. Blades glabrous, or almost so, triangular, ternate; three divisions pinnate-pinnatifid; pinnules oblong, blunt. Sori small, situated near the margin.

The oak fern is fairly common in Canada from the Atlantic to the far West. Its small, delicate, triangular blades horizontal to the ground and its bright lime to yellow green color are distinctive. Some researchers see it as a miniature bracken, but bracken is too coarse and its tissue too thick for such a comparison. The oak fern is particularly striking in early spring, when the trees are beginning to leaf out. The fronds unfold early in this species and they are smooth or with only an occasional gland.

Cytology: $n = 80$ (Britton 1953*; Cody and Mulligan 1982*) eastern Canada; ca. $4\times$ (R.L. Taylor and Brockman 1966*), British Columbia, which is the same chromosome number for this taxon as in Europe.

Habitat: Cool rocky woods, swamp margins, and shaded slopes.

Aspidiaceae

Fig. 121 *Gymnocarpium dryopteris* ssp. *dryopteris*; (a) frond, 1/3 ×; (b) fertile pinnule, 1 1/2 ×.

Range: Circumpolar; in North America from Newfoundland to British Columbia, the Yukon and Alaska, south to Virginia, Michigan, and Wisconsin.

Remarks: The species makes attractive patches in moist hollows in open woods and is generally abundant over most of forested Canada, particularly in boreal woods.

1.1 **Gymnocarpium dryopteris** (L.) Newm. ssp. **disjunctum** (Rupr.)
    Sarvela
    Dryopteris disjuncta (Rupr.) Morton
    western oak fern
Map 120.

Differs from ssp. *dryopteris* in being more robust (to 50 cm long). Blades tripinnate rather than bipinnate.

This subspecies has few characters to delineate it from ssp. *dryopteris*, other than those given above. It is a basic diploid entity, and if the definition of a biological species is strictly applied, ssp. *disjunctum* should be accorded specific rank. We agree with W.H. Wagner (1966b) that this would be a mistake because of our lack of knowledge of the small-spored plants in Alaska and neighboring USSR. Also, it should be noted that both northern populations of this subspecies and those at higher altitudes would then agree with the description of ssp. *dryopteris*.

Cytology: $n = 40$ (R.L. Taylor and Mulligan 1968*). This is a basic diploid entity, $x = 40$.

Habitat: Moist woods and rocky slopes.

Range: British Columbia, Washington, Oregon, Idaho, and Alaska; Sakhalin Island and Kamchatka.

Remarks: Plants that grow under optimum conditions such as those in the MacMillan Memorial Grove, Vancouver Island, B.C., are easy to identify as to subspecies. Plants from higher elevations are quite another matter. One sometimes hears the generalization that polyploids are larger. In this case, the basic diploid is larger.

2. **Gymnocarpium jessoense** (Koidz.) Koidz. ssp. **parvulum** Sarvela
    G. continentale (Petrov) Pojak
    Nahanni oak fern
Fig. 122 (a) frond; (b) fertile pinnule. Map 121.

                                      Aspidiaceae

Fig. 122 *Gymnocarpium jessoense* ssp. *parvulum*; (a) frond, 1/2 ×; (b) fertile pinnule, 3 ×.

Aspidiaceae

Fronds up to 30 cm long, arising singly from a slender forking blackish rhizome. Blades glandular, narrowly triangular, bipinnate-pinnatifid; proximal basal pinnules usually only slightly longer than the corresponding upper pinnules; lobes of pinnules of central part of basal pinnae oblique to the rachis or curved. Sori small, situated near the margin.

This species and subspecies are new to Canada's flora. The subspecies was described by Sarvela (1978), and the type chosen came from below Virginia Falls, in Nahanni National Park, District of Mackenzie.

The blade and rachis are *glandular*, and so previous workers have associated this subspecies with *G. robertianum*. It differs from that species in being a smaller, more slender species of cool, moist, calcareous cliffs. The pinnae are usually curved upwards, and the pinnules have a definite curve outwards, rather than being perpendicular to their axis (Sarvela et al. 1981). Subspecies *jessoense* has a widespread distribution, entirely in Eurasia (Sarvela 1978).

Cytology: $n = 80$ (Sarvela et al. 1981*).

Habitat: Limestone or basic rock cliffs and moist, rocky woods.

Range: Upper Great Lakes in Ontario to British Columbia and Alaska, south to Minnesota and Wisconsin; northern Eurasia. The records for this species in Atlantic Canada are from very old sheets and should be verified from new collections.

Remarks: We have called this new subspecies the Nahanni oak fern, to highlight the fact that the type was collected there and because it has a generally northern distribution on cool, moist, calcareous sites.

3.   **Gymnocarpium robertianum** (Hoffm.) Newm.
     G. dryopteris (L.) Newm. var. *pumilum* (DC.) Boivin
     Dryopteris robertiana (Hoffm.) C. Chr.
     Carpogymnia robertiana (Hoffm.) Löve & Löve
     limestone oak fern
Fig. 123, frond. Map 122.

Fronds up to 40 cm long, arising singly from a slender blackish rhizome. Blades glandular, triangular, bipinnate-pinnatifid; proximal basal pinnules of the lowermost pair of pinnae usually much longer than the corresponding upper pinnule; lobes or pinnules in the central part of the basal pinnae at right angles to the rachis. Sori small, situated near the margin.

This species has been recognized for a long time and, for North America at least, has always been considered a rare plant. Boivin (1962) almost alone believes that it does not merit specific rank. The pronounced glandularity of the blades, including the upper surface, and the long triangular shape of the blade with the pinnules at right angles are features to notice.

Fig. 123 *Gymnocarpium robertianum*; frond, 1/2 ×.

Aspidiaceae

Habitat: Moist calcareous ledges, limestone paving, cliffs, and rocky woods.

Range: Newfoundland to Ontario, south to Minnesota; Europe.

Remarks: This species was already rare in North America before *G. jessoense* ssp. *parvulum* was segregated; consequently, even fewer localities are known now for this species and they are all in eastern Canada. Rare in Ontario (Argus and White 1977).

Hybrids of *Gymnocarpium*

W.H. Wagner (1966*b*) was the first to draw attention to "the apparent cross of *G. dryopteris* and *G. robertianum*," and he named this new apomictic species *G. heterosporum*. The plants were triploid. Sarvela (1978) gave this species the hybrid designation *G.* × *heterosporum* W.H. Wagner and said the cross was *G. jessoense* × *robertianum*. Sarvela (1978) also described a new hybrid, *G.* × *intermedium* (*G. dryopteris* × *jessoense*). Studies by Pryer (1981) indicate that *G.* × *intermedium* is a frequent hybrid at sites where *G. jessoense* ssp. *parvulum* grows, but instead of being tetraploid as one would expect, it is triploid. Accordingly, she concluded that one parent was *G. dryopteris* ssp. *disjunctum* (2*x*) and the other was *G. jessoense* ssp. *parvulum* (4*x*). All *Gymnocarpium* hybrids are recognized by their mostly aborted spores. However, few large spherical spores that do germinate are produced, and we believe that the hybrids are able to propagate themselves apomictically in this way, although no one has raised mature plants yet. If we are right, it should be noted that these hybrids are then much more difficult to study than *Dryopteris* hybrids, because they could have been formed far away from their present location and at quite a different time. *Gymnocarpium* × *intermedium* is found in very large colonies north of Lake Superior. It can greatly outnumber *G. jessoense* ssp. *parvulum* at some locations.

Another hybrid of *Gymnocarpium* was described by Sarvela (1980) as *G. dryopteris* ssp. × *brittonianum*. This is considered to be *G. dryopteris* ssp. *disjunctum* × ssp. *dryopteris*. It is an easy hybrid to identify and seems to be of frequent occurrence. Its hybrid nature is evident again by its mostly aborted spores, and it has a larger stature than ssp. *dryopteris*. The parents are glabrous, as is the hybrid. Again, Pryer (1981) believes that the few large spores that are present are able to perpetuate this hybrid as an apomictic species. It has been identified as occurring from coast to coast in Canada (Sarvela 1980; Pryer 1981).

One additional hybrid might be mentioned, *G.* × *achriosporum* Sarvela (*G. dryopteris* × *robertianum*). Hybrids of *G. robertianum* seem to be of rare occurrence. The type is referable to Swedish material, and only two collections in North America have been ascribed to this combination. These are from Chicoutimi and Gaspé, both in Quebec (Sarvela 1981). No material has been examined cytologically, but since both parents are tetraploid, the hybrid should be tetraploid.

Gymnocarpium × *heterosporum* is known only from the type material in North America, but was ascribed also to one locality in Finland (Sarvela 1978). It is not known in Canada (Sarvela 1980).

7. *Thelypteris* Schmidel

Small to medium-sized ferns with more or less pubescent fronds arising from a slender (stout in *T. limbosperma*) rhizome. Veins free, simple, or forking. Sori small, round. Indusia reniform or horseshoe-shaped, attached at the sinus.

It is difficult to estimate the number of species in the genus *Thelypteris* at this time because of the large number of segregate genera that are being recognized in both the Old World and the New World. The number in the world with affinities to this genus is large (approximately 800). We have recognized five species in the genus for Canada. Other researchers might include these species in as many as four genera. The five Canadian species, however, have features in common that can be identified by the amateur. They all form upright colonies or patches rather than neat individual "shuttlecocks," and none is as finely divided as *Dennstaedtia punctilobula* or *Athyrium filix-femina*, for example. All, except *T. limbosperma*, are roughly in the same size range of 20–60 cm and all are deciduous. In general, the blades are quite soft and thin in texture, unless in the full sun, and the plants wilt rapidly when picked.

There is one species in western Canada, *T. nevadensis*, another in both the West and isolated in the East, *T. limbosperma*, and three in the East, *T. noveboracensis*, *T. palustris*, and *T. simulata*.

An indication of the diversity of Canada's five species is illustrated by their chromosome numbers, which range from 27 to 64. In other words, the situation is quite unlike the one for *Dryopteris*, where the basic diploids are all derived from the number 41.

A. Lower pinnae gradually decreasing in size; lowermost pinnae often very much decreased.
    B. Rhizome short and stout . . . . 1. *T. limbosperma*
    B. Rhizome slender and elongate.
        C. Pinnules ciliate (eastern) . . . . . . 3. *T. noveboracensis*
        C. Pinnules not ciliate (western) . . . . . 2. *T. nevadensis*
A. Lower pinnae only slightly if at all smaller.
    D. Fronds dimorphic; lateral veins of pinnules of sterile fronds mostly forking; glandular dots on undersurface of pinnules absent . . . . 4. *T. palustris* var. *pubescens*

D. Fronds similar; lateral veins of pinnules of sterile fronds not
forked; glandular dots on undersurface of pinnules present
. . . . 5. *T. simulata*

1. **Thelypteris limbosperma** (All.) H.P. Fuchs
T. oreopteris (Ehrh.) Slosson
Dryopteris oreopteris (Ehrh.) Maxon
Oreopteris limbosperma (All.) Holub
mountain fern
Fig. 124 (a) frond; (b) fertile pinnule; (c) sorus. Map 123.

Fronds up to 1 m in length, tufted at the end of a short thick more
or less ascending rhizome. Stipes and rachis scaly. Blades lanceolate,
elongate, abruptly acuminate; lower pinnae reduced, pinnate-
pinnatifid, triangular; middle and upper pinnae linear-lanceolate,
tapering at the tip; pinnules oblong, obliquely set, blunt or subacute,
entire or somewhat wavy margined; margins of pinnules slightly
inrolled. Sori situated near the margins. Indusia glandular.
Thelypteris limbosperma was transferred to *Oreopteris
limbosperma* by Holub (1969). We are aware of the shuffling of the
genera and of the new interpretations that are occurring (Holtum
1971). For Canada's limited flora, however, it seems more sensible to
group our few species into a common genus at this time.
This is a taller species than the others, with leaf margins often
recurved and finely hyaline-papillose. It is aromatic when crushed,
and therefore in Great Britain it is referred to as the lemon-scented
fern. (Jermy et al. 1978)

Cytology: $n = 34$ (T.M.C. Taylor and Lang 1963*) as in Europe.

Habitat: Margins of creeks and runnels in rocky woods, outcrops, and
crevices of cliffs to at least 700 m.

Range: Coastal in Alaska and British Columbia, but occurs inland in
the Cascade Mountains, Wash.; disjunct in Gros Morne National Park
in western Newfoundland (Bouchard and Hay 1976); Eurasia.

Remarks: The discovery of this fern in Gros Morne National Park,
Nfld., was most interesting, even if it poses a problem regarding
whether it is a western disjunct species there or whether it has its
affinities with the populations in Europe (Bouchard et al. 1977).

2. **Thelypteris nevadensis** (Baker) Clute
T. oregana (C. Chr.) St. John
Dryopteris nevadensis (Baker) Underw.
Fig. 125 (a) fronds; (b) fertile pinna. Map 124.

Aspidiaceae

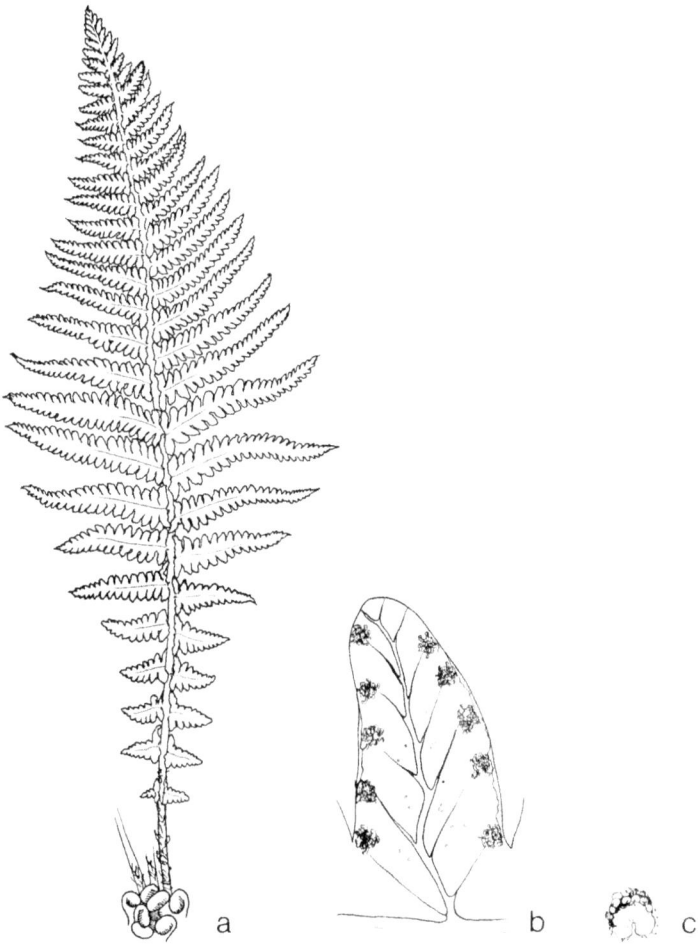

Fig. 124 *Thelypteris limbosperma*; (a) frond, 1/3 × ; (b) fertile pinnule, 8 × ; (c) sorus, 8 × .

Fig. 125 *Thelypteris nevadensis*; (a) fronds, 1/2 × ; (b) fertile pinna, 2 × .

Aspidiaceae

Fronds 20–60 cm long or longer, tufted at the end of the slender, horizontal rhizome. Blades elliptic-lanceolate, 5–12 cm wide, long attenuate; lower pinnae much reduced, pinnate-pinnatifid; pinnae linear to linear-lanceolate, acuminate to caudate, somewhat hairy on the midveins; pinnules oblong, blunt, obliquely set, entire or slightly toothed, resin-dotted on the lower surface. Sori round, situated near the middle. Indusia horseshoe-shaped, glandular.

This species can readily be distinguished from the only other western species, *T. limbosperma*, by its more delicate fronds and slender rhizome.

Cytology: $n = 27$ (T.M.C. Taylor and Lang 1963 as $n = 26$-27^*; A.R. Smith 1971; A.F. Tryon and R.M. Tryon 1974).

Habitat: Rocky banks of streams.

Range: In British Columbia known in a single locality (Sooke River, Vancouver Island), southward in the foothills and middle altitudes to central California.

Remarks: This species is very closely related to *T. noveboracensis*. A comparative study that included the Asiatic vicariads (A.F. Tryon and R.M. Tryon 1974), together with attempts at hybridization, would be very interesting. It was placed in *Parathelypteris* by Ching (1963).

3. **Thelypteris noveboracensis** (L.) Nieuwl.
    Dryopteris noveboracensis (L.) Gray
    New York fern
Fig. 126 (a) fronds; (b) fertile pinnule. Map 125.

Fronds 25–55 cm long, arising from a slender rhizome. Blades elliptic to elliptic-lanceolate, 9–15 cm wide; lower pinnae very reduced, pinnate-pinnatifid; pinnae oblong to oblong-lanceolate, somewhat hairy on the rachis and veins; pinnules oblong, somewhat blunt. Sori round, situated near the margin. Indusia glandular-ciliate.

This is a characteristic fern of sandy, acid glades and roadsides in southeastern Canada. The light green patches are soon easy to identify by noting the double taper to the blades of a frond. The last characteristic should be used with caution, however, because some small plants of the lady fern have this same general shape. The New York fern has round sori, whereas those of *Athyrium filix-femina* are elongated.

Cytology: $n = 27$ (Britton 1964*; A.F. Tryon and R.M. Tryon 1973, 1974).

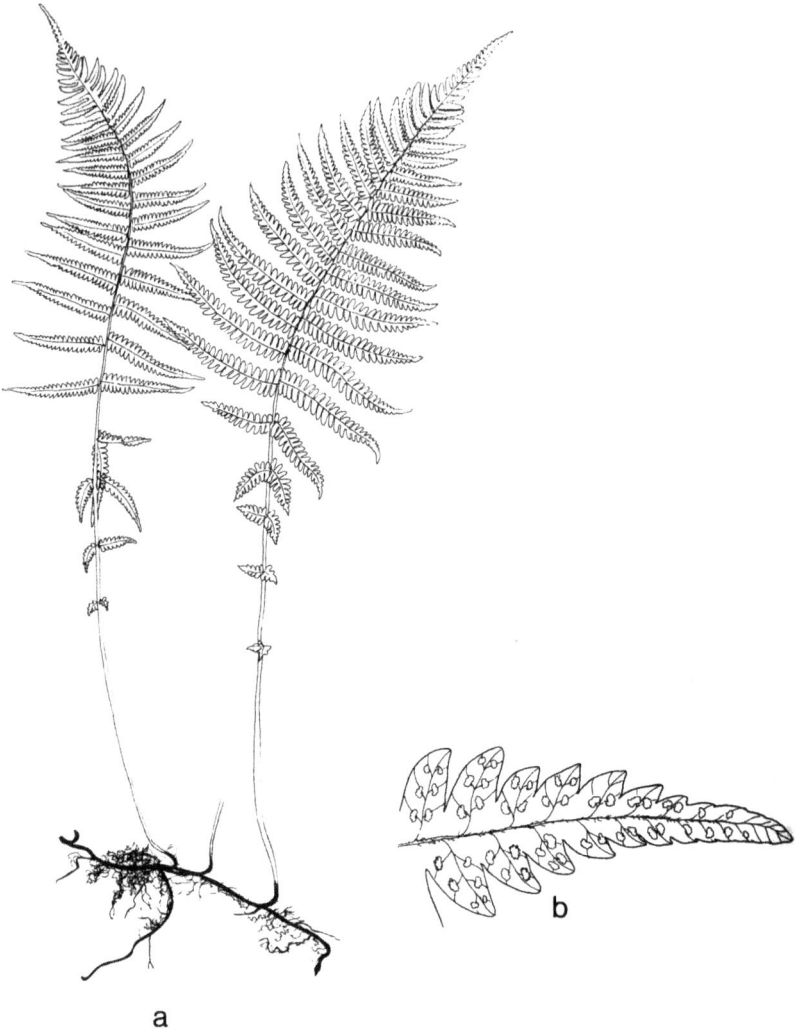

Fig. 126 *Thelypteris noveboracensis*; (a) fronds, 1/3 × ; (b) fertile pinna, 5 × .

Habitat: Moist woods, thickets, and swamps, chiefly in moderately acid soil.

Range: Eastern North America, Newfoundland to Ontario, Michigan, and Illinois, south to Georgia, Alabama, Mississippi, and Arkansas.

Remarks: For most ferns (*Osmunda* excepted), this species has a low chromosome number. It is the lowest number for the thelypterids. See also remarks under *T. nevadensis.*

4. **Thelypteris palustris** Schott var. **pubescens** (Lawson) Fern.
   T. thelypterioides sensu Holub
   Dryopteris thelypteris (L.) Gray var. *pubescens* (Lawson) Nakai
   marsh fern
Fig. 127 (*a*) fronds; (*b*) fertile pinna. Map 126.

Fronds 20–60 cm long or longer, arising from an elongate rhizome. Blades lanceolate, 7–15 cm wide, pinnate-pinnatifid; pinnae linear-lanceolate; pinnules oblong, blunt, dimorphic. Fertile fronds usually longer than the sterile fronds; pinnules somewhat thicker and inrolled. Rachis and blade minutely pubescent and sometimes glandular. Sori round, situated about halfway between the margin and midvein, at maturity partly covered by the inrolled margin. Lateral veins of pinnules of sterile fronds mostly forking.

Cytology: $n = 35$ (Britton 1953*; A.F. Tryon and R.M. Tryon 1973). This is the same chromosome number as for the species in Europe.

Habitat: Marshes, swamps, wet thickets, bog margins, and ditches.

Range: Eastern North America, southern Newfoundland to southeastern Manitoba, south to Georgia, Tennessee, and Oklahoma, Japan, and northeastern Asia; var. *palustris* occurs in Eurasia.

Remarks: This is a common fern in much of southeastern Canada. It is not a particularly graceful species and is so soft and fragile that it is often rather windblown and distorted—so much so, that some researchers consider it either weed-like or an immature form of a larger species. The reflexed margins of the segments of the blades of the fertile fronds suggested to early writers the shapes of snuff boxes.
   The marsh fern is suffering the vicissitudes of nomenclatural changes. If the Canadian plant is considered as a variety or subspecies of the species in Europe, it may be considered as var. *pubescens* (Lawson) Fernald. Recently, workers in Europe have taken up the name *T. thelypterioides* (Michx.) Holub, but A.F. Tryon et al. (1980)

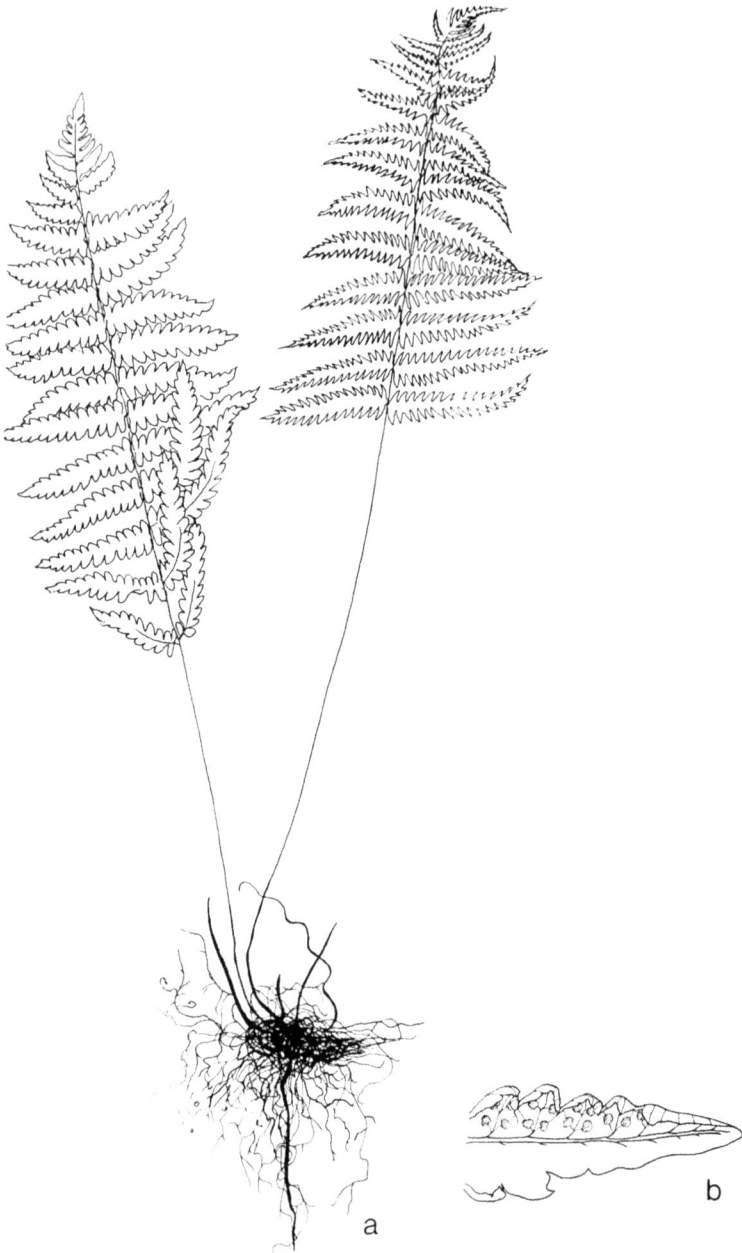

Fig. 127 *Thelypteris palustris* var. *pubescens*; (a) fronds, 1/2 × ; (b) fertile pinna, 3 × .

Aspidiaceae

say that this is a mistake, and that the familiar name *T. palustris* should be used. Variety *pubescens* is rare in Newfoundland and Manitoba.

5. **Thelypteris simulata** (Davenp.) Nieuwl.
Dryopteris simulata Davenp.
Massachusetts fern
Fig. 128 (*a*) frond; (*b*) fertile pinnule. Map 127.

Fronds 20–50 cm long or longer, arising from an elongate rhizome. Blades oblong-lanceolate, 7–15 cm wide, tapering at the tip, pinnate-pinnatifid; pinnae oblong-lanceolate; fertile pinnae long-acuminate; pinnules oblong, obtuse, flat, or sometimes slightly inrolled. Upper surface of pinnules strigose. Indusia and lower surface of the pinnules glandular dotted.

This species is sufficiently similar to both the New York fern and the marsh fern to cause identification problems. In the Massachusetts fern the lower pinnae are noticeably narrowed at their base next to the rachis (pinched in) and are usually retrorse (angled down). The veins in both the fertile and sterile fronds are unbranched. The marsh fern has veins in the sterile fronds mostly forking.

Cytology: $n = 64$ (A.F. Tryon and R.M. Tryon 1973).

Habitat: Moist woods and boggy thickets in intensely acid situations often in association with *Sphagnum*.

Range: Eastern North America from southwestern Nova Scotia, south to Virginia, and extending locally inland to southwestern Quebec, southeastern Ontario, central New York, and Maryland; also disjunct in the Driftless Area of Wisconsin (mapped by A.F. Tryon and R.M. Tryon 1973).

Remarks: At one time *T. simulata* was considered as a possible hybrid derivative of *T. noveboracensis* and *T. palustris*, but as pointed out by A.F. Tryon and R.M. Tryon (1973), there are vicariads in Asia, and all the species of the world of the same general affinity should be studied and carefully compared. Löve and Löve (1976) created the genus *Wagneriopteris* for this species, but this seems premature. *Thelypteris simulata* is rare in Quebec and Ontario.

8. **Phegopteris** Fée

Small to medium-sized ferns with more or less pubescent fronds arising singly from elongate, slender, horizontal, scaly rhizomes. Veins free, simple, or forking. Sori small, round. Indusium absent.

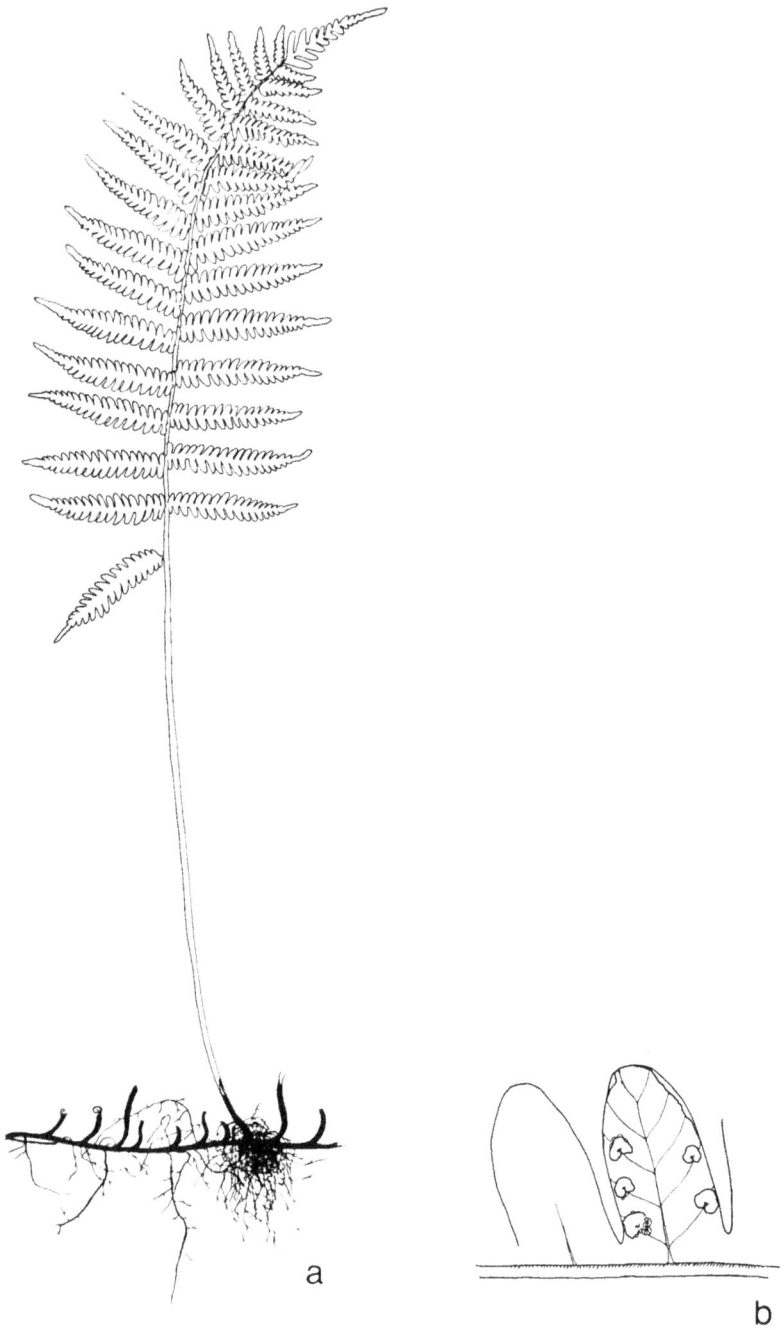

Fig. 128   *Thelypteris simulata*; (a) frond, 1/3 × ; (b) fertile pinnule, 3 × .

Aspidiaceae

This genus has progressed from being part of *Dryopteris* s.l., in Christensen's time, to a *Thelypteris*, and now to the segregate genus *Phegopteris*, which comprises only three or four species in the world. The basic chromosome number *x* is 30.

A. Wings of rachis extending down to the lowest pinna-like divisions; blades broadly triangular ...... 1. *P. hexagonoptera*
A. Wings of rachis not extending down to the lowest pinnae; blades narrowly triangular; lowest pair of pinnae usually projected downward and forward .... 2. *P. connectilis*

1. **Phegopteris hexagonoptera** (Michx.) Fée
   Dryopteris hexagonoptera (Michx.) Christens.
   Thelypteris hexagonoptera (Michx.) Weatherby
   broad beech fern
Fig. 129 (a) frond; (b) fertile pinnule. Map 128.

Fronds 30–60 cm long or longer. Blades broadly triangular, 15–30 cm wide or wider, about as broad as long, tapering to the top, pinnate-pinnatifid; middle and upper pinna-like divisions lanceolate; lower pinna-like divisions unequally ovate to lanceolate-ovate, not projected forward; all divisions connected by a wing; segments, particularly those of the lower pinnae, often deeply pinnatifid. Stipe naked except at the base; rachis not chaffy or with almost colorless scales. Rachis and veins minutely glandular puberulent. Sori small, near the margin.

Although the other species in Canada is called *P. connectilis*, it is *P. hexagonoptera* that has all the divisions of the blade connected to the rachis, including the basal pair. In shape, the blade is more broadly triangular in the broad, or southern, beech fern than in the long beech fern. The shape of the basal segments is unlike that in *P. connectilis*, being widest in the middle and lobed again rather than entire.

Cytology: $n = 30$ (Mulligan and Cody 1979*). A sexual basic diploid.

Habitat: Rich often rocky woods and wooded slopes.

Range: Southwestern Quebec and southern Ontario, south to Florida and Texas.

Remarks: This species is very local even within its Canadian distribution. It would seem to be perhaps Canada's only fern with a strictly Carolinian distribution, compared with the distribution of *Sassafras*, or black walnut, although some researchers might question even that observation for the Quebec populations. *Phegopteris hexagonoptera* is rare in Ontario (Argus and White 1977) and Quebec (Vincent 1981).

Aspidiaceae   249

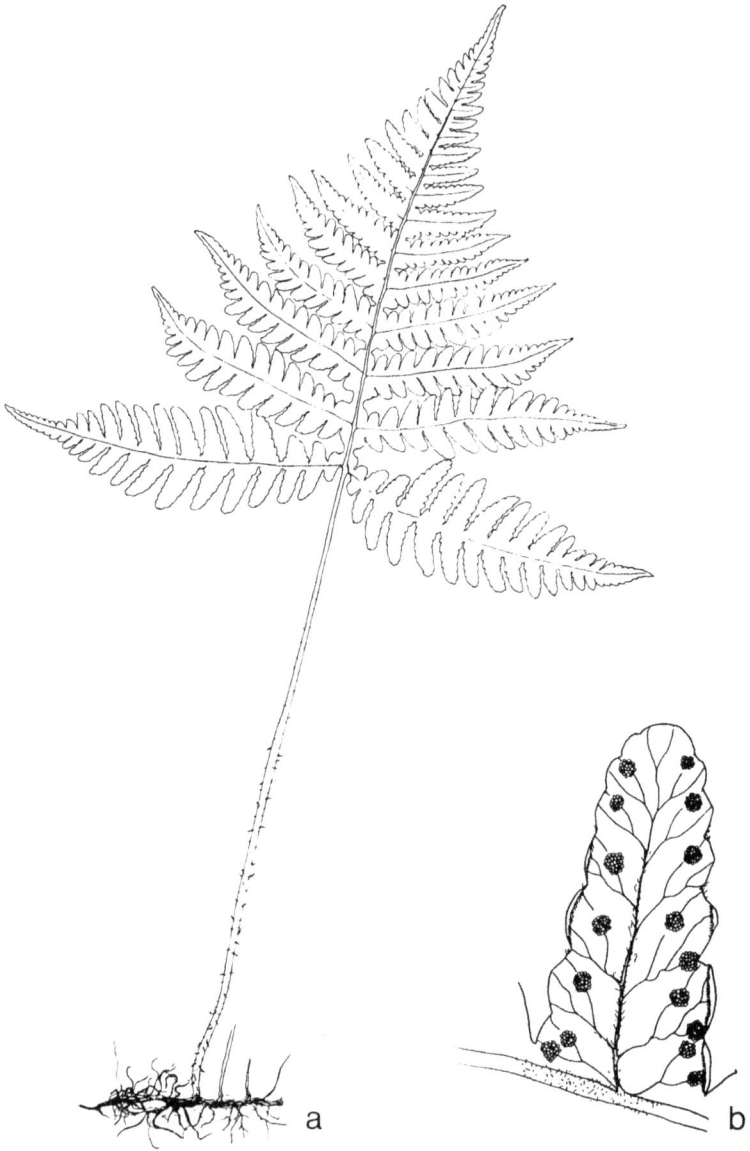

Fig. 129 *Phegopteris hexagonoptera*; (a) frond, 1/3 × ; (b) fertile pinnule, 3 × .

Aspidiaceae

2. **Phegopteris connectilis** (Michx.) Watt
   P. polypodioides Fée
   Dryopteris phegopteris (L.) Christens.
   Thelypteris phegopteris (L.) Slosson
   long beech fern
   Fig. 130 (*a*) frond; (*b*) portion of fertile pinnule. Map 129.

Fronds 15–35 cm long or longer. Blades triangular, longer than broad, 8–20 cm long or longer, 6–16 cm wide or wider, pinnate-pinnatifid, tapering to the tip; pinna-like divisions lance-acuminate, with all but the lower pair (usually projected downward and forward) connected by a wing. Pinnules oblong, rounded at the tip, more or less hairy on both faces and on the rachis. Stipe and rachis with brown scales. Sori small, situated near the margin.

Our more common representative, the long, or northern, beech fern, was known as *P. polypodoides* for a number of years, but is now called *P. connectilis* (Michx.) Watt. This species is known from coast to coast in Canada and is soon familiar to amateurs in our northern woods. The distinctive triangular blades with only the base pair of pinna-segments free, and these projecting downward and forward, are good field characters.

Cytology: "*n*" = 2*n* = 90 (Mulligan and Cody 1979*). The species reproduces apogamously, and the chromosome number in the spores is identical to that of the somatic cells of the mature plant. Because the basic chromosome number for the genus is 30, this species is a triploid. There are reports for a diploid sexual race of limited range in Japan (Mulligan and Cody 1979), although for most of the distribution there and elsewhere in the world, the number is 2*n* = 90.

Habitat: In Ontario often in soil on banks of rivulets or creeks. In Newfoundland and British Columbia on moist rocky hillsides, appearing from under large boulders or on moist rocky ledges.

Range: Circumpolar; in North America, Greenland, Labrador, and Newfoundland to Alaska, south to North Carolina, Michigan, Iowa, and Oregon.

Remarks: This is an attractive species of interesting form. It is not as abundant as some species, but is of broad occurrence. *Phegopteris connectilis* is rare in the District of Mackenzie (Cody 1979), the Yukon (Douglas et al. 1981), Manitoba (White and Johnson 1980), Saskatchewan (Maher et al. 1979), and Alberta (Argus and White 1978).

Hybrids of *Phegopteris*

Mulligan et al. (1972) reported the occurrence in Quebec of a hybrid between *P. connectilis* and *P. hexagonoptera*. Mulligan and

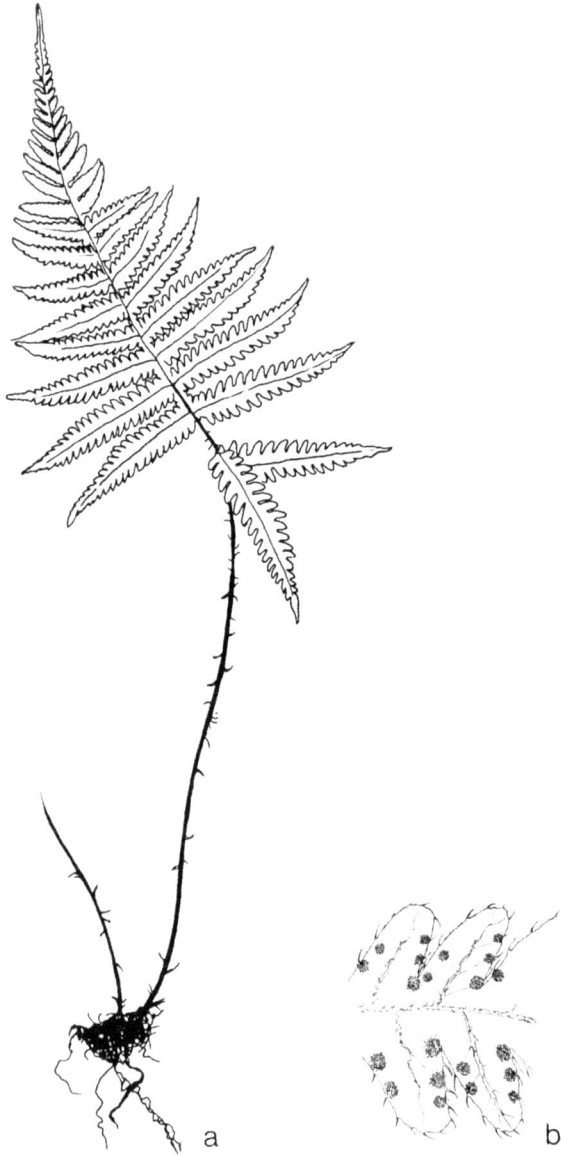

Fig. 130 *Phegopteris connectilis*; (a) frond, 2/3 × ; (b) portion of fertile pinnule, 6 × .

Aspidiaceae

Cody (1979) reported the chromosome numbers of 10 collections of this hybrid from six different localities in Quebec, New Brunswick, and Nova Scotia. The plants were tetraploid, based on $x = 30$, and had "n" $= 2n = 120$ chromosomes, i.e., the spores and the somatic cells both had 120 chromosomes, and the plants were apogamous and fertile. They elected to refer this entity to *P. connectilis* because the tetraploid hybrid was "not, however, always distinguishable from triploid plants of *P. connectilis*."

9. *Cystopteris* Bernh. bladder fern

Delicate medium-sized ferns, with bipinnate to tripinnate or ternate fronds arising from short creeping rhizomes. Veins free. Indusium hoodshaped, thin, and withering, attached at one side and arching over the rounded sori.

The genus *Cystopteris* occurs worldwide and is a common element of temperate floras in both the northern and southern hemispheres. Blasdell (1963), who studied the species of the world, recognized 10 species, five varieties, and six hybrids. The genus is noted for much variation and cytogenetic complexity, with much polyploidy (Lovis 1977). The basic chromosome number is 42, and diploids to octaploids are known.

In the Canadian flora there is the highly distinctive northern and alpine *C. montana*; *C. bulbifera*, a basic diploid of southeastern and central North America; the very widespread and abundant *C. fragilis* with its variety, *mackayii*; a diploid segregate species of *C. fragilis*, now known as *C. protrusa*; and a derived allohexaploid species, *C. laurentiana*. We are recognizing five species, but because of all the biosystematic work on this genus that is in progress, it seems inevitable that further segregate species will be recognized. These will be derived from the "*C. fragilis* complex."

A. Fronds ternate .... 1. **C. montana**
A. Fronds bipinnate to tripinnate.
    B. Fronds lanceolate and usually long-attenuate, often bearing bulblets beneath; veins mostly ending in a notch . . .. 2. **C. bulbifera**
    B. Fronds lanceolate only, without bulblets.
        C. Fronds scattered along a creeping rhizome .... 3. **C. protrusa**
        C. Fronds tufted from a short creeping rhizome.
            D. Indusium glandular; veins ending both in teeth and sinuses .... 5. **C. laurentiana**
            D. Indusium not glandular; veins usually ending in the teeth .... 4. **C. fragilis**

1. **Cystopteris montana** (Lam.) Bernh.
   mountain bladder fern
Fig. 131 (a) frond; (b) fertile pinnule. Map 130.

Fronds 40 cm long or longer, arising singly from a widely creeping slender rhizome. Stipes usually longer than the blade, sparsely chaffy. Blades ternate, broadly deltoid-ovate; two lower divisions somewhat narrower and shorter than the upper; pinnae bipinnate to tripinnate; ultimate segments ovate, rounded, often cleft at the apex. Veins ending in the sinuses between minute teeth. Similar in aspect to *Gymnocarpium dryopteris*, but more finely dissected.

Cytology: $n = 84$ (Britton 1964*). This is a tetraploid and has the same number as reported for Europe. Blasdell (1963) suggests from spore sizes that diploids exist.

Habitat: Cool moist woods and rocky slopes, mainly in calcareous places.

Range: Circumpolar; in North America in southern Greenland, Quebec, north shore of Lake Superior, Alaska, British Columbia, southwestern District of Mackenzie and western Alberta, and the mountains of Colorado.

Remarks: This is a most attractive species. In general aspect it might remind one of an oak fern, but the degree of dissection of the blade is so fine that it is sometimes called a lace fern. It is widely distributed, but of rare occurrence, and is considered a find when one finally sees it. *Cystopteris montana* is rare in the District of Mackenzie (Cody 1979), Ontario (Argus and White 1977), and Saskatchewan (Maher et al. 1979).

2. **Cystopteris bulbifera** (L.) Bernh.
   Filix bulbifera (L.) Und.
   bulblet fern
Fig. 132 (a) frond; (b) fertile pinna. Map 131.

Fronds 30–80 cm long or longer, from a short stout rhizome. Blades lanceolate and usually long-attenuate, 6–15 cm wide or wider at the base; sterile blades usually shorter, bipinnate; pinnules oblong, obtuse, pinnatifid to lobed. Veins mostly ending in a notch or sinus. Dark green bulblets often borne on the underside of the rachis and pinnules. Indusium minutely glandular.
    Key field characters are the fine dissections, the veins ending in notches, the bulblets on the axes, and the glandular indusia. It is a distinctive species.

Fig. 131 *Cystopteris montana*; (a) frond, 1/3 × ; (b) fertile pinnule, 7 × .

Fig. 132   *Cystopteris bulbifera*; (a) frond, 1/3 × ; (b) fertile pinnule, 8 × .

Aspidiaceae

Cytology: $n = 42$ (Britton 1953*). This is a basic diploid species.

Habitat: Chiefly on calcareous rocks, in shaded ravines, and in moist woods.

Range: Newfoundland to Ontario, Minnesota, and South Dakota, south to Georgia and Texas. Doubtfully recorded from southeastern Manitoba (Scoggan 1957).

Remarks: The long, graceful, arching fronds of mature plants are distinctive. Smaller, shorter, and more triangular fronds must be identified with care. In early spring, the bright maroon-colored stipes are striking.

3. **Cystopteris protrusa** (Weath.) Blasdell
   C. fragilis (L.) Bernh. var. *protrusa* Weath.
   Fig. 133, sterile and fertile fronds. Map 132.

Fronds 20–45 cm long or longer, scattered along a creeping rhizome; rhizome projecting beyond the current year's fronds. Stipes greenish, or straw-colored, or pale brown. Blades lanceolate, 13–25 cm long, 5–10 cm wide; sterile blades usually shorter, bipinnate; pinnules sharply toothed, ovate-lanceolate to deltoid-ovate; lower pinnules tapering to a stalk-like base. Veins mostly ending in a tooth or on the unnotched margin. Indusium up to 0.5 mm long, shallowly or not at all toothed at the apex.

Cystopteris protrusa may be distinguished from *C. fragilis* var. *mackayii*, with which it might be confused, by the long internodes on the rhizome, the greenish or straw-colored stipes, the softer and more easily wilting blade, which is more ample and more feathery, and the lower pinnules, which taper to a stalk-like base.

Cytology: $n = 42$ (Britton unpublished*).

Habitat: In the shade of deciduous trees on rich river-bottom benches.

Range: In Canada known from only two stations in southwestern-most Ontario (Carolinian Zone); Ontario to Wisconsin, south to Georgia and Mississippi. Further investigation of similar habitats should yield more stations in Ontario.

Remarks: This basic diploid species has only recently been identified in the Canadian flora, although its occurrence in northern Michigan (W.H. Wagner and Hagenah 1956) had suggested that it should be present in southwestern Ontario in similar sites.

Aspidiaceae

Fig. 133 *Cystopteris protrusa*; sterile and fertile fronds, 1/2 ×.

Aspidiaceae

4. **Cystopteris fragilis** (L.) Bernh. var. *fragilis*
    Filix fragilis (L.) Und.
    fragile fern
Fig. 134, frond. Map 133.

Fronds 10–35 cm long or longer, tufted from short creeping rhizomes. Blades lanceolate, 3–8 cm wide or wider near the base, bipinnate; pinnae pinnatifid to lobed, and at least the basal pinnules varying from orbicular to triangular and rounded to the base. Veins mostly ending in a tooth or on the unnotched margin. Indusium up to 1 mm long and more or less cleft at the apex.

A taxon in which the spores are rugose rather than echinate has been called *C. dickieana* Sim or has been treated as a subspecies, a variety, or a form of *C. fragilis*. This taxon is in need of further investigation (Hagenah 1961).

Field characters are the translucent stipe, the veiny thin blades with the veins going to the very tips of the teeth, and a smooth rachis and indusium.

This highly variable species is known to have at least two levels of ploidy (4x and 6x) after *C. protrusa* (2x) has been segregated. There is sufficient interest in this wide-ranging taxon, together with a great deal of biosystematic work in progress, that one can expect further segregate species to be described (Lovis 1977).

Cytology: $n = 84$ (Britton 1953*, eastern; Cody and Mulligan 1982*, western). Both tetraploids and hexaploids are known in Europe.

Habitat: Sheltered crevices in cliffs, moist banks, and wooded talus slopes.

Range: Circumpolar; in North America from Greenland to Alaska, south to Virginia, Texas, and California.

Remarks: This is an abundant, attractive species, which one expects to find on most moist, shady cliffs. It is present on both acidic and basic rocks.

4.1 **Cystopteris fragilis** (L.) Bernh. var. **mackayii** Lawson
Fig. 135 (*a*) frond; (*b*) portion of fertile pinna. Map 134.

Similar to var. *fragilis*, but with the pinnules oblong to nearly lanceolate and evenly wedge-shaped at the base, and the indusium about 0.5 mm long and shallowly or not at all toothed at the apex.

This quite frequent variety of *C. fragilis* has a rather constant morphology. Blasdell (1963) relegated the variety to a *Cystopteris diaphana* × *fragilis* complex, thereby eliminating *C. fragilis* from

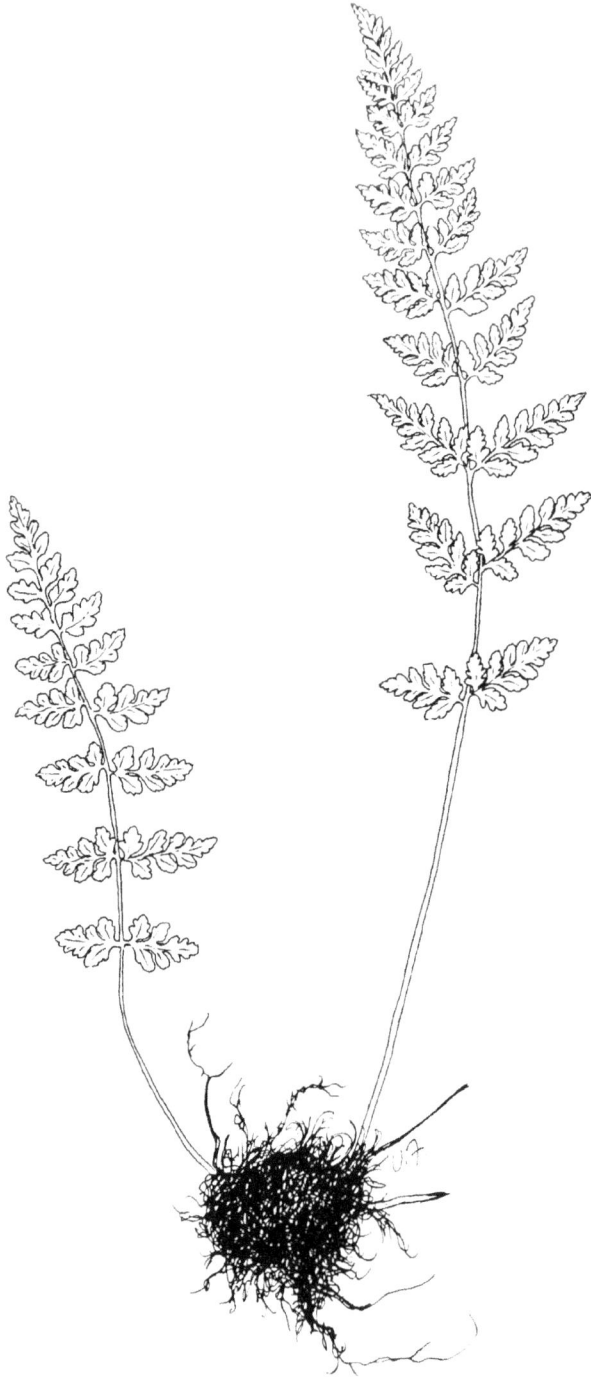

Fig. 134 *Cystopteris fragilis* var. *fragilis*; fronds, 2/3 ×.

Aspidiaceae

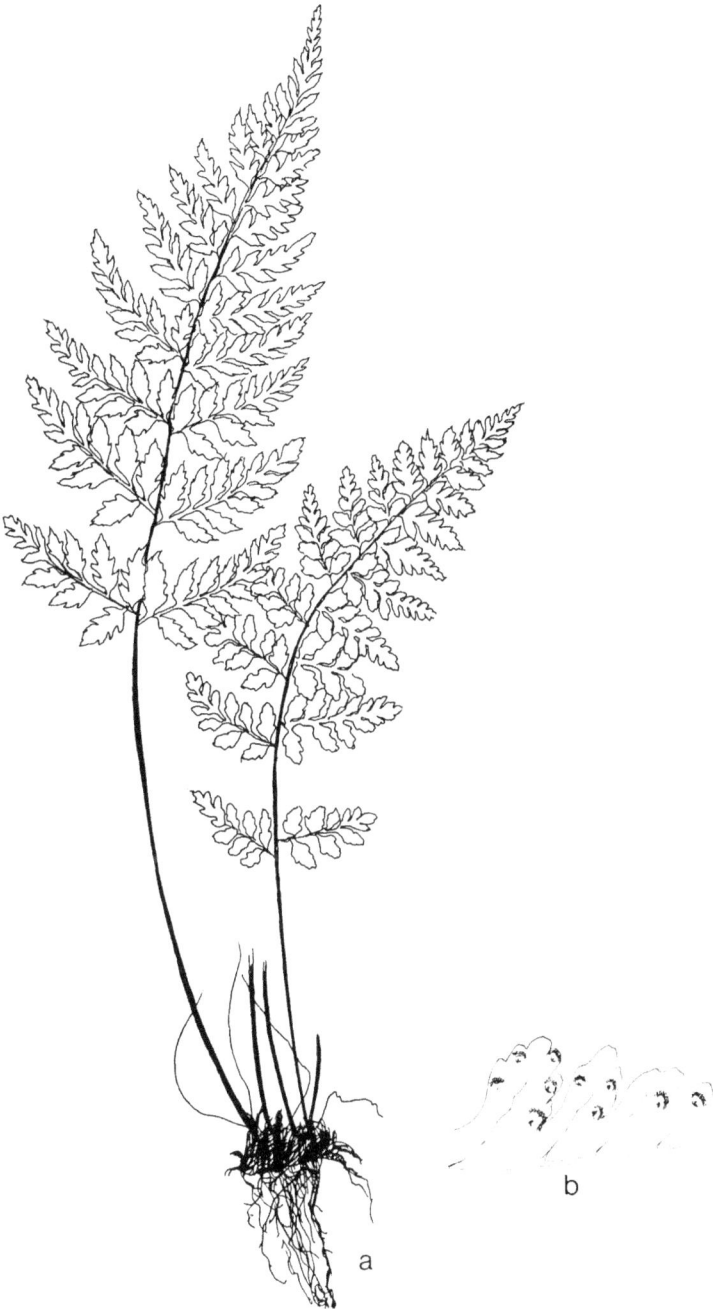

Fig. 135 *Cystopteris fragilis* var. *mackayii*; (a) fronds, 2/3 ×; (b) portion of fertile pinna, 4 ×.

Newfoundland, New Brunswick, and Nova Scotia. A new interpretation has recently appeared that will require a great deal of biosystematic study to validate. Lellinger (1981) suggests that this taxon is *C. tenuis* (Michx.) Desv., which has arisen from a cross of *C. protrusa* (2x) × *reevesiana* (2x), the latter a new species from Utah and Arizona (Lellinger 1981).

Cytology: $n = 84$ (Britton unpublished*).

Habitat: Found in habitats similar to var. *fragilis*, but more often on banks, on rotted logs, and in moist glades on soil.

Range: Nova Scotia to Ontario and Michigan, south to Virginia and Missouri. In Canada it is a more southern and eastern variety than is *C. fragilis* var. *fragilis*.

5.   **Cystopteris laurentiana** (Weath.) Blasdell
     C. fragilis (L.) Bernh. var. *laurentiana* Weath.
     Laurentian fragile fern
Fig. 136, frond. Map 135.

Fronds tufted from a short creeping rhizome, up to 60 cm long or longer. Stipes light brown to red-tinged. Blades ovate-attenuate, up to 34 cm long and 13 cm wide; sterile blades usually shorter, tripinnatifid; pinnules with the veins ending both in teeth and emarginations. Indusium up to 1 mm in diameter, minutely glandular.
This species combines the attributes of its presumed parents, *C. fragilis* var. *fragilis* (4x) and *C. bulbifera* (2x). It is usually an upright, vigorous plant of greater stature than a small or medium-sized *C. fragilis*. The veins go both to the points and to the sinuses, the indusia are glandular, and the spores are larger than those of either parent.

Cytology: $n = 126$ (Britton 1974*).

Habitat: Calcareous rock or slopes.

Range: Newfoundland, Nova Scotia, New Brunswick, Quebec, and Ontario, west to Wisconsin. Reported as highly localized.

Remarks: This species of hybrid origin is comparable to *Dryopteris clintoniana* and *Polystichum setigerum*, which are also considered to be allohexaploids. The spores are large, regular, and freely produced. The chromosome number and the morphology of the plants are consistent with the presumed parentage.
     Cystopteris laurentiana is usually considered a rare plant, but too often is merely lumped into *C. fragilis* s.l. We found it to be quite

abundant on Manitoulin Island, Ont. It was reported by Argus and White (1977) as rare in Ontario and by Maher et al. (1978) as rare in Nova Scotia.

Hybrids of *Cystopteris*

We know of no confirmed hybrids for Canada, after *C. laurentiana* and some of Blasdell's *C. diaphana* complexes are removed from consideration. Now that *C. protrusa* has been confirmed for the Canadian flora, *C. bulbifera* × *protrusa* is a further possibility (Cranfill 1980), as are also various interspecific crosses such as *C. fragilis* × *montana*, known in central Europe but not in North America.

10. *Athyrium* Roth

Rather large ferns with large rootstocks; fronds pinnate to tripinnatifid. Veins either simple or somewhat forked. Sori curved or straight, borne along the veins. Indusium attached on one side of the sorus or lacking.

This is a large genus of about 300 species, mainly of tropical distribution. At one time all the species were included in the genus *Asplenium*. More recently, there has been an attempt to split off such small genera as *Homalosorus* and *Lunathyrium* (Löve et al. 1977), both of which have the same basic number, $x = 40$, as *Athyrium*. Pending further study, we are following Copeland (1947) in maintaining all our species in the genus *Athyrium*.

A.   Fronds pinnate or bipinnatifid.
    B.   Fronds pinnate .... 3. ***A. pycnocarpon***
    B.   Fronds deeply bipinnatifid .... 4. ***A. thelypterioides***
A.   Fronds bipinnate to tripinnatifid.
    C.   Ultimate segments of frond broad and close together; indusium curved or horseshoe-shaped ....
    .... 2. ***A. filix-femina***
    C.   Ultimate segments of frond very narrow and distant; indusium lacking, with sori roundish ....
    .... 1. ***A. alpestre*** ssp. ***americanum***

1.   ***Athyrium alpestre*** (Hoppe) Rylands ssp. ***americanum*** (Butters) Lellinger
    A. alpestre (Hoppe) Rylands var. *americanum* Butters
    A. alpestre (Hoppe) Rylands var. *gaspense* Fern.
    A. distentifolium Tausch. ssp. *americanum* (Butters) Hultén
    A. distentifolium Tausch. var. *americanum* (Butters) Boivin
    A. americanum (Butters) Maxon
Fig. 137 (*a*) frond; (*b*) fertile pinnule. Map 136.

Fig. 136   *Cystopteris laurentiana*; frond, 1/3 ×.

Aspidiaceae

Fig. 137 *Athyrium alpestre* ssp. *americanum*; (a) frond, 1/2 × ; (b) fertile pinnule, 5×.

Aspidiaceae

Fronds up to 80 cm long, forming large clumps from stout erect or somewhat decumbent scaly rhizomes. Stipe short, sparsely scaly. Blades glabrous, subcoriaceous, linear to oblong-lanceolate, acuminate, twice pinnate-pinnatifid; pinnae narrowly deltoid, gradually acuminate; pinnules oblong-lanceolate to narrowly triangular; ultimate segments narrow and distant. Sori roundish, lacking indusia.

The subalpine–alpine habitat of this species, its narrow and more distant pinnules, and the absence of an indusium separate this fern from the more common and widespread lowland *A. filix-femina*.

Cytology: $n = 40$ (Taylor and Lang 1963*).

Habitat: Moist, open, rocky subalpine slopes and alpine meadows.

Range: The species is circumpolar; ssp. *americanum* occurs in Greenland, Newfoundland, and Gaspé Peninsula, Que., and in western North America from southeastern Alaska through western and southern British Columbia to California, Nevada, and Colorado.

Remarks: This is a wide-ranging species, comprising several subspecies. Fernald (1928) described the plant from the Tabletop Range, in the Gaspé Peninsula, as var. *gaspense*, but more recent authors have included it in the North American ssp. *americanum* (Scoggan 1978).

2. **Athyrium filix-femina** (L.) Roth
   lady fern
Fig. 138 (*a*) frond; (*b*) fertile pinnule. Map 137 (var. *cyclosorum*).
Fig. 139 (*a*) frond; (*b*) fertile pinnule. Map 138 (var. *michauxii*).

Fronds up to 2 m long, tufted and erect-spreading from stout chaffy erect or ascending rhizomes. Stipes brittle, scaly near the base. Blades narrowly to broadly lanceolate, bipinnate to tripinnate; pinnae lanceolate, acuminate to attenuate; pinnules somewhat lobed to deeply toothed, blunt, or acute at the tip. Sori oblong to horseshoe-shaped; indusia often toothed, ciliate, and attached by their inner side to a veinlet.

Athyrium filix-femina is a very variable fern throughout its circumpolar range. In western North America var. *cyclosorum* (Ledeb.) Moore (var. *sitchense* Rupr.) is a tall coarse fern with subcoriaceous fronds, and its ultimate segments have broad, bluntish teeth. In eastern North America var. *michauxii* (Spreng.) Farw. is shorter and more delicate; on the basis of differences in the frond, the following forms of var. *michauxii* might be recognized, but in all cases they gradate to one another (Butters 1917).

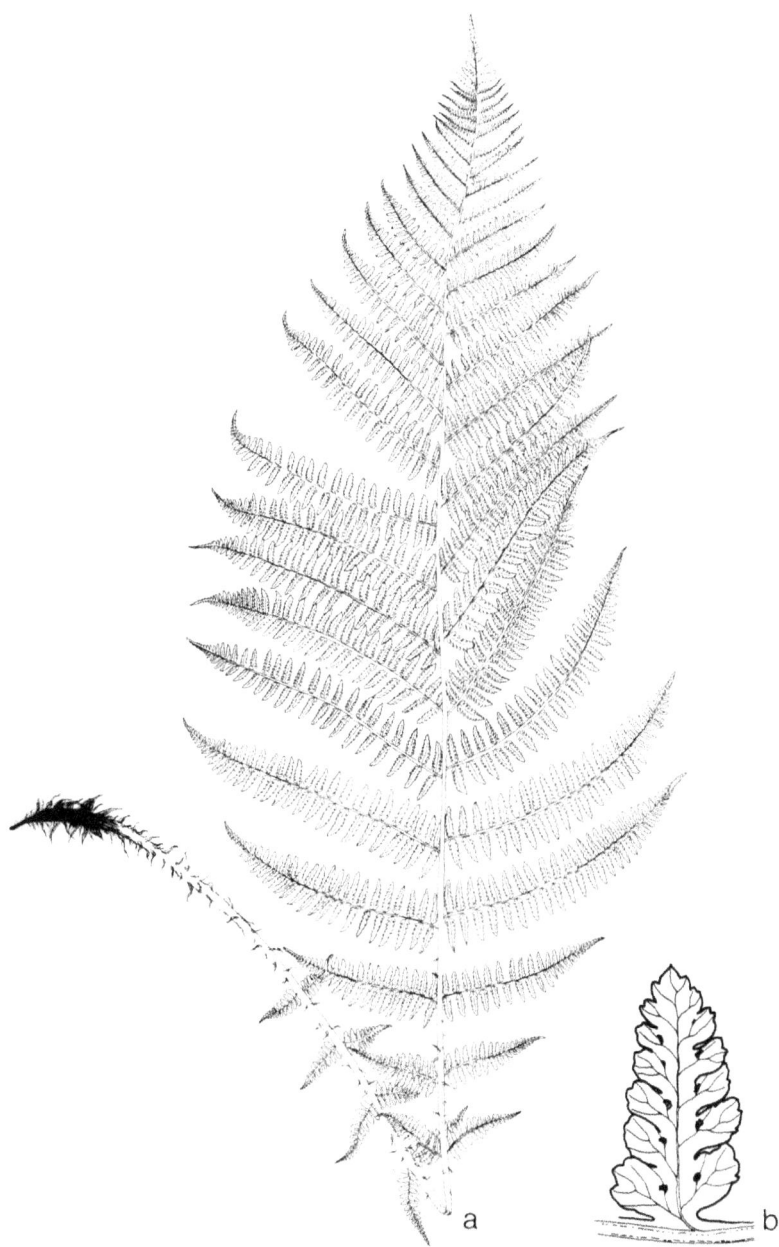

Fig. 138 *Athyrium filix-femina* var. *cyclosorum*; (a) frond, 1/3 ×; (b) fertile pinnule, 3 ×.

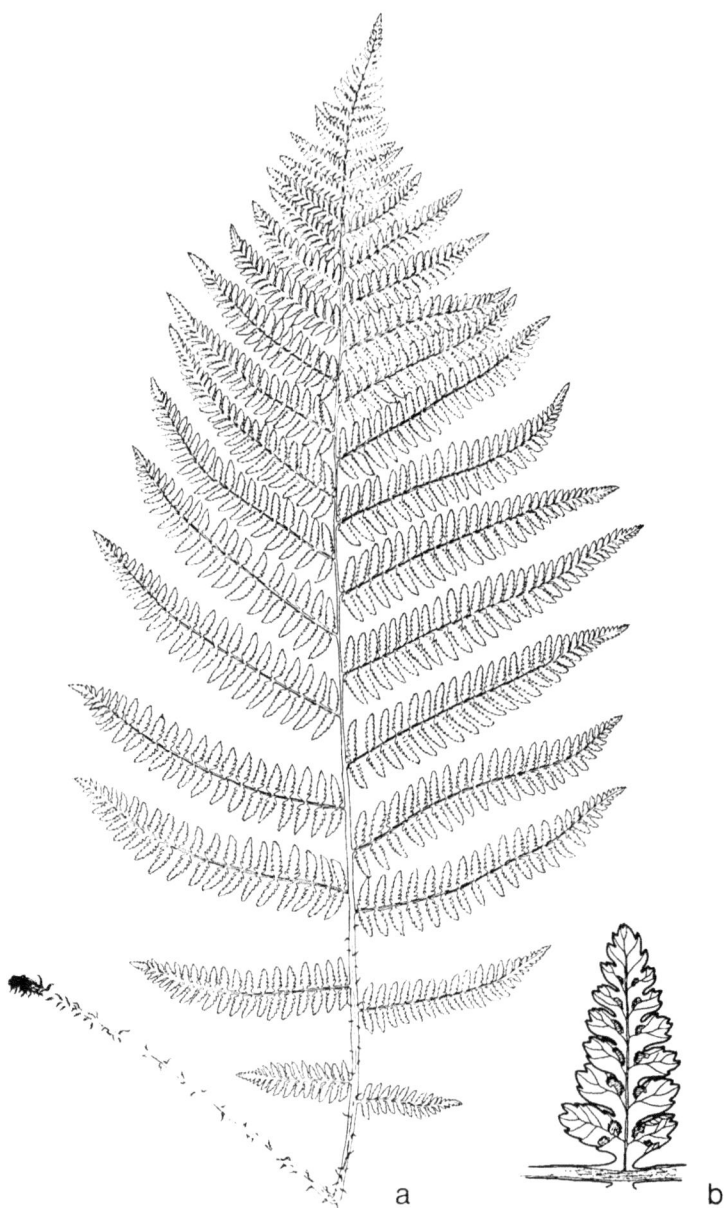

Fig. 139 *Athyrium filix-femina* var. *michauxii*; (a) frond, 1/3 × ; (b) fertile pinnule, 3 × .

Forma *michauxii* has fronds dimorphic; fertile fronds contracted. Pinnae 5–12 cm long. Pinnules 7–12 mm long, rounded, and only shallowly lobed. Sori usually confluent at maturity.

Forma *elatius* (Link) Clute has fronds dimorphic; fertile fronds contracted. Pinnae 10–20 cm long. Pinnules 12–25 mm long, pinnatifid, acutish. Lower sori often strongly curved or horseshoe-shaped; sori usually becoming confluent at maturity.

Forma *rubellum* (Gilbert) Farw. has fronds not dimorphic, larger than the two preceding forms. Pinnules strongly toothed or pinnatifid. Sori separate at maturity.

The lady fern might possibly be confused with some segregates of the *Dryopteris carthusiana* complex, from which it can readily be distinguished by its elongate, sometimes curved (rather than round) sori, which are covered by an indusium attached on one side.

Cytology: var. *cyclosorum*, n = 40 (Mulligan and Cody 1968*); var. *michauxii*, n = 40 (Britton 1953*; Cody and Mulligan 1982*).

Habitat: Meadows to open thickets, moist woods, and even swamps.

Range: The species is circumpolar; var. *cyclosorum* occurs in Alaska, south to California; var. *michauxii* occurs in Labrador and Newfoundland to northern Saskatchewan, south to Pennsylvania, Ohio, Wisconsin, and Iowa.

Remarks: This species can be almost weedy in low-lying overgrown meadows and roadside ditches. The var. *sitchense*, a synonym of var. *cyclosorum*, was recorded by Fernald (1950) from northern Newfoundland and the Gaspé Peninsula, Que. The various forms of these plants are here included in var. *michauxii*. The group is in need of further study.

3.   **Athyrium pycnocarpon** (Spreng.) Tidestr.
      A. angustifolium (Michx.) Milde
      Diplazium pycnocarpon (Spreng.) Brown
      Homalosorus pycnocarpon (Spreng.) Small ex Pichi Sermolli
      narrow-leaved spleenwort
Fig. 140 (*a*) sterile and fertile fronds; (*b*) portion of fertile pinna; (*c*) venation of pinna. Map 139.

Fronds up to 80 cm long or longer, forming a crown at the end of a stout horizontal rhizome. Sterile blades lanceolate, 8–16 cm wide, simply pinnate; pinnae long-acuminate, rounded to truncate at the base, membranous; fertile pinnae lance-linear. Sori linear, situated on the veins in crowded rows between the midrib and the margin. Indusium opening along one side.

Aspidiaceae   269

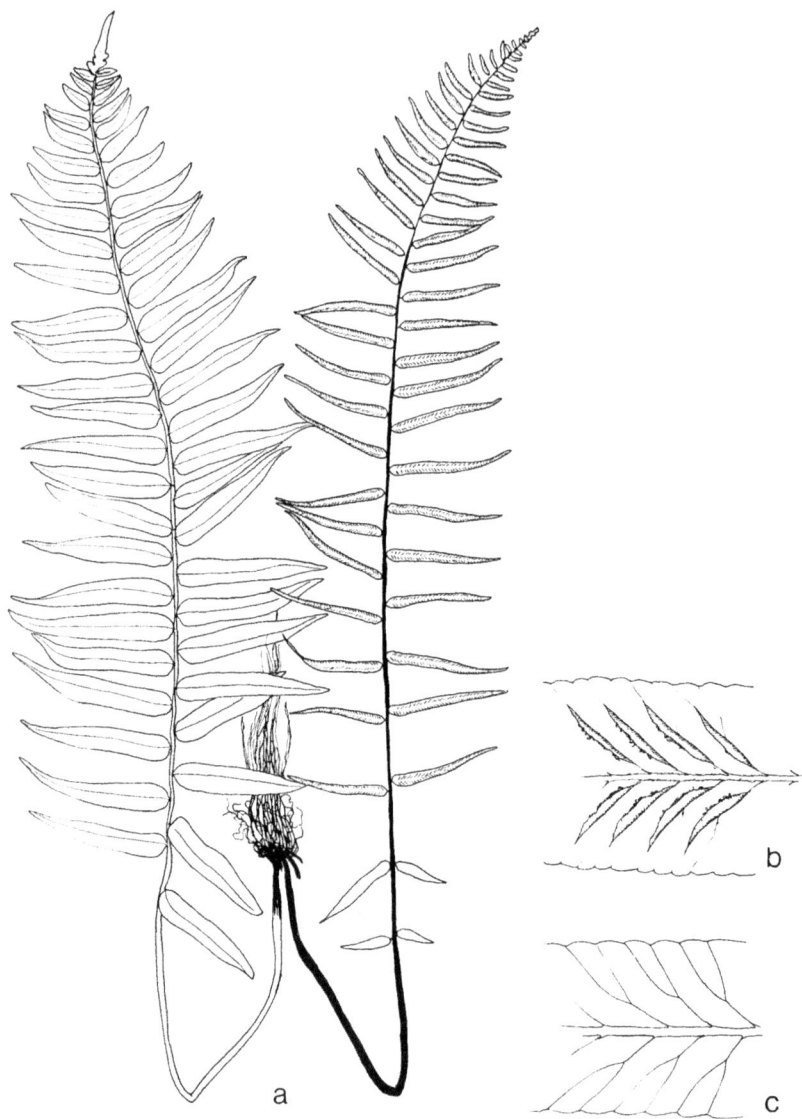

Fig. 140 *Athyrium pycnocarpon*; (a) sterile and fertile fronds, 1/3 ×; (b) portion of fertile pinna, 4 ×; (c) venation of pinna, 2 ×.

Aspidiaceae

The sterile fronds of the narrow-leaved spleenwort are somewhat similar to the Christmas fern, but are much more delicate and thus easily differentiated from the evergreen fronds of the latter.

Cytology: $n = 40$ (Britton 1964*; Cody and Mulligan 1982*).

Habitat: Deep, rich, moist woods and ravines.

Range: Southern Quebec and Ontario, south and west to Georgia, Alabama, Minnesota, Iowa, and Kansas.

Remarks: This species is recorded as rare in Ontario by Argus and White (1977), and it certainly is rare in Quebec also.

4. **Athyrium thelypterioides** (Michx.) Desv.
    A. acrostichoides (Sw.) Diels
    Diplazium thelypterioides (Michx.) Presl
    D. acrostichoides (Sw.) Butters
    Lunathyrium acrostichoides (Sw.) Ching.
    silvery spleenwort
Fig. 141 (a) frond; (b) fertile pinnule. Map 140.

Fronds up to 100 cm long or longer, forming a crown at the end of the horizontal rhizome. Blades lanceolate to elliptic-lanceolate, 8–22 cm wide, pinnate-pinnatifid; pinnae long-tapering; segments oblong, blunt, finely toothed. Sori straight or slightly curved, situated on the veins between the midrib and margin. Indusia becoming silvery at maturity, opening on one side or, if double, opening on both sides.

The silvery spleenwort can readily be distinguished by the straight or slightly curved sori, which toward maturity fill the underside of the fertile fronds and become silvery in color.

Cytology: $n = 40$ (Britton 1964*; Cody and Mulligan 1982*).

Habitat: Rich woods, stream banks, shaded slopes, and rarely in open thickets.

Range: Nova Scotia to southern Ontario, south and west to Georgia, Alabama, Michigan, Wisconsin, and Minnesota.

Remarks: *Athyrium thelypterioides* is often associated with other ferns of rich woodlands, such as *Dryopteris goldiana*. A report by Macoun (1890) of this species growing at Current River, Lake Superior (Thunder Bay, Ont.) was repeated by Scoggan (1978), but no substantiating specimen has been found, nor have recent collectors gathered it in that area.

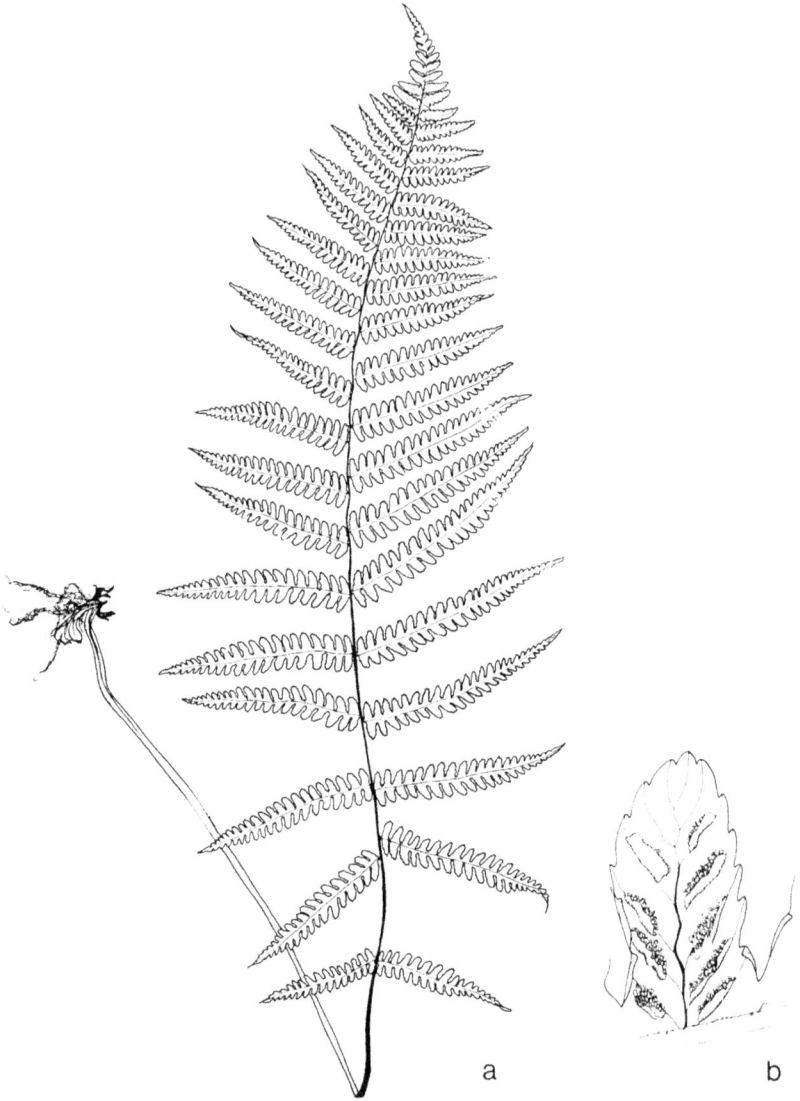

Fig. 141 *Athyrium thelypterioides*; (a) frond, 1/3 × ; (b) fertile pinnule, 5 × .

Aspidiaceae

11. BLECHNACEAE

Ferns coarse, tufted from a short creeping rhizome or scattered along a widely creeping rhizome. Fronds similar or dimorphic, pinnatifid, pinnate, or bipinnatifid. Veins anastomosing to form costal or costular areolae, and then free to the margin or forming additional areolae. Sori elongate. Indusium opening on the costal side. A small family of terrestrial or occasionally climbing ferns.

A. Fronds evergreen, dimorphic, pinnate . . . . 1. *Blechnum*
A. Fronds deciduous, dimorphic or similar, pinnatifid or bipinnatifid . . . . 2. **Woodwardia**

1. *Blechnum* L.

1. ***Blechnum spicant*** (L.) Roth
   B. spicant (L.) Roth ssp. *nipponicum* (Kunze) Löve & Löve
   B. doodioides Hook.
   Lomaria spicant (L.) Desv.
   Struthiopteris spicant (L.) Weiss
   deer fern
Fig. 142 (*a*) sterile and fertile fronds; (*b*) portion of fertile pinna. Map 141.

Fronds dimorphic, tufted from a short creeping rhizome. Sterile fronds evergreen, 10–40 cm long, spreading and appressed to the ground; stipes short; blades linear-oblanceolate, pinnate; pinnae oblong or linear-oblong, blunt or somewhat pointed, becoming much reduced towards the base. Fertile fronds fewer, upright, deciduous, much longer than the sterile; pinnae linear, narrower than the sterile. Veins simple or branched. Sporangia confluent, parallel to the midrib. Indusium continuous, brown-hyaline, attached close to the margin.

The genus *Blechnum* numbers about 200 species, mainly of south-temperate and tropical distribution. The species are both terrestrial and epiphytic. Only one species, *B. spicant*, of wide distribution, is represented in our area. As stated by Hitchcock et al. (1969), the American plants tend to be a little more robust than the European ones, but are scarcely separable taxonomically. Löve and Löve (1966*b*, 1968) did, however, refer the plants of the Pacific region to ssp. *nipponicum*. If these were treated at specific levels, then the name to be applied would probably be *B. doodioides*, the type of which was collected in British Columbia.

Cytology: $n = 34$ (Cody and Mulligan 1982*).

Fig. 142 *Blechnum spicant*; (a) sterile and fertile fronds, 1/3 ×; (b) portion of fertile pinna, 3 ×.

Blechnaceae

Habitat: Wet woods and clearings near the Pacific coast, and also inland in the Revelstoke region.

Range: Interruptedly circumpolar; in North America from coastal Alaska to California.

Remarks: A common fern in coastal British Columbia, *B. spicant* is found in the same habitat as *Polystichum munitum.* In the Queen Charlotte Islands, it is found in almost every habitat, from sea level to the tree line (Calder and Taylor 1968).

2. *Woodwardia* Sm.

Coarse ferns, with horizontal widely creeping or stout ascending rhizomes.   Fronds similar or dimorphic, pinnatifid or bipinnatifid. Veins anastomosing to form costal or costular areolae, then free to the margin or forming additional areolae.   Sori linear or oblong, parallel to the midveins, borne along the veinlets, which form the outer side of the first row of areolae.   Indusium persistent, opening on the side adjacent to the midrib.

The species of this small genus are found in bogs and wet shady places.   Some authors (Cranfill 1980) have separated *Woodwardia areolata* from the genus as a monotypic species in the genus *Lorinseria*, based on the marked dimorphism and other characters.   In a treatment of the Florida chain ferns, Lucansky (1981) states that comparative anatomical data support the placement in the genus *Woodwardia* of the three species found in that state, *W. virginica, W. radicans*, and *W. areolata.*   Cody (1963) reported on the genus in Canada.

A.   Fronds dimorphic, pinnatifid . . . . 1. **W. areolata**
A.   Sterile and fertile fronds similar or nearly so, pinnate-pinnatifid.
    B.   Fronds 0.7–1.3 m long; pinnules 0.6–1.5 cm long; sori usually confluent when mature . . . . 3. **W. virginica**
    B.   Fronds 1–2 m long; pinnules 2–6 cm long; sori usually separate at maturity . . . . 2. **W. fimbriata**

1.   **Woodwardia areolata** (L.) Moore
    Lorinseria areolata (L.) Presl
    W. angustifolia J.E. Smith
    netted chain fern
Fig. 143 (*a*) sterile and fertile fronds; (*b*) portion of fertile pinna; (*c*) venation.   Map 142.

Fronds dimorphic, pinnatifid, arising from a slender branching rhizome.   Sterile fronds 10–70 cm long; stipes greenish yellow; blades

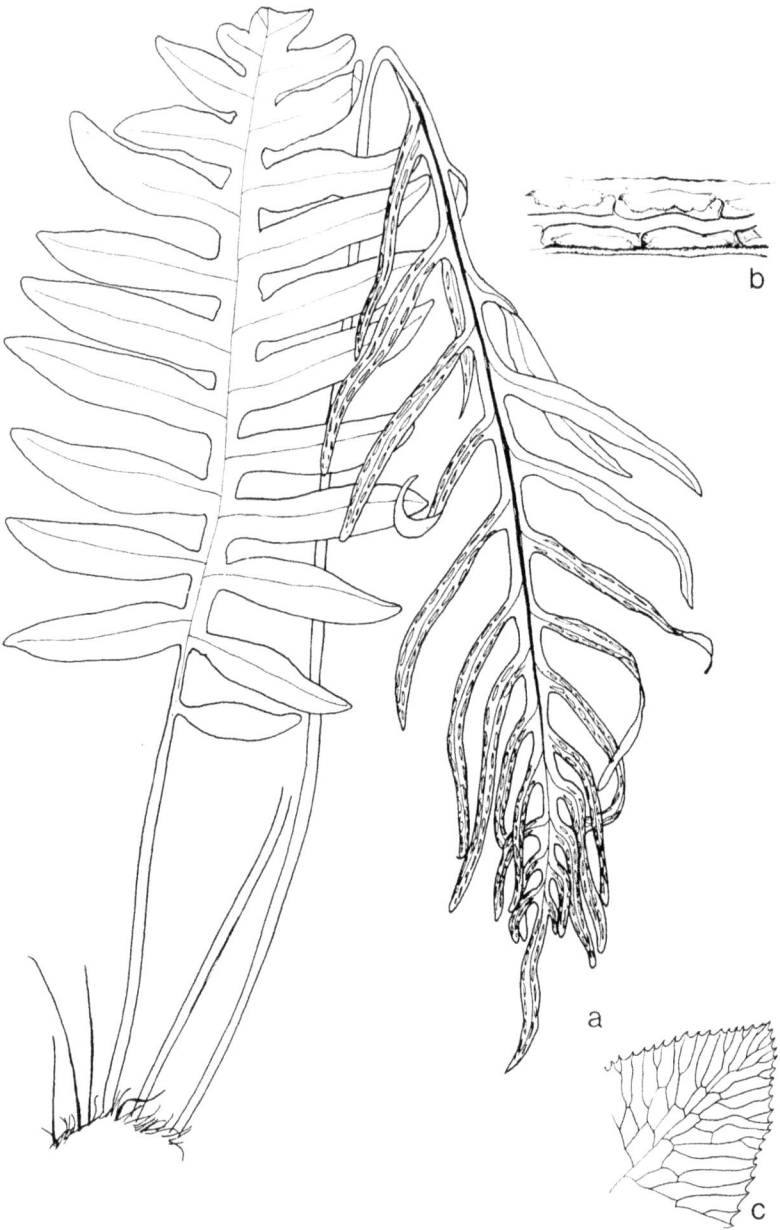

Fig. 143 *Woodwardia areolata*; (a) sterile and fertile fronds, 1/2 × ; (b) portion of fertile pinna, 2 × ; (c) venation, 2 × .

Blechnaceae

oblong-lanceolate to ovate, 7–35 cm long or longer; lanceolate divisions united at the base by a broad wing, with margins finely toothed; venation in several rows of areolae, then open to the margin. Fertile fronds taller than the sterile; stipes darker; blades narrower; divisions narrowly linear and almost distinct. Sori in a single row on each side of the secondary midrib.

The sterile fronds of this species might at first glance be confused with those of the common and sometimes weedy *Onoclea sensibilis*, sensitive fern. *Woodwardia areolata* can easily be distinguished from that species by its minutely serrate rather than entire margins and by its basal pinnae, which are alternate rather than subopposite.

Cytology: $n = 35$ (Cody and Mulligan 1982*).

Habitat: Mediacid situations along streams and among cobblestones on beaches.

Range: In Canada known only in Yarmouth, Shelbourne, and Queens counties in Nova Scotia; in the United States on or near the coastal plain to Florida and Texas and sparingly inland to Missouri and Oklahoma.

Remarks: *Woodwardia areolata* is rare in Nova Scotia (Maher et al. 1968).

2.   **Woodwardia fimbriata** J.E. Smith
     W. chamissoi Brack.
     W. paradoxa Wright
     W. radicans (L.) Smith var. *americana* Hook.
Fig. 144 (*a*) portion of frond; (*b*) fertile pinnule. Map 143.

Fronds up to 70 cm long or longer (longer to the south), evergreen, forming a crown at the end of the stout and widely creeping or more or less erect rhizome. Stipes equaling the blades, conspicuously chaffy at the base. Blades linear-oblong to oblong-lanceolate, pinnate-pinnatifid, acuminate at the tip and narrowed towards the base; pinnae linear-oblong, obliquely pinnatifid; ultimate segments narrowly triangular to linear, acuminate; venation in a single row of areolae on either side of the midvein, then free and sometimes forking to the sharply and closely serrulate teeth. Sori in shallow pits in the areolae on either side of the secondary midrib.

The elongated sori on either side of the secondary midrib and the tall stature of this fern readily set it off from other fern species in British Columbia.

Cytology: Manton and Sledge (1954) reported a count of $n = 34$ based on material from Ceylon (Sri Lanka), but it is rather doubtful if the species was *W. fimbriata*.

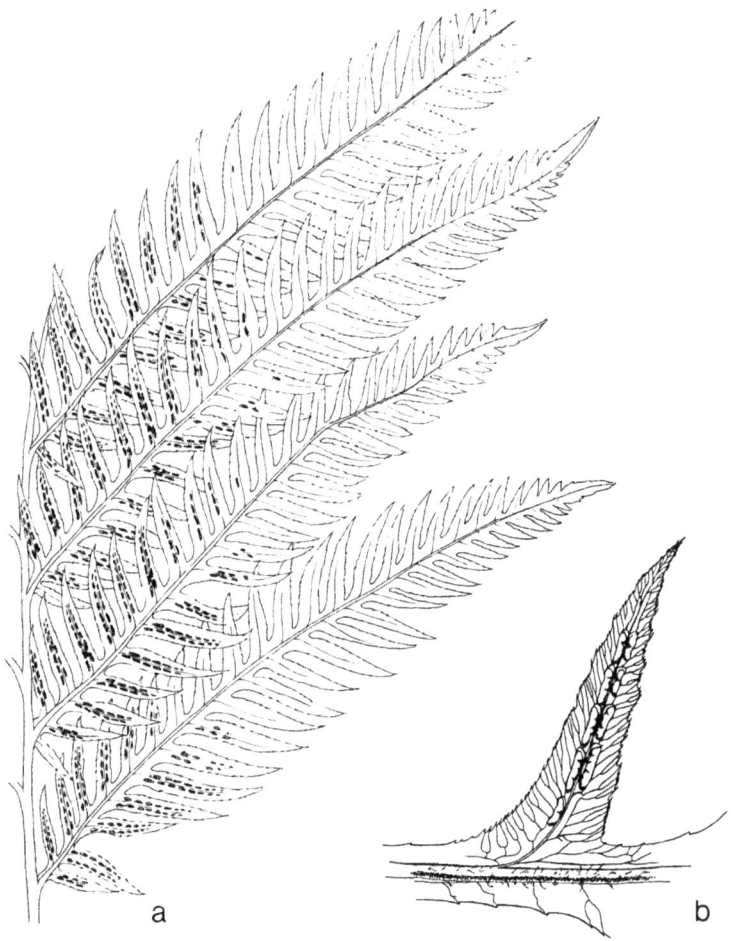

Fig. 144 *Woodwardia fimbriata*; (a) portion of frond, 1/3 × ; (b) fertile pinnule, 1 ×.

Blechnaceae

Habitat: Damp and boggy woodland and banks.

Range: In Canada known only from coastal British Columbia: Lasqueti Island, Texada Island, and Saanich Arm on Vancouver Island; south in the United States to California and Arizona.

Remarks: This species will undoubtedly be placed on the list of rare plants of British Columbia.

3.   **Woodwardia virginica** (L.) Sm.
   Anchistea virginica (L.) Presl
   Virginian chain fern
Fig. 145 (a) portion of frond; (b) fertile pinnules. Map 144.

Fronds 60–100 cm long, scattered along the creeping rhizome. Stipes long, lustrous.   Blades oblong-lanceolate, 10–30 cm wide, pinnate-pinnatifid; pinnae linear-lanceolate; pinnules oblong, obtuse, with finely serrulate margins.   Veins united to form a single series of areolae next to the midrib of both the pinnae and the pinnules, then free to the margin.   Sori oblong, usually becoming confluent at maturity, one to each areole.
   The distinctive chains of sori on the areolae adjacent to the midrib set this fern off from all other fern species in eastern Canada.

Cytology: $n = 35$ (Britton 1964*).

Habitat: Swampy woods, boggy shores, cobbly lakeshores, and peat bogs.

Range: Eastern North America, Nova Scotia to Ontario, south to Florida and Texas.

Remarks: *Woodwardia virginica* was originally placed in the genus *Blechnum* by Linnaeus.   Presl (1851) later transferred it to the genus *Anchistea*, where some researchers still prefer to retain it, although only the glandular indusia and the presence of a single row of areoles distinguish this species from other species of *Woodwardia*.

Blechnaceae   279

Fig. 145 *Woodwardia virginica*; (a) portion of frond, 1/3×; (b) fertile pinnules, 1 1/2×.

Blechnaceae

12. ASPLENIACEAE

Mostly small ferns with fronds firm, simple, pinnate, or bipinnate. Veins free or forking. Sori elongate, occurring along the veinlets. Indusia attached to the veinlets.
The family consists of 600–700 species worldwide and is dominated by the very large genus *Asplenium*. In all, there are fewer than 14 genera, and the other genera have very few species. *Asplenium* has been extensively studied in Europe, the United States, and New Zealand, and chromosome numbers for over 140 species are known. Lovis, who followed Manton at Leeds, has made many artificial crosses and, together with Meyer in Berlin, has analyzed the European species in great detail (Lovis 1977; see also Reichstein 1981). There are 38 European species of which 18 are tetraploids and of these, 11 are genomic or alloploids and six are considered to be autoploids (Lovis 1977). Evans (1970) reviewed studies on *Asplenium*, with special emphasis on those in the southern Appalachians. Ten species and 10 hybrids are included in the discussion, and *Camptosorus* is such an integral part of the crossing diagram and is involved in so many hybrid combinations that it was evident that it should be considered as an entire-leaved *Asplenium* rather than a separate genus. Recent work from Europe (Jermy et al. 1978) has included the genera *Phyllitis* and *Ceterach* in *Asplenium* as well.
In the Canadian flora, there are only six species to consider, but these are all attractive species that are always admired when seen in the wild. Most are found in rock crevices and usually in shady and mossy places, so that they create a very pleasing tapestry of different shades of green.

A.　Fronds simple, commonly auricled at the base.
　　B.　Fronds long-caudate, sometimes rooting at the tip . . . .
　　. . . . . 2. **Camptosorus**
　　B.　Fronds oblong, not attenuate or rooting at the tip . . . .
　　. . . . . 3. **Phyllitis**
A.　Fronds pinnate or bipinnate . . . . 1. **Asplenium**

1. *Asplenium* L.

Small ferns, usually of rocky places, from small compact rhizomes. Veins free. Sori linear, oblique. Indusium usually membranous, attached lengthwise along one side of the sorus.
A very large genus with a broad distribution on all continents. In our flora we have only four species of distinctive morphology. One, *A. viride*, is a northerner, and the other three are cool temperate plants. *Asplenium platyneuron* has almost all its distribution south of Canada, and *A. ruta-muraria* has very few stations in Canada.

Our species are either diploid or tetraploid, but polyploidy is impressive in some species in the world. Twelve-ploid and 16-ploid species are known based on $x = 36$ (Lovis 1977).

A. Fronds dimorphic; fertile fronds upright, much taller than the spreading sterile ones .... 1. *A. platyneuron*
A. Fronds similar.
    B. Rachis purplish black, lustrous .... 3. **A. trichomanes**
    B. Rachis green.
        C. Blade linear, simply pinnate .... 4. **A. viride**
        C. Blade deltoid-ovate, bipinnate .... 2. **A. ruta-muraria**

1. **Asplenium platyneuron** (L.) Oakes
    A. ebeneum Ait.
    ebony spleenwort
Fig. 146 (a) sterile and fertile fronds; (b) fertile pinnules. Map 145.

Fronds dimorphic, tufted from a short rhizome. Fertile fronds stiff and upright, 20–40 cm long, 2.5–4.0 cm wide, gradually tapering to the base; pinnae linear-oblong or basal pinnae triangular, auricled, widely separated and alternated; rachis lustrous, chestnut purple. Sterile fronds shorter, spreading, and prostrate, with oblong approximate pinnae. Sori linear-oblong, situated on the veins, nearer the midvein than the margin.

This species is easily identified by the stiff, upright fertile fronds. It is not often associated with the other species of *Asplenium* in our flora, and so the probability of hybridization is reduced.

Cytology: $n = 36$ (W.H. Wagner 1973a).

Habitat: In partial shade in open woods, grown-over areas and clearings, often in moss or in very shallow soil over rocks.

Range: Southwestern Quebec and southern Ontario, south to Florida and Texas and west to Iowa and Kansas in recent years.

Remarks: W.H. Wagner and Johnson (1981) have documented the recent spread of *A. platyneuron* in parts of Ontario and adjacent Quebec and have provided a wealth of biological data on the species. These recently produced plants are found in very atypical habitats and may prove to be short-lived. We have seen small colonies in young pine plantations on sand; in beds of *Equisetum hyemale* on moist sandy slopes, i.e., acidic locations; and on rocks and in moss on limestone paving, both presumably of high pH. The species was considered to be rare in Ontario by Argus and White (1977), but now seems to be too widespread to be so regarded.

Aspleniaceae

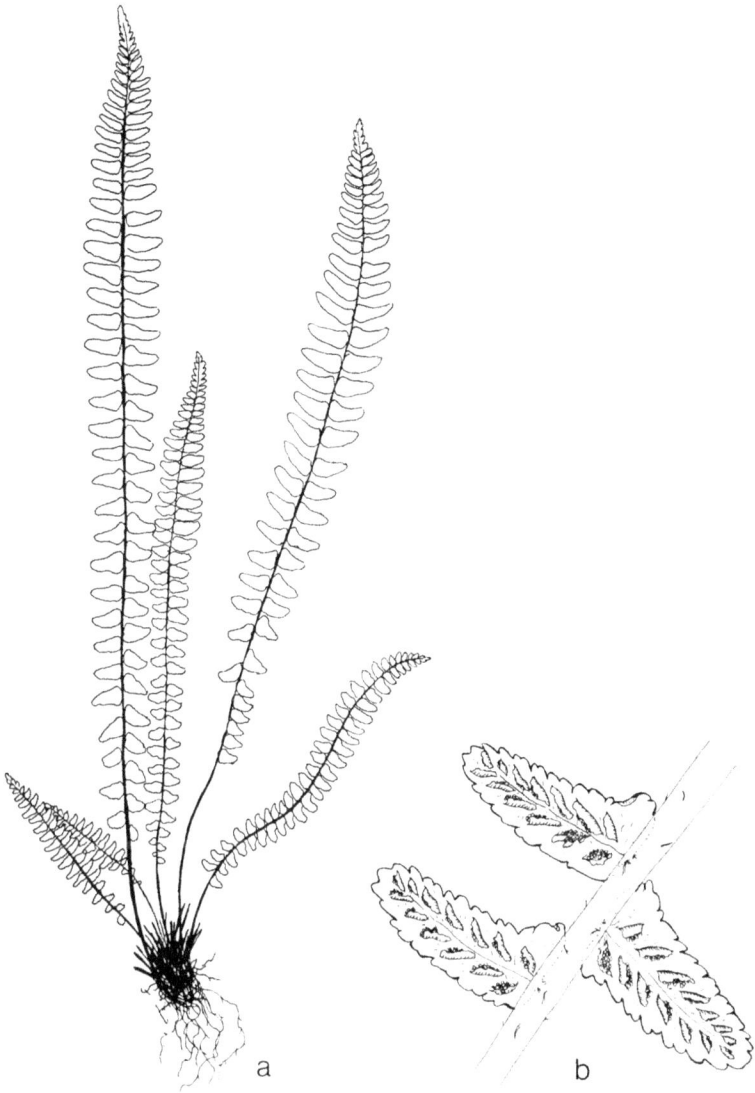

Fig. 146 *Asplenium platyneuron*; (a) sterile and fertile fronds, 1/2 × ; (b) fertile pinnules, 3 × .

2. **Asplenium ruta-muraria** L.
   A. *cryptolepis* Fern.
   wall-rue
Fig. 147 (a) fronds; (b) fertile pinnules. Map 146.

Fronds 3–9 cm long, tufted from a short rhizome. Stipes naked or minutely scaly at the base, green. Blades deltoid-ovate, 1–6 cm long, bipinnate; pinnae and pinnules mostly alternate, petioled; ultimate segments rhombic or obovate, long-cuneate at the base, with the broadly rounded apex crenately toothed. Veins flabellate, simple or forked. Sori few, linear-oblong, usually not confluent in age.

This is a very diminutive fern that is rarely abundant in our area. One is fortunate to locate a small plant after inspecting literally thousands of large dolomitic talus boulders without success.

Cytology: $n = 72$ (W.H. Wagner and F.S. Wagner 1966). This has the same chromosome number as the European plants. Lovis (1977) says that ssp. *ruta-muraria* is an autotetraploid of ssp. *dolomiticum* ($n = 36$) in Europe.

Habitat: Sunny or shaded crevices of limestone cliffs.

Range: Manitoulin Island and Bruce Peninsula, Ont., southern Quebec, in the United States from Vermont to Michigan, south to Alabama and Missouri.

Remarks: The distribution in Ontario is considered in detail by Soper (1955). The species is on the rare plant list for Ontario (Argus and White 1977). This is such a common and abundant plant of stone walls in Great Britain that it is a surprise for field workers from Europe to find that the plant is considered such a rarity in Canada. Frère Louis-Alphonse in 1951, 1952, and 1953 collected plants from "crevasses dans un rocher" (crevices in a rock) at Baie Missisquoi in southern Quebec near the United States border (specimens at MT). B. Boivin has annotated these specimens "var. *ruta-muraria* et vraisemblement planté," (var. *ruta-muraria* and presumably planted) but in view of the fact that the plant is known in nearby Vermont, this seems doubtful.

Variety *cryptolepis* (Fern.) Massey was considered to differ from the European ssp. *ruta-muraria* by hidden rhizome scales (Fernald 1928). We are in agreement with Wherry (1961) that the variety should be dropped.

Fig. 147 *Asplenium ruta-muraria*; (a) fronds, 1 × ; (b) fertile pinnules, 3 × .

3. **Asplenium trichomanes** L.
maidenhair spleenwort
Fig. 148 (a) fronds; (b) fertile pinnae. Map 147.

Fronds 6–20 cm long or longer, forming a dense tuft from a compact rhizome. Stipe and rachis purple brown; old rachises persistent. Blades linear, pinnate; pinnae usually opposite or subopposite, oval, rounded to cuneate at the inequilateral base and slightly toothed on the sides and at the blunt apex. Sori linear, situated on the veins between the midrib and the margin.

Cytology: $n = 36$ (Britton 1964*; Cody and Mulligan 1982*); $n = 72$ (Britton 1953*; Cody and Mulligan 1982*).

Habitat: Sheltered rock crevices.

Range: Circumpolar; in North America from western Newfoundland to Ontario, British Columbia, and south to Georgia, Alabama, Arkansas, Oklahoma, and Arizona.

Remarks: This extremely attractive species has received a great deal of attention from the Leeds school (Lovis 1977). Early work showed that there were both diploids and tetraploids in the complex, the former on more acidic rocks than the latter. Extensive analysis showed that the tetraploid was an autotetraploid, and names were selected to indicate this relationship. The basic diploid became *Asplenium trichomanes* ssp. *trichomanes* and the autotetraploid *A. trichomanes* ssp. *quadrivalens* D.E. Meyer emend Lovis. Later, another diploid was found in Europe on limestone and was described as *A. trichomanes* ssp. *inexpectans* Lovis. This work has been followed with interest in North America, although authors have been reluctant to utilize the subspecific names from Europe without further study of North American plants. It has been known since 1953 (Britton 1953) that both diploids and tetraploids occur in North America and that soil preferences exist for these (diploid on granite and tetraploid on limestone). The complexity of the situation was highlighted by D.H. Wagner and W.H. Wagner (1966), who reported a colony in Virginia consisting of one diploid, three triploids, and 85 tetraploids, and no superficial differences in morphology.
      For those who wish to delineate ssp. *trichomanes* from ssp. *quadrivalens*, one should study the note by Lovis in the *Atlas of Ferns of the British Isles* (Jermy et al. 1978), Jermy and Page (1980), and Moran (1982).
      Arriving at names for a simple recognition of diploids and tetraploids in North America has been greatly complicated by Löve (Löve et al. 1977), who decided that *Asplenium trichomanes* L. ssp. *trichomanes* is the tetraploid, not the diploid, as treated by Lovis

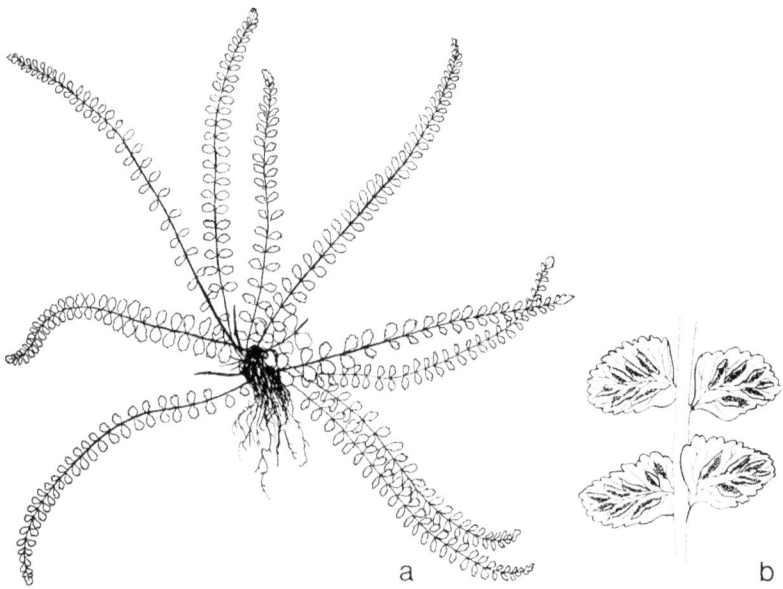

Fig. 148   *Asplenium trichomanes*; (a) fronds, 1/2 × ; (b) fertile pinnae, 3 × .

(1964), and that the diploid should be known as *A. melanocaulon* Willd. It would be a great pity if the taxonomic rules forced us to use two entirely different specific names for plants that have so many features in common. We side with Lovis (1964) "that when a Linnean name covers a species complex, this name should be retained by that segregate with which Linnaeus would have been most familiar, and which would therefore most likely have represented to him the most typical form of his species."

4.   **Asplenium viride** Huds.
     green spleenwort
Fig. 149 (*a*) fronds; (*b*) fertile pinnae. Map 148.

Fronds 2–14 cm long, tufted from a short rhizome. Stipes darkened below, green above. Rachis green. Blade linear to linear-lanceolate, pinnate; pinnae rounded or rhomboid-ovate, crenate. Sori elongate, borne near the indistinct midrib, becoming confluent at maturity.

This northern species, with its bright green stipes, poses no identification problems unless extremely small, poorly developed fronds from subarctic sites are encountered and are subsequently confused with *Woodsia glabella*. One should check for an elongate sorus, with the indusium attached on one side, for positive identification of *A. viride*.

Cytology: $n = 36$ (Taylor and Lang 1963*; Britton 1964*).

Habitat: Among talus and in usually protected crevices of limestone or basic rocks in shady locations.

Range: Circumpolar; in North America from Newfoundland to Alaska, south to New York, Colorado, Utah, and Washington.

Remarks: A distinctive species usually less common on the Niagara Escarpment than is *Asplenium trichomanes*, with which it is sometimes associated. One might expect hybrids to be frequent in localities where the two species grow intermixed, but although that hybrid is reported as a very rare one in Europe, none are known for North America. Considered to be rare in Ontario by Argus and White (1977).

Hybrids of *Asplenium*

Since there are four species of *Asplenium* in Canada and two levels of ploidy ($2x$ and $4x$), the theoretical expectations with five entities and no barriers to hybridization are (4–3–2–1), or 10 possible primary crosses. Five of these are known in nature or have been

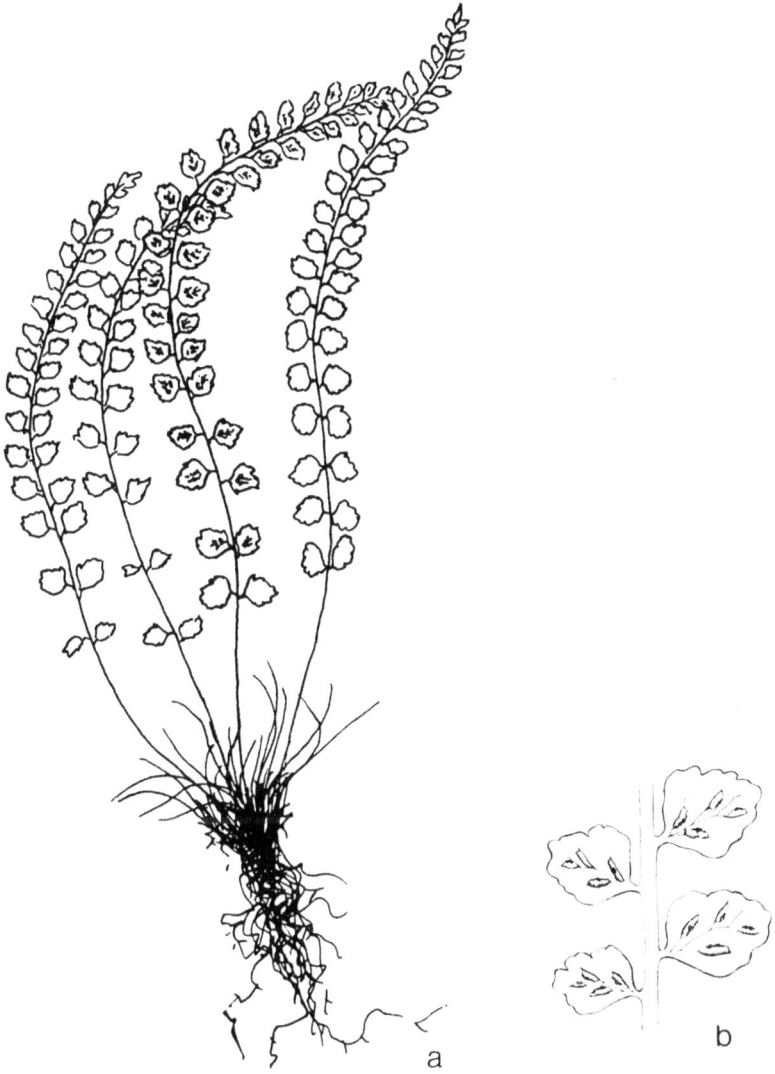

Fig. 149 *Asplenium viride*; (a) fronds, 1 × ; (b) fertile pinnae, 3 × .

produced artificially and are recorded in the literature (Knobloch 1976). Reichstein (1981) considers four known in Europe. We know of none in Canada, although five *A. trichomanes* hybrids are known in the United States and all are very rare (Moran 1982).

What was once considered even more unusual — intergeneric crosses between *Camptosorus* and various species of *Asplenium* — are well known in the United States. In theory, each of the five taxa of *Asplenium* could cross with *Camptosorus*, and so there would be five possible combinations (Knobloch 1976). None has been found in Canada so far.

It is worth noting that one can see evolution in action when one compares the primary hybrid of the two diploids, *Asplenium platyneuron* (PP) × *Camptosorus rhizophyllus* (RR). The result is a sterile diploid (P)(R) known as Scott's spleenwort, or *Asplenium* × *ebenoides*. If an occasional spore should be produced with all the chromosomes of this hybrid, it gives rise after fertilization to a new fertile allotetraploid with the constitution PPRR and would be known as *A. ebenoides*, a derived species.

Excluded report

Asplenium marinum L.

Lawson (1889) stated, "There are Nova Scotian specimens in the Kew Herbarium, referred to in Hooker & Baker's *Synopsis Filicum*, second edition, 1883, but this fern has not recently been found in Canada. It grows around the shores of Western Europe, and extends from Orkney, the British Isles, Canaries and Azores, to St. Vincent and South Brazil." A search of the herbaria at Kew, Edinburgh, and Glasgow failed to turn up the specimens. Presumably they have been revised to some other species.

2. *Camptosorus* Link

1. *Camptosorus rhizophyllus* (L.) Link
   Asplenium rhizophyllum L.
   walking fern
   Fig. 150 (*a*) fronds; (*b*) portion of fertile frond. Map 149.

Fronds 5–30 cm long or longer, clustered at the end of the erect or ascending scaly rhizome. Blades evergreen, entire, 1–3 cm wide at the cordate or auriculate base, usually tapering to a long caudate tip. Veins reticulate. Sori elongate, scattered along the veins. Indusium attached on one side of the sorus. Tips of the arching blades often rooting to form new plants, hence the name walking fern.

Aspleniaceae

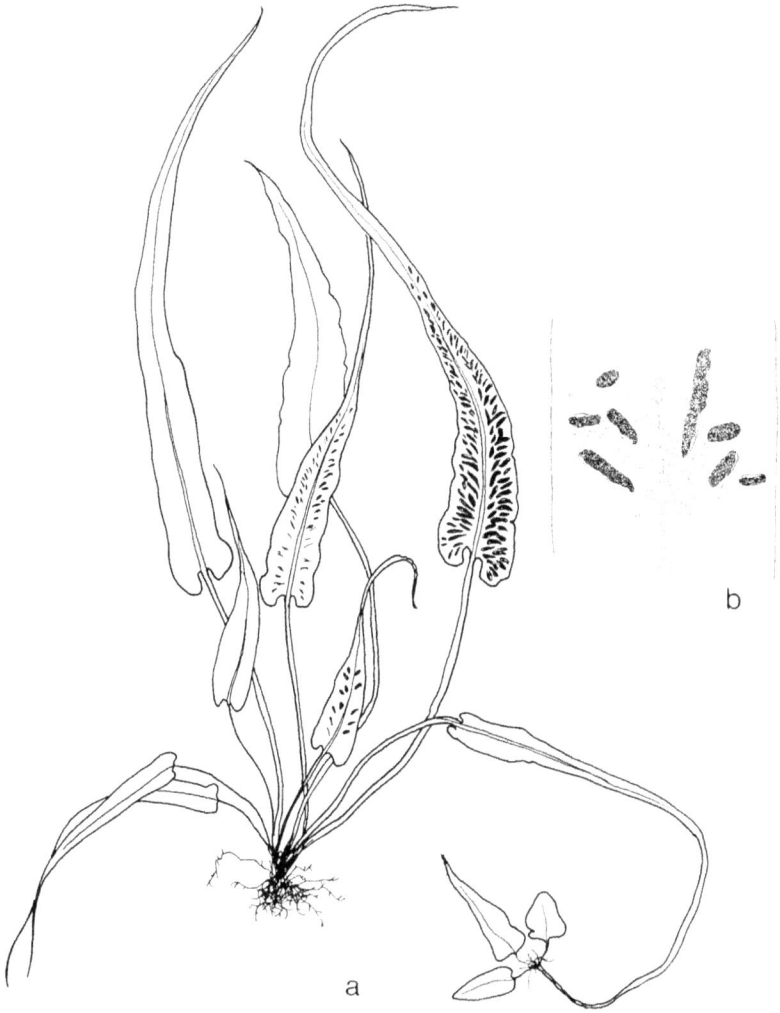

Fig. 151 *Phyllitis scolopendrium* var. *americana*; (a) fronds, 1/3 ×; (b) portion of fertile frond, 4 ×.

This fern is most distinctive and is much admired by rock garden enthusiasts. European visitors are always eager to see the species in the wild and are often initially disappointed in its stature, expecting it to be as long as a hart's-tongue.

Cytology: $n = 36$ (W.H. Wagner et al. 1970).
Habitat: Damp, sheltered, often mossy rocks and stony banks, preferring limestone.

Range: Southwestern Quebec and southern Ontario to Minnesota, south in the United States to Georgia and Mississippi, and west to Oklahoma and Kansas.

Remarks: It is impossible to refute the argument that this fern is in reality an entire-leaved *Asplenium*, based on its propensity to cross with *Asplenium* species. We have retained it here largely for sentimental reasons, because it is well known by this name and to highlight the fact that it has a counterpart in Asia, *C. sibiricus* Rupr. These represent a pair of interesting species, treated as subspecies by Löve (Löve et al. 1977), and are the only members of the genus.

3. *Phyllitis* Hill

1. **Phyllitis scolopendrium** (L.) Newm. var. **americanum** Fern.
   Phyllitis fernaldiana Löve
   Phyllitis japonica Kom. ssp. *americana* (Fern.) Löve & Löve
   Scolopendrium vulgare auth.
   hart's-tongue
   Fig. 151 (a) fronds; (b) portion of fertile frond. Map 150.

Fronds 15–40 cm long or longer, from a short caudex; blade simple, oblong-lingulate or strap-like, deeply cordate-auriculate at the base, and tapered to a point at the tip. Stipe short, clothed with narrow, curling, long-caudate scales. Veins free, forking. Sori narrowly oblong, nearly at right angles to the midrib, and located on either side of adjacent veinlets so that the sori appear to have double indusia opening along the middle.

This is a very striking fern that reminds one of tropical plants— to some it is even reminiscent of banana leaves. It is an exciting experience to see a large number of these plants on the rich, cool, shady limestone talus of the Niagara Escarpment. Identification is easy, but there is some disagreement as to what name should be applied to North American plants because of their cytology. Canada's plants are tetraploid ($n = 72$, Britton 1953), as are those of Japan, whereas the European ones are diploid. Emmott (1964) made a number of crosses between European, American, and Japanese plants,

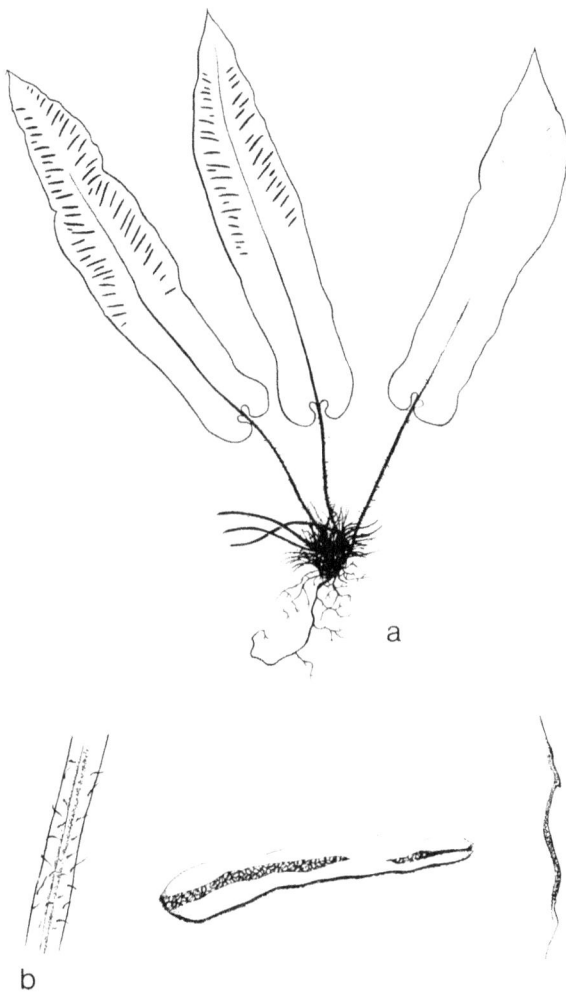

Fig. 151 *Phyllitis scolopendrium* var. *americana*; (a) fronds, 1/3×; (b) portion of fertile frond, 4×.

but the degree of relatedness between these widely separated populations was not clear from her results. Löve, initially impressed by the fact that the North American plants were tetraploid, suggested the name *Phyllitis fernaldiana* Löve, but later believed that the two tetraploids should be called *P. japonica* Kom., with the North American known as *P. japonica* ssp. *americana* (Fern.) Löve and Löve. Further experimental study is needed, and so we have retained the familiar name *Phyllitis scolopendrium*, even though we are aware that some Europeans now refer this species to *Asplenium* rather than to *Phyllitis*.

Cytology: $n = 72$ (Cody and Mulligan 1982*).

Habitat: Rich rocky woodland adjacent to the Niagara Escarpment, often associated with *Polystichum lonchitis*.

Range: In Ontario known from Bruce, Dufferin, Gray, Halton, Peel, Simcoe, and Welland counties. It may have been introduced in Welland County. An early report from Woodstock, N.B., has not been substantiated by recent collections and may also have represented an introduction. In the United States, hart's-tongue is known only from a few widely scattered localities: central New York, Tennessee, and Michigan.

Remarks: Because North American plants are tetraploid, it is now possible to reinvestigate plants in isolated sinkholes, such as those in Tennessee and Alabama, to determine if the plants are native or were introduced from Europe. A small colony of small plants discovered on western Vancouver Island consists of diploid plants. These are considered to have been introduced.
 Most of the extant North American plants occur in a small area of Ontario. A special effort should be made to protect them. It is a mistake to move them to gardens, because they are adapted to very special requirements of substrate, drainage, moisture, and degree of shade, which cannot be duplicated in the average garden. The species is rare in all of North America, including Canada and Ontario (Argus and White 1977).

13. POLYPODIACEAE

1. *Polypodium* L.

Small to medium-sized ferns from widely creeping branching chaffy rhizomes. Fronds jointed to the rhizome, evergreen, not reduced toward the base. Veins once or twice branched or anastomosing. Sori round or oval, naked, in rows on either side of the midrib, with or without glandular paraphyses among the sporangia.

The genus *Polypodium* has many tropical representatives and over 100 species in the world. The basic chromosome number (x) for almost all of these is firmly established as 37, although there are very few species with either 36 or 35. Ploidy levels range from $2x$ to $6x$ in the genus (Lovis 1977). In north-temperate regions, there are few species to consider, but these have proved to be quite complex. In Britain, *P. vulgare* s.l. has proved to be a complex of three levels of ploidy: diploid, tetraploid, and hexaploid, each with its own morphological and ecological characteristics. These are now known as *P. australe*, *P. vulgare* s.s., and *P. interjectum*, respectively (T.G. Walker 1979). Interestingly, *P. vulgare*, which is proven as an allotetraploid, does not have the genomes of *P. australe* in its constitution, and it was necessary to look elsewhere for the diploid progenitors of this long-established species. Cytogenetic analysis has suggested that two of our species, *P. glycyrrhiza* and the diploid *P. virginianum*, might be the ancestors of the European *P. vulgare*. Accordingly, in spite of attempts through the years to keep *P. virginianum* in North America clearly separate from *P. vulgare* in Europe, it would seem that the *P. vulgare* complex must be approached on a worldwide basis. *Polypodium fauriei* Christ in Asia should not be forgotten.

In eastern Canada, there is only *P. virginianum* to consider, although it is clear that there are both diploids and tetraploids to delineate and a very common sterile triploid hybrid that arises from hybridization between these two.

In western Canada, there are four other species to consider. One, *P. scouleri* $x = n = 37$, is most distinctive in our flora, and is apparently quite remote from the ancestry of the others. There are two other basic diploids in the West, the widespread coastal *P. glycyrrhiza* (GG) and the more montane *P. amorphum* (AA). There is also a presumptive allotetraploid, *P. hesperium* (AAGG), considered to have arisen from the doubled hybrid of *P. amorphum* and *P. glycyrrhiza* (Lang 1971).

These polypodies are most often found on rock surfaces, e.g., talus boulders, rocky ledges, and cliffs, although in the coastal forests of British Columbia, *P. glycyrrhiza* may be seen quite far up on trunks and branches of maple trees, suggesting epiphytes of the tropics. The leathery evergreen fronds of all our species, with little dissection of the blade, make it easy for the amateur to identify the genus.

A. Blade stiffly coriaceous; veins anastomosing ...... 1. **P. scouleri**
A. Blade herbaceous to membranous; veins free.
   B. Segments of fronds usually 3 cm long or longer, with the tips acute to attenuate .... 2. **P. glycyrrhiza**
   B. Segments of fronds usually less than 3 cm long, with the tips obtuse to acute.
      C. Sori round; rhizome scales with a darker central stripe.
         D. Blades abruptly tipped; pinnae oblong, blunt or somewhat acute (eastern and northern) ....
         .... 5. **P. virginianum**
         D. Blades tapering to the tip; pinnae somewhat oval (western British Columbia) ....
         .... 3. **P. amorphum**
      C. Sori oval; rhizome scales concolorous ....
      .... 4. **P. hesperium**

1. **Polypodium scouleri** Hook. & Grev.
Fig. 152, fronds. Map 151.

Fronds 15–40 cm long or longer, from a sparsely scaly stout rhizome. Stipes stiff, shorter than the blade, and with a few deciduous scales. Blades thick and leathery, deltoid-ovate, pinnate to pinnatifid; segments linear to linear-oblong, rounded at the apex. Rachis and midveins of pinnae with deciduous scales. Veins anastomosing. Sori round, large, near the midvein, and situated on the upper pinnae.
The very thick and leathery fronds, with blunt divisions and lobes, make this species easy to identify at a glance.

Cytology: $n = 37$ (Cody and Mulligan 1982*).

Habitat: Near the coast and often reached by the salt spray, on banks, cliffs, and tree trunks.

Range: Southern British Columbia, south to southern California.

Remarks: This species has others that resemble it in California and is part of the *P. californicum* complex. Various hybrids are known in that region.

2. **Polypodium glycyrrhiza** D.C. Eat.
P. vulgare L. var. *occidentale* Hook.
P. vulgare L. ssp. *occidentale* (Hook.) Hultén
licorice fern
Fig. 153 (*a*) fronds; (*b*) fertile pinna. Map 152.

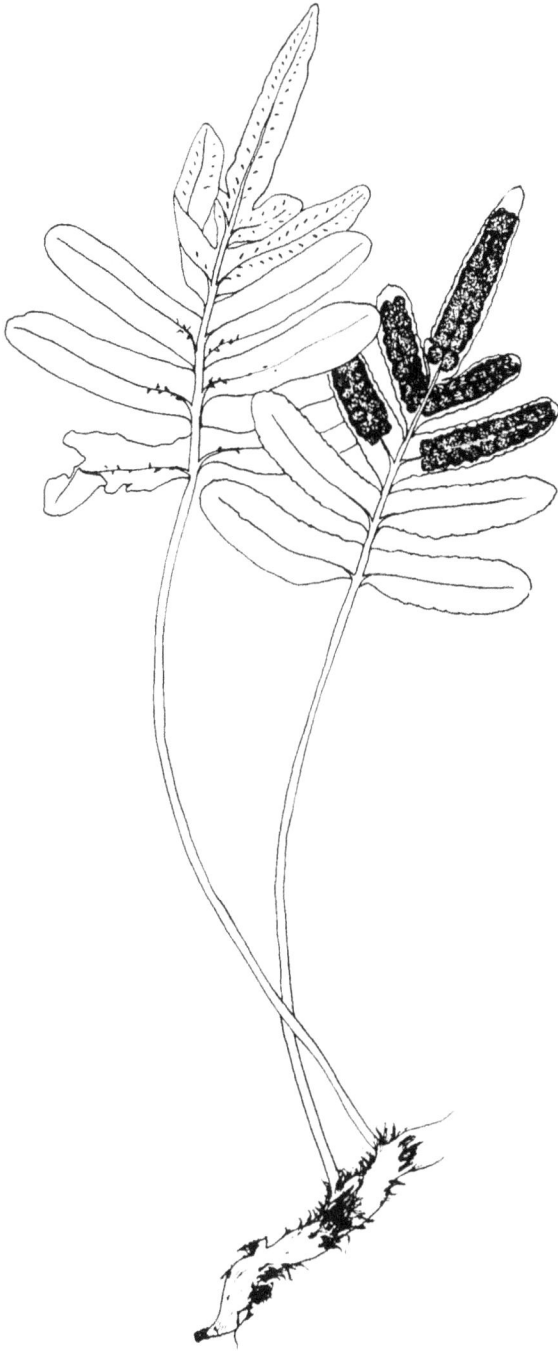

Fig. 152 *Polypodium scouleri*; fronds, 1/3 ×.

Fig. 153 *Polypodium glycyrrhiza*; (a) frond, 1/2 × ; (b) fertile pinna, 1 1/2 ×.

Polypodiaceae

Fronds up to 60 cm long or longer from the relatively thick licorice-tasting rhizome. Scales light brown or straw-colored, cordate to peltate, ovate, often with a capillary tip. Blades oblong to ovate; segments oblong-attenuate, with acute to acuminate tips, with finely serrate margins, and pubescent along the midveins on the lower surface. Sori on the upper segments round or occasionally somewhat oval, about equidistant between the margin and midvein. Paraphyses absent.

The long, acute tips to the segments and the licorice taste of the rhizome are useful field characters.

Cytology: $n = 37$ (Lang 1971*). A basic diploid ancestral species.

Habitat: On tree trunks, mossy logs, and moist banks at low elevations near the coast.

Range: Kamchatka, the Aleutian Islands, and through the Alaskan Panhandle, south to central California.

Remarks: This basic diploid species forms great sheets over rocky banks. It is conspicuous along several highways in coastal British Columbia.

3. **Polypodium amorphum** Suksdorf
   P. montense F.A. Lang
Fig. 154, fronds. Map 153.

Fronds up to 30 cm long from a thin acrid rhizome. Scales dark brown to chestnut, usually with a darker centre, narrowly ovate, often constricted near the base and with a long capillary tip. Blades oblong, up to 20 cm in length; segments oblong to obovate, with obtuse or rarely acute tips; margins entire to crenulate. Veins free. Sori round, near the margins of the segments, and rarely with a few glandular paraphyses.

This basic ancestral diploid species was segregated from *P. hesperium*, by Lang (1969), as *P. montense* Lang. Morton (1970) pointed out that *P. amorphum* Suks., even if monstrous in form, had priority; it apparently was extirpated at the type locality. Unfortunately, T.M.C. Taylor (1970) has *P. amorphum* listed as a synonym of *P. hesperium* Maxon, whereas now *P. amorphum* is the accepted name for what was *P. montense*.

This species is more montane than *P. glycyrrhiza*, has rounded segments, round sori, and an acrid-tasting rhizome. It resembles closely some of the plants of the tetraploids of *P. virginianum*.

Cytology: $n = 37$ (Lang 1969, 1971*). A basic diploid ancestral species.

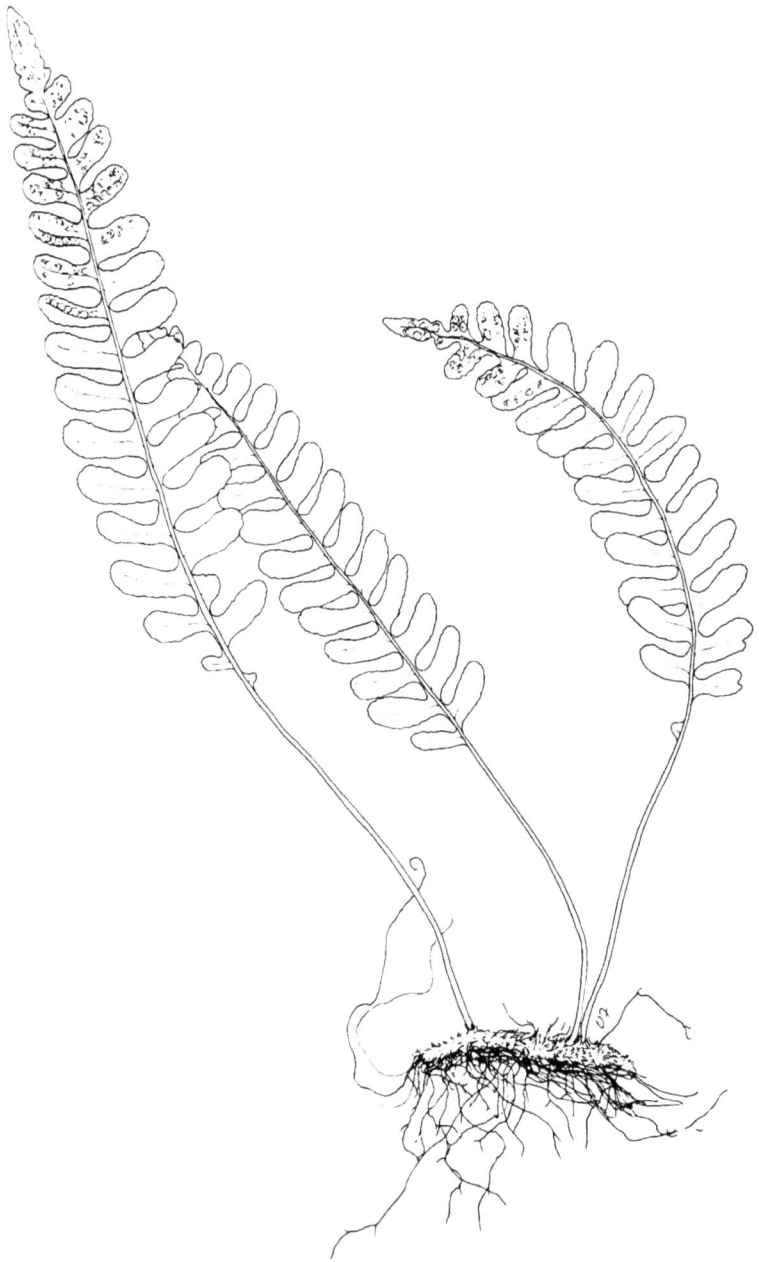

Fig. 154 *Polypodium amorphum*; fronds, 1/2 ×.

Polypodiaceae

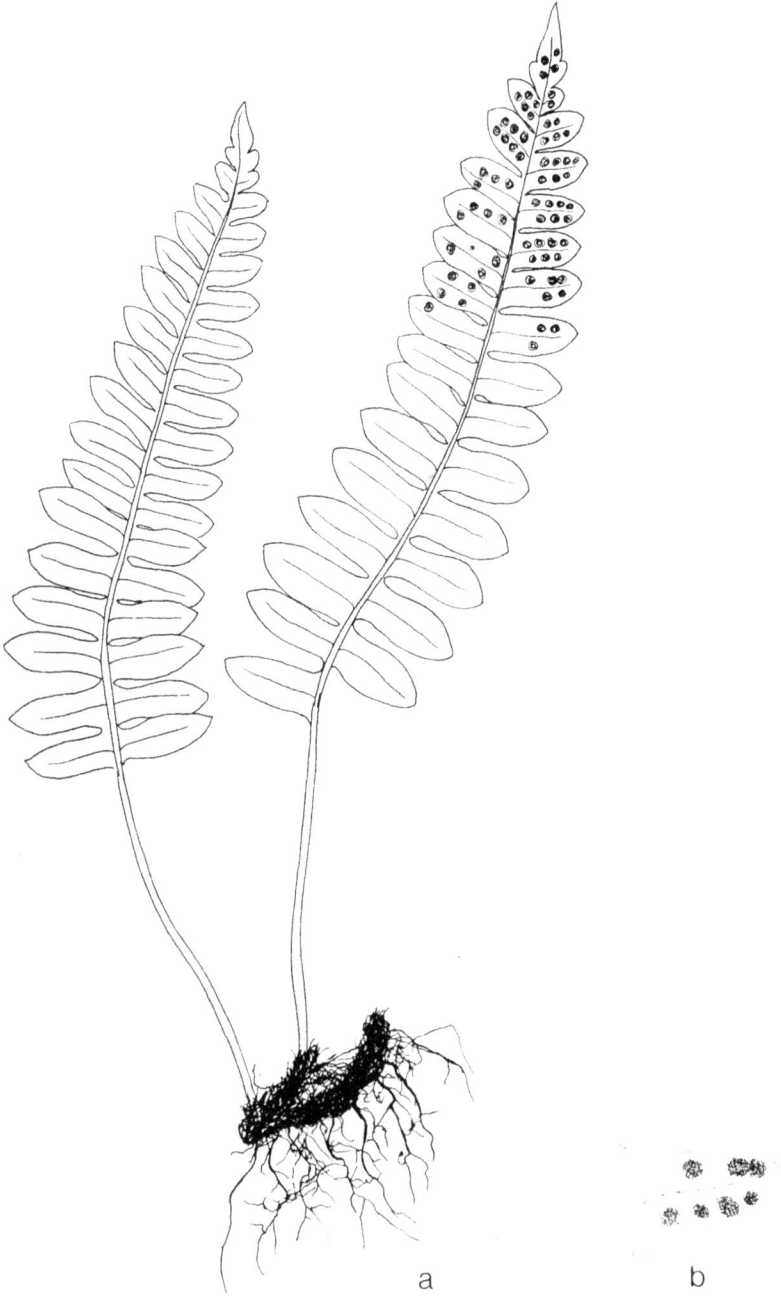

Fig. 155 *Polypodium hesperium*; (a) fronds, 2/3 × ; (b) fertile pinna, 1 1/2 × .

Polypodiaceae

Habitat: Rock crevices in the mountains.

Range: Coastal Mountains in British Columbia, south to Oregon and the Sierra Nevada, Calif.

Remarks: The rhizome is acrid to the taste.

4. **Polypodium hesperium** Maxon
   P. vulgare L. var. *columbianum* Gilbert
   P. vulgare L. ssp. *columbianum* (Gilbert) Hultén
   Fig. 155 (a) fronds; (b) fertile pinna. Map 154.

Fronds up to 35 cm long or longer from a rather thick, licorice-tasting rhizome. Scales chestnut brown, lance-ovate, more or less crenate-serrate. Blades oblong, up to 20 cm in length; segments oblong, obtuse to acute at the tip; margins entire to serrate. Sori on the upper segments oval, about equidistant between the margin and midvein. Paraphyses common.

This species is considered to be an allotetraploid derivative of *P. amorphum* and *P. glycyrrhiza*. From the former, it has the markedly rounded lobes of the segments and from the latter it has inherited the licorice taste of the rhizomes. The medial sori, which are oval, are good field characters.

Cytology: $n = 74$ (Lang 1971*). Allotetraploid of constitution AAGG.

Habitat: Rocky slopes and crevices.

Range: Southern British Columbia, south to Arizona, New Mexico, and Baja California, Mexico, and east to the Black Hills of South Dakota.

Remarks: Although Lang had two of the three hybrid combinations needed to analyze this species (*P. amorphum* × *hesperium* and *P. glycyrrhiza* × *hesperium*), and the morphological data are fairly consistent, the direct proof of the allotetraploid origin of *P. hesperium* is lacking (Lovis 1977). Artificial crosses that take 6–8 years to produce fertile plants or the analysis of a natural hybrid of *P. amorphum* × *glycyrrhiza* are needed for certain proof. *Polypodium hesperium* is rare in Alberta (Argus and White 1978).

5. **Polypodium virginianum** L.
   P. vulgare L. ssp. *virginianum* (L.) Hultén
   rock polypody
   Fig. 156, fronds (diploid). Fig 157 (a) fronds; (b) portion of fertile pinna (tetraploid). Map 155 (s.l.).

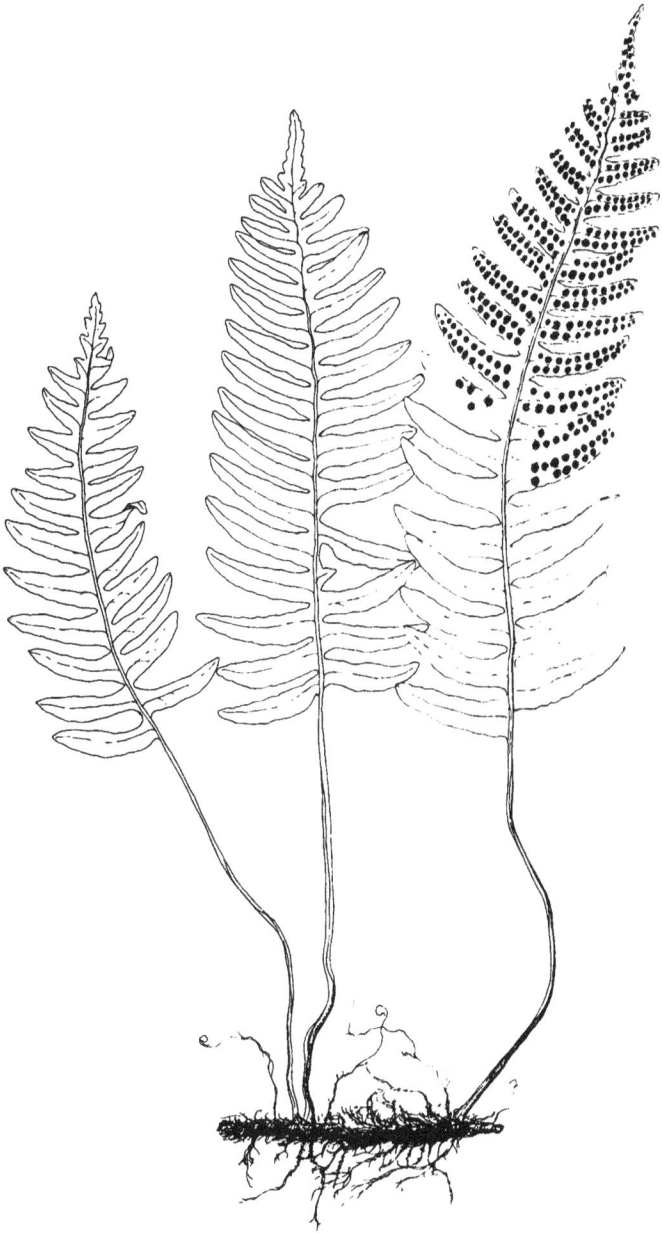

Fig. 156 *Polypodium virginianum*; fronds (diploid), 1/3 × .

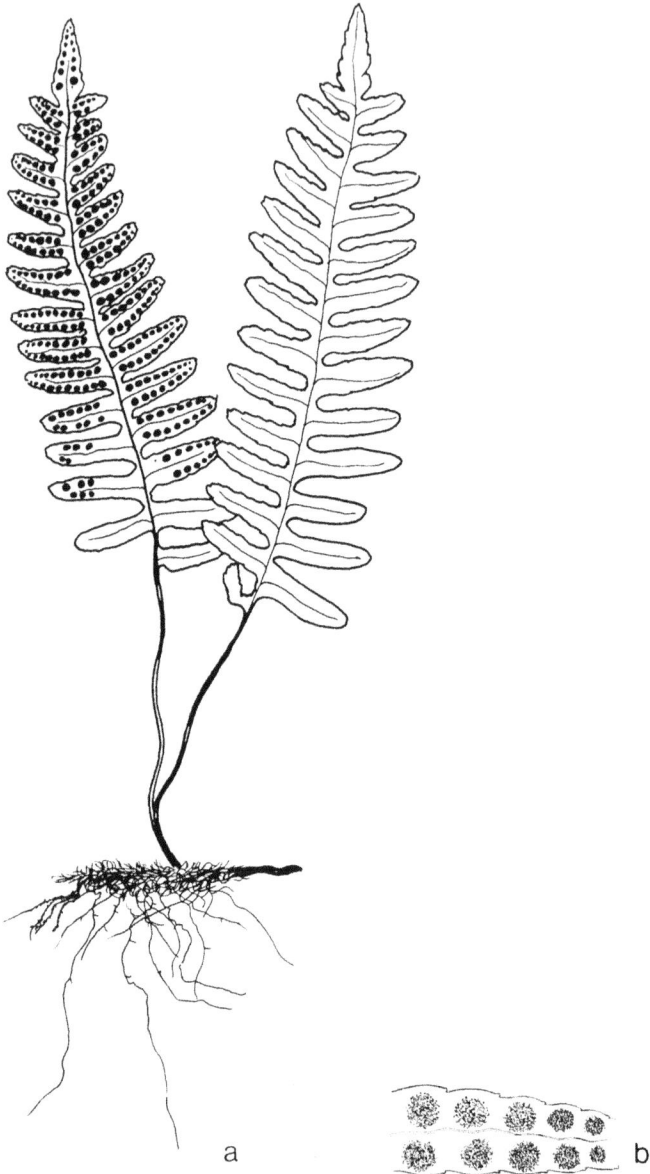

Fig. 157 *Polypodium virginianum*; (a) fronds, 2/3 ×; (b) portion of fertile pinna (tetraploid), 2 ×.

Polypodiaceae

Fronds up to 35 cm long from a creeping somewhat acrid rhizome. Scales often with a dark central stripe, deeply cordate. Blades oblong-lanceolate; segments linear-oblong to deltoid, entire to remotely dentate, blunt or acutish at the tip. Veins free. Sori round, midway between the midvein and margin, and occurring on the upper segments. Paraphyses present.

In eastern Canada *Polypodium virginianum* is easily identified by its small evergreen fronds of distinctive morphology and its usually large colonies on rocky talus, capping boulders, climbing rocky slopes, and ledges. The species is more complex than a casual glance would indicate. It is known that there are both diploids ($n = 37$) and tetraploids ($n = 74$) in eastern Canada, and that the triploid hybrid is quite common (Evans 1970). The diploids have blades that are more deltoid in shape, with acute apices of the segments, and the tetraploids have a more oblong blade (narrowed at the base), with rounded apices of the segments. There are a number of other morphological differences that are considered in detail by Kott and Britton (1982*b*) and Cranfill (1980).

Cytology: $n = 37$ and 74 (Britton 1953*; Löve and Löve 1976*; Cody and Mulligan 1982*).

Habitat: In shallow humus on rocks, in crevices, on woodland banks, and rarely on mossy stumps and in crotches of trees.

Range: Newfoundland to central Alaska and northeastern British Columbia, south in the east to Georgia, Alabama, Tennessee, and Arkansas. The tetraploid is more widespread in Canada to the north and west, whereas the diploid is more abundant in the southeast. Most plants in Ontario are tetraploids. The diploid has a distribution in Canada similar to that of *Dryopteris campyloptera*.

Remarks: Löve and Löve (1977) named the tetraploid *P. vinlandicum* without lectotypification of *P. virginianum* L. The type of the former is given as *P. virginianum* L. var. *americanum* Hooker. It is unfortunate to designate such old, poorly studied material as the type for a new species, when we would like to know as much as possible about the type material, e.g., chromosome number, number of paraphyses, spore morphology, and chromatography. This is impossible with old herbarium specimens of uncertain provenance, which sometimes contain mixtures.

The frequent triploid hybrid, with mostly aborted spores, is thought to be able to reproduce by occasional spores that have an unreduced chromosome number (Evans 1970).

It is premature to describe a new species before *P. virginianum* is lectotypified and before the relationship of the tetraploid to the diploid race as well as to the diploid *P. amorphum* is clarified by experimental

Hybrids of *Polypodium*

Polypodium amorphum × *glycyrrhiza* has not yet been reported in Canada.

Polypodium glycyrrhiza × *hesperium* has been reported in three localities in British Columbia: Kaske Creek, approximately 100 km east of Prince Rupert; Alexandra Bridge, Fraser River; and Green River, Pemberton.

Polypodium amorphum × *hesperium* has been reported in Alexandra Bridge, Fraser River, B.C.

The triploid hybrid between the two cytotypes of *P. virginianum* is frequent in eastern Canada (Kott and Britton 1982*b*). It is more widespread in Ontario than is the diploid.

14. MARSILEACEAE

Marsileaceae, which contains only three genera, is characterized by the presence of sporocarps. It is not closely related to other fern families, except perhaps Salvineaceae. About 75 species, mainly of Old World distribution, are in the genus *Marsilea*.

1. *Marsilea* L. water-clover

Plants aquatic, perennial, herbaceous, submersed or emersed, growing from widely creeping rhizomes. Leaves alternate, long-petioled; blade divided into four clover-like leaflets. Sori embedded in a gelatinous sheath in sporocarps; sporocarps hard, ovoid, pedunculate, two-loculate, bearing 2 more or less conspicuous teeth near the base. Microspores numerous, but megaspores only one to a sporangium.

As is apparent by the common name, the four-foliate blade of this genus mimics some members of the family Leguminosae.

A. Leaflets broadly obovate-cuneate, glabrous; sporocarps often in pairs borne on a long stalk arising from the stipe above its base .... 2. **M. quadrifolia**

A. Leaflets broadly cuneate, sparsely appressed-pubescent; sporocarps borne singly on short stalks from the rhizome or base of the stipe .... 1. **M. vestita**

1. ***Marsilea vestita*** Hook. & Grev.
    M. mucronata A.Br.
    hairy pepperwort
Fig. 158, fronds and sporocarp. Map 156.

Leaves with petioles up to 10 cm long or longer, tufted or scattered from an elongate rhizome. Leaflets to 1 cm long, broadly cuneate, sparsely appressed pubescent. Sporocarps borne singly on short stalks from the rhizome or from the base of the stipe.

Cytology: No recent reports seen.

Habitat: Shallow ponds, ditches, sloughs, marshy places, and quiet streams, often stranded later in the season.

Range: Southern Saskatchewan to southern British Columbia, south to California, Texas, and Arkansas.

Remarks: The species is rare in British Columbia. It was collected at the Indian reserve in the vicinity of Kamloops, B.C., by John Macoun also known at Goose Lake near Vernon, B.C. The hard sporocarps may allow the species to survive dry periods. Indeed, Bloom (1955, 1961) has found that as long as the sporocarps remain dry, they could be autoclaved for up to 15 minutes and still show excellent "germination." Scoggan (1978) stated that there is no confirmation for the report by Burman (1909) of this species occurring in Manitoba. There is, however, a specimen in WIN labeled "Western Manitoba," which presumably is the basis of the report, but the exact locality, date of collection, and collector are unknown. The species should be searched for in western Manitoba.

2. **Marsilea quadrifolia** L.
   water-clover
Fig. 159 (a) fronds and sporocarps; (b) venation. Map 157.

More robust than *M. vestita*. Leaves with petioles up to 20 cm long, scattered from the elongated rhizome. Leaflets up to 2 cm long, obovate-cuneate, glabrous. Sporocarps often in pairs, borne on a long stalk arising from the stipe above its base.

Cytology: $n = 20$ (Mehra and Loyal 1959).

Habitat: Shallow water of slow-moving streams.

Range: In Ontario, only in Nanticoke Creek in Haldimand and Norfolk counties and Mississauga in Peel County; in the United States from New England to Iowa and Kentucky; naturalized in Europe.

Remarks: This species has been sold for use in pools and aquaria whence it escapes to slow-moving streams and ponds.

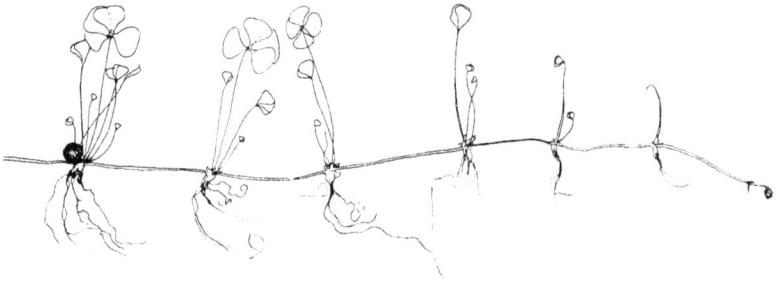

Fig. 158 *Marsilea vestita*; fronds and sporocarp, 1/2 ×.

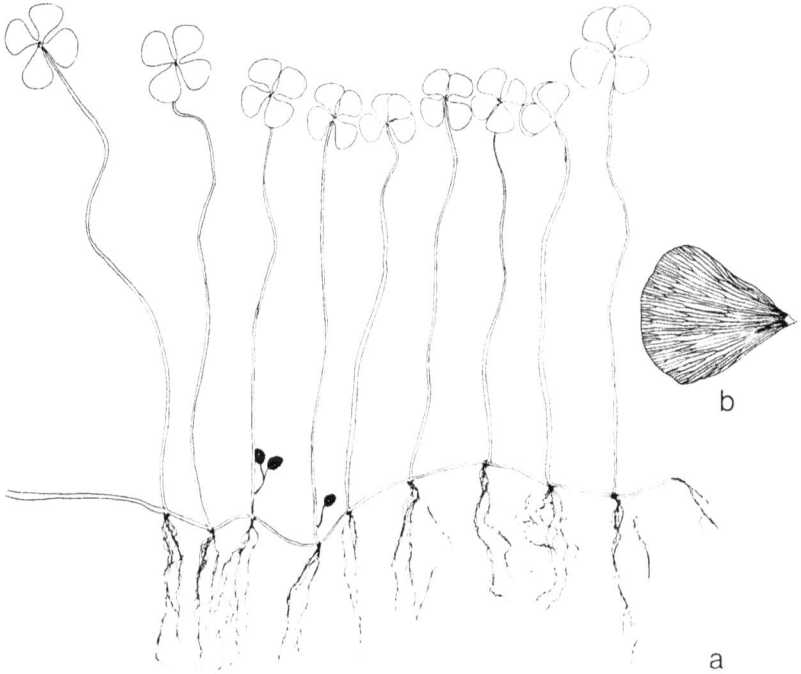

Fig. 159 *Marsilea quadrifolia*; (a) fronds and sporocarps, 1/2 ×; (b) venation, 2 ×.

15. SALVINIACEAE

Salviniaceae is a family of only two genera, *Azolla* and *Salvinia*; it is sometimes separated into two families, with a total of about 16 species that occur primarily in tropical regions around the world.

1. *Azolla* Lam. mosquito fern

Ferns small, annual, aquatic, free-floating, with unbranched thread-like roots. Plants compact, dichotomously branched, forming small mats. Leaves usually crowded, two-lobed; upper lobe green or often reddish later in the season; lower lobe usually larger than the upper, only one cell thick, and mostly without chlorophyll. Sporocarps in pairs in the leaf axils, each enclosed in an indusium. Microsporocarps containing numerous microsporangia, each of which produces masses of microspores, which when released exhibit peculiar barb-tipped hairs (glochidia). Megasporocarps smaller, acorn-shaped, each containing a single megasporangium with a single megaspore.

The New World species of *Azolla* were treated by Svenson (1944); species of *Azolla* are often difficult to distinguish because of the absence of sporocarps and the need to use a microscope to examine the glochidia; Svenson's report of *A. filiculoides* for Alaska was not taken up by Hultén (1967).

A. Plants less than 1 cm in diameter; leaves about 0.5 mm long; glochidia without cross walls .... 1. **A. caroliniana**
A. Plants 1 cm in diameter or larger; leaves about 0.7 mm long or longer; glochidia with cross walls .... 2. **A. mexicana**

1. **Azolla caroliniana** Willd.
Map 158.

Plants 0.5–1.0 cm in diameter. Upper leaf-lobes 0.5–0.6 mm long, smooth, not closely imbricate. Glochidia without cross walls. Megaspores unknown.

Cytology: $n = 24$ (Tschermak-Woess and Dolezal-Janisch 1959).

Habitat: Quiet waters.

Range: Eastern United States.

Remarks: A collection by Judge Logie in 1862 from Hamilton Beach, Ont., is now in the National Museum of Natural Sciences, Ottawa (Lawson Herbarium). The species has not been collected in that area

Fig. 160 *Azolla mexicana*; crowded leaves and sporocarps, 9 ×.

since and has probably been extirpated. Soper (1949) noted that this *Azolla* had been collected in 1934 from the American side of the Niagara frontier region, and Pursh (1814) reported it from Lake Ontario.

2. **Azolla mexicana** Presl
   A. caroliniana auth. non Willd.
   A. filiculoides sensu Scoggan (1978)
Fig. 160, crowded leaves and sporocarps. Map 159.

Plants 1.0–1.5 cm in diameter. Leaves crowded and overlapping; upper leaf-lobes 0.7 mm long, papillose. Glochidia many-septate. Lower part of megaspores pitted.

Cytology: None.

Habitat: Ponds and slow-moving streams.

Range: Southern British Columbia, south to northern South America, east to Missouri, Illinois, and Wisconsin.

Remarks: In British Columbia known only in the area adjacent to Sicamous and Salmon Arm, where it was collected as recently as 1976. Scoggan (1978) was of the opinion that the species was introduced in this area.

DISTRIBUTION MAPS

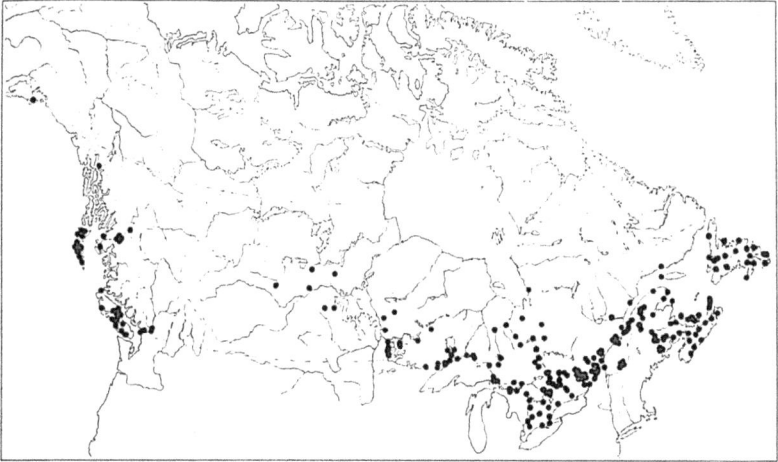

Map 1 *Lycopodium clavatum* var. *clavatum*

Map 2 *Lycopodium clavatum* var. *monostachyon*

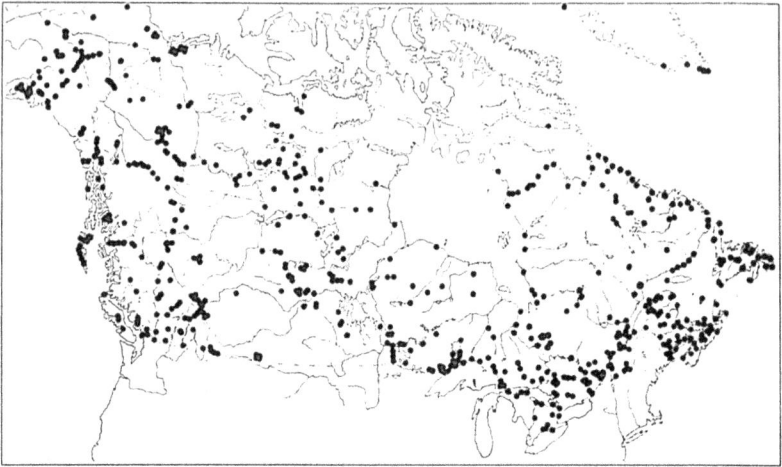

Map 3 *Lycopodium annotinum*

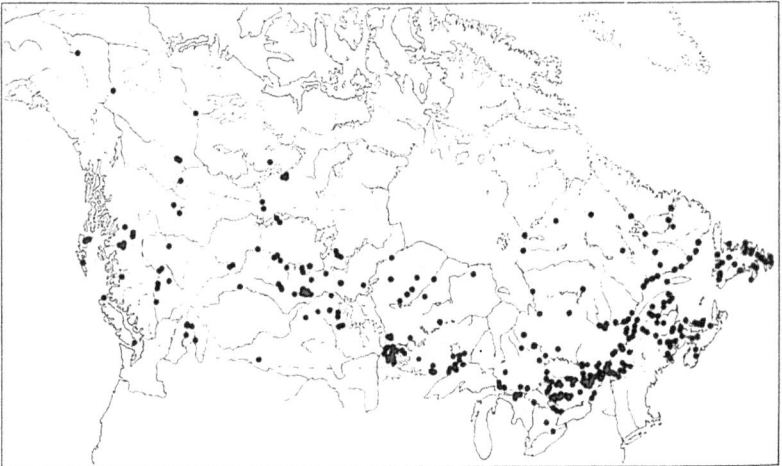

Map 4 *Lycopodium dendroideum*

Maps

Map 5 *Lycopodium obscurum* var. *obscurum*

Map 6 *Lycopodium obscurum* var. *isophyllum*

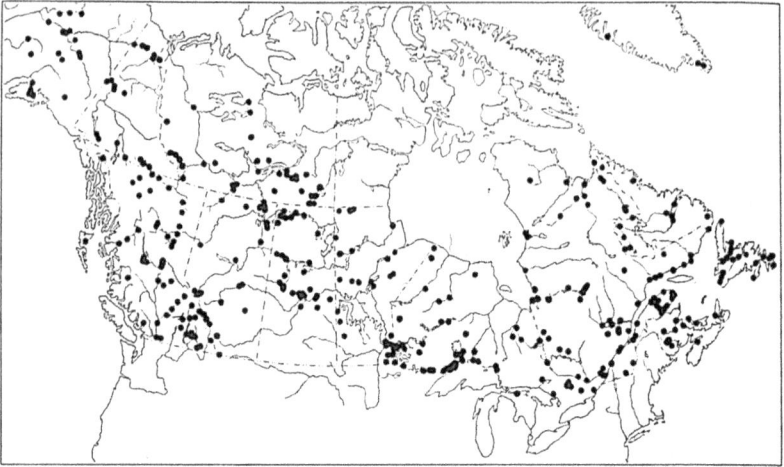

Map 7 *Lycopodium complanatum*

Map 8 *Lycopodium digitatum*

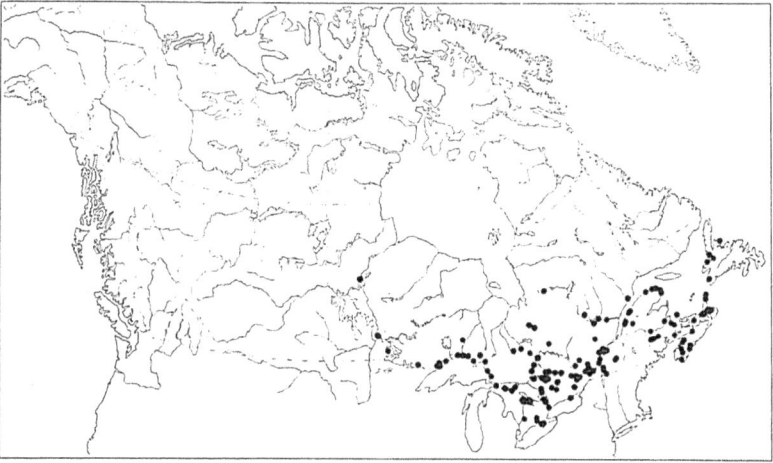

Map 9 *Lycopodium tristachyum*

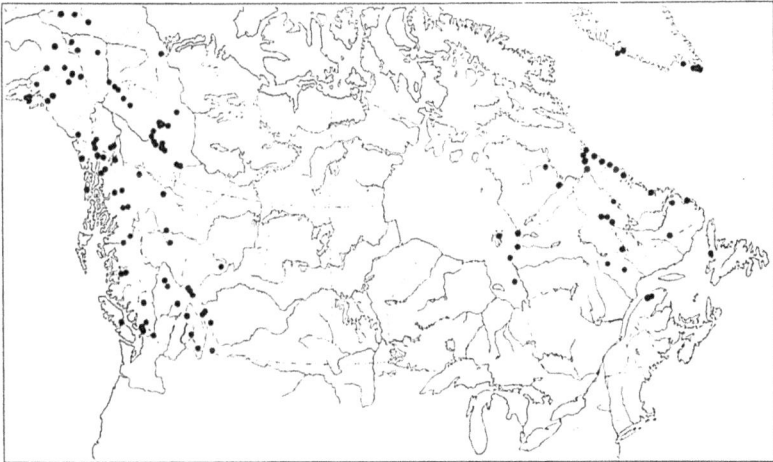

Map 10 *Lycopodium alpinum*

Map 11 *Lycopodium sitchense*

Map 12 *Lycopodium sabinifolium*

Maps

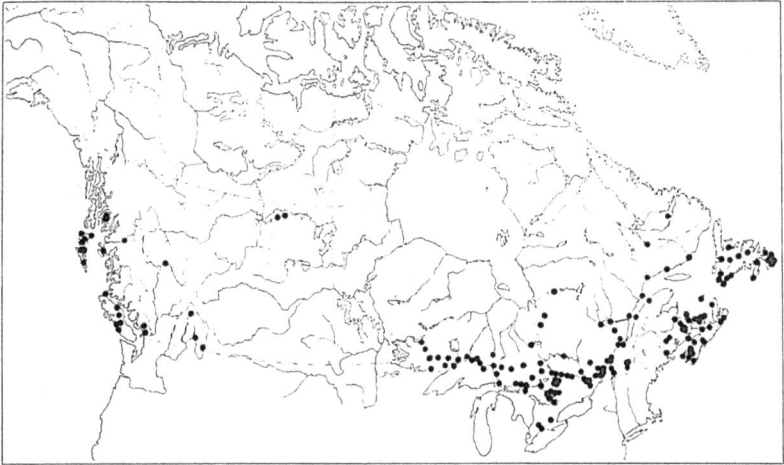

Map 13 *Lycopodium inundatum* var. *inundatum*

Map 14 *Lycopodium inundatum* var. *bigelovii*

Map 15 *Lycopodium lucidulum*

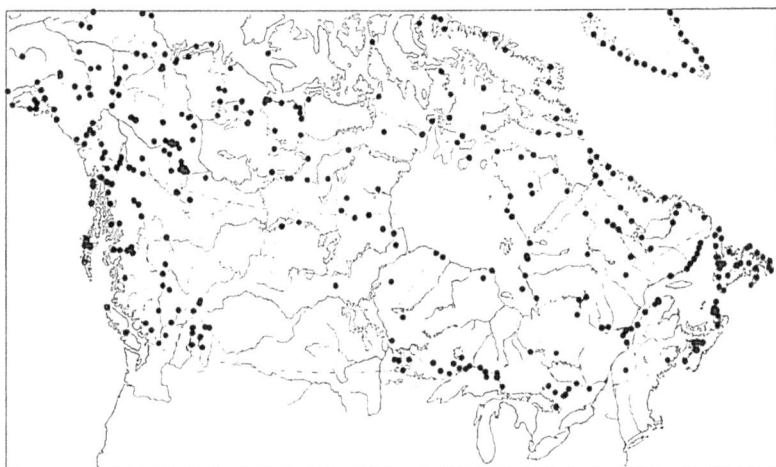

Map 16 *Lycopodium selago* ssp. *selago*

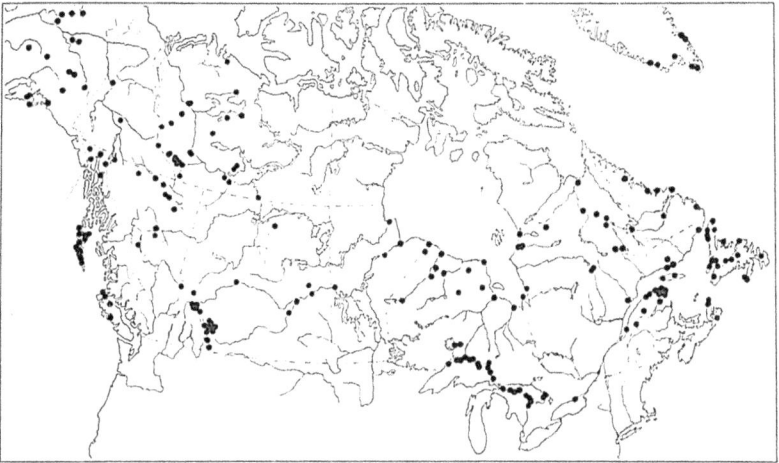

Map 17 *Selaginella selaginoides*

Map 18 *Selaginella apoda*

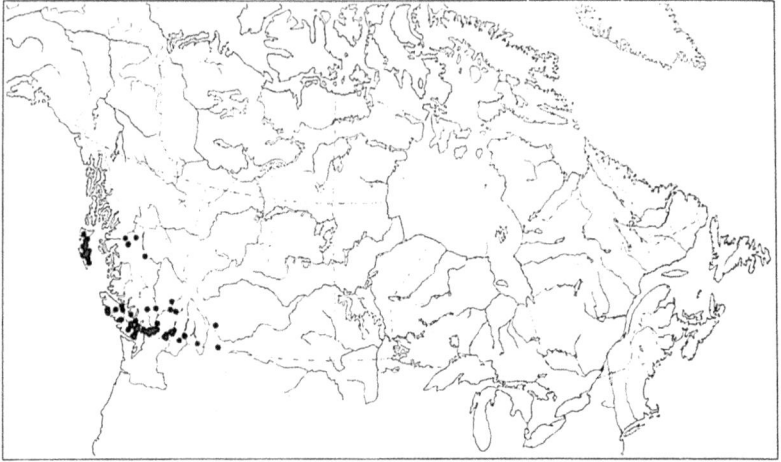

Map 19 *Selaginella wallacei*

Map 20 *Selaginella oregana*

Maps

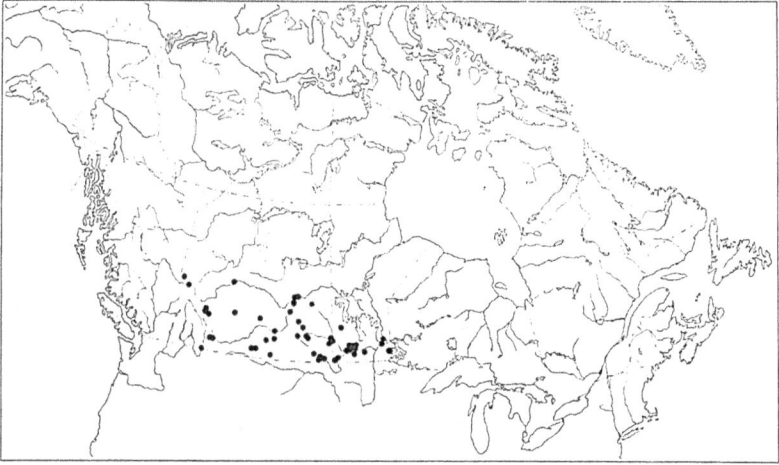

Map 21 *Selaginella densa*

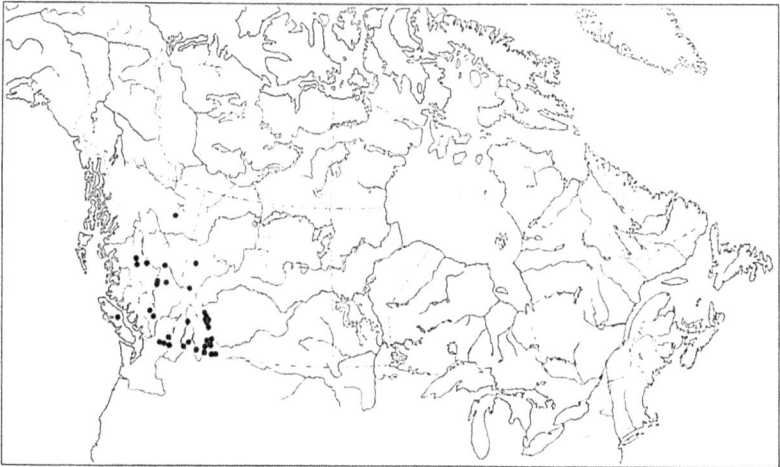

Map 22 *Selaginella densa* var. *scopulorum*

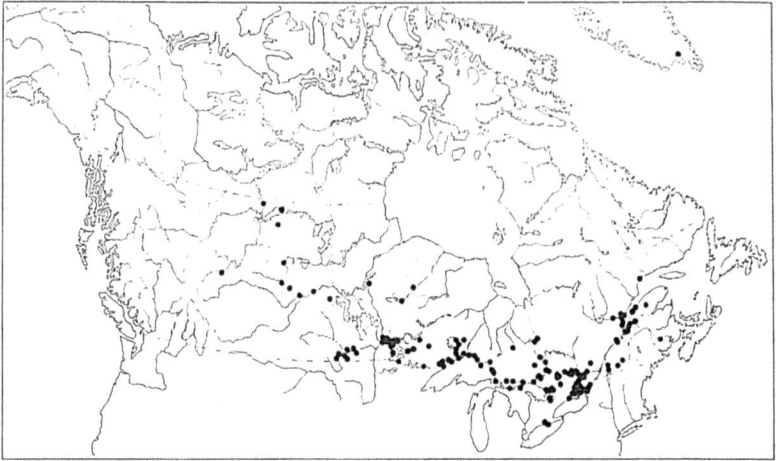

Map 23 *Selaginella rupestris*

Map 24 *Selaginella sibirica*

324

Maps

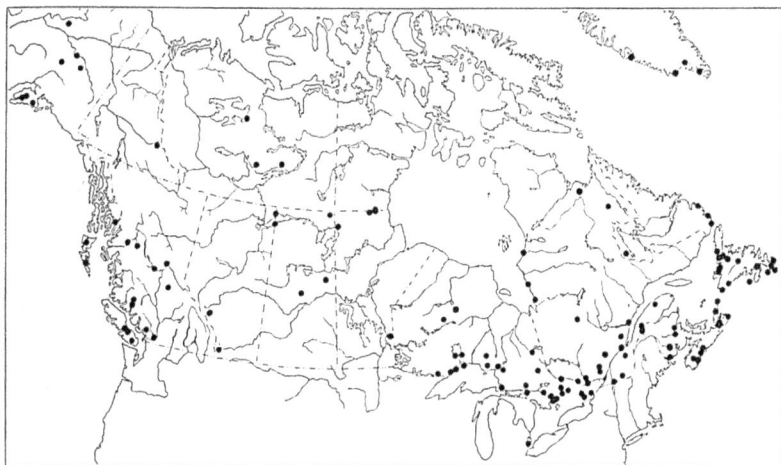

Map 25 *Isoetes echinospora*

Map 26 *Isoetes maritima*

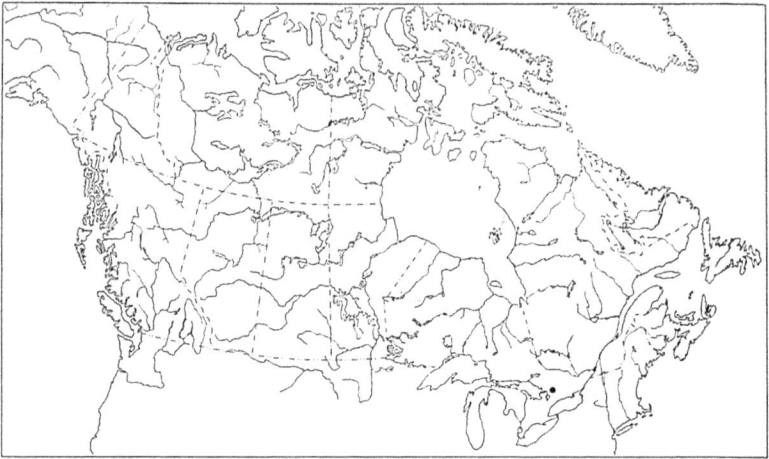

Map 27 *Isoetes eatonii*

Map 28 *Isoetes riparia*

Maps

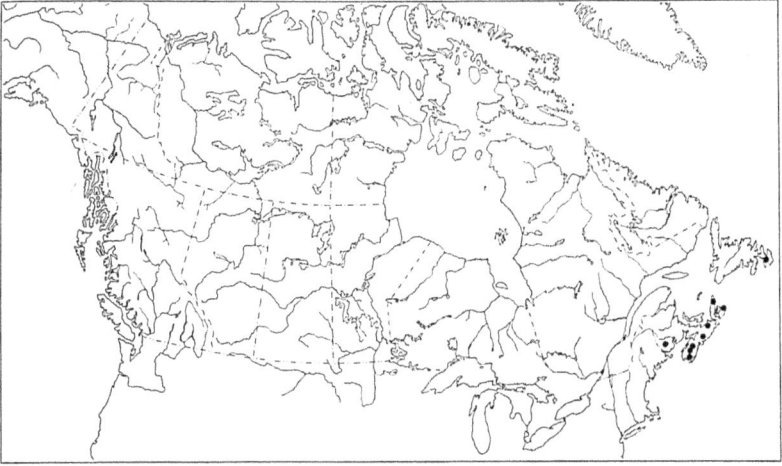

Map 29 *Isoetes acadiensis*

Map 30 *Isoetes hieroglyphica*

Map 31 *Isoetes tuckermanii*

Map 32 *Isoetes macrospora*

Maps

Map 33 *Isoetes nuttallii*

Map 34 *Isoetes howellii*

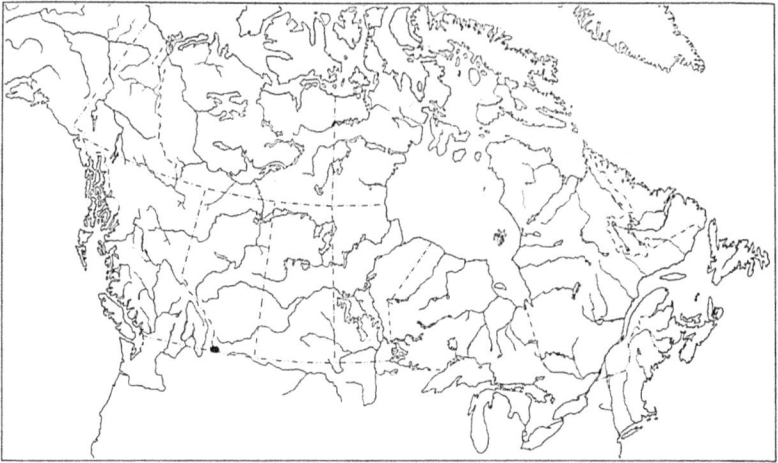

Map 35 *Isoetes bolanderi*

Map 36 *Isoetes occidentalis*

Maps

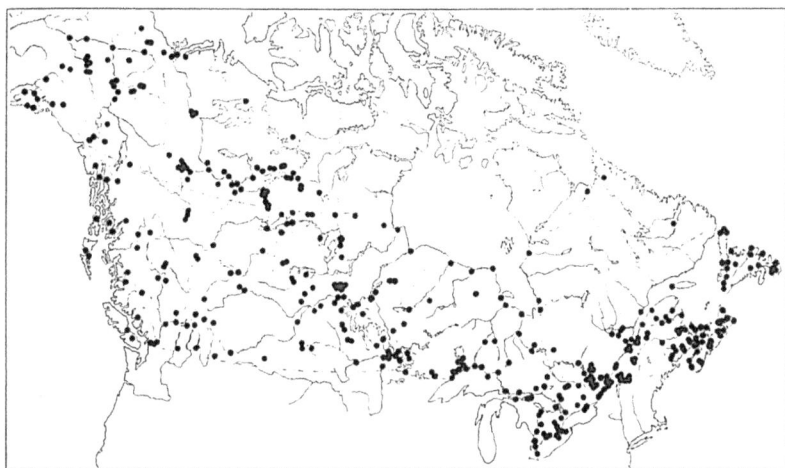

Map 37   *Equisetum fluviatile*

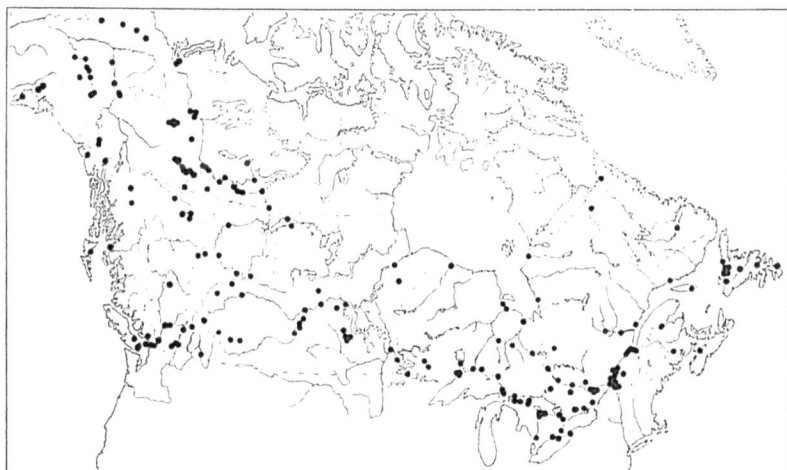

Map 38   *Equisetum palustre*

Map 39 *Equisetum telmateia* ssp. *braunii*

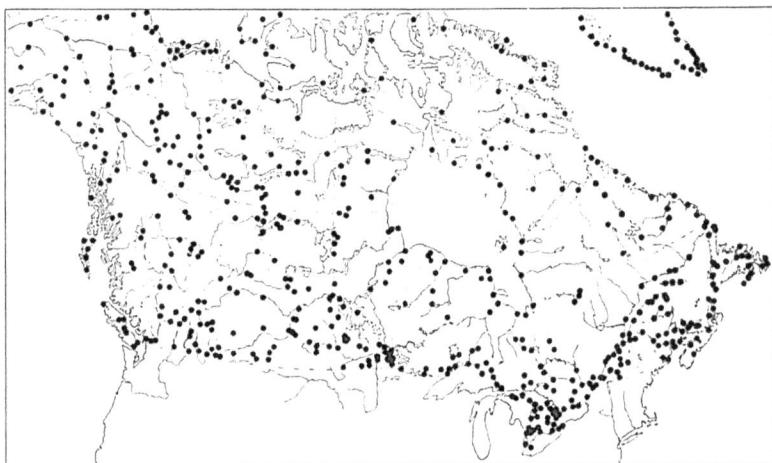

Map 40 *Equisetum arvense*

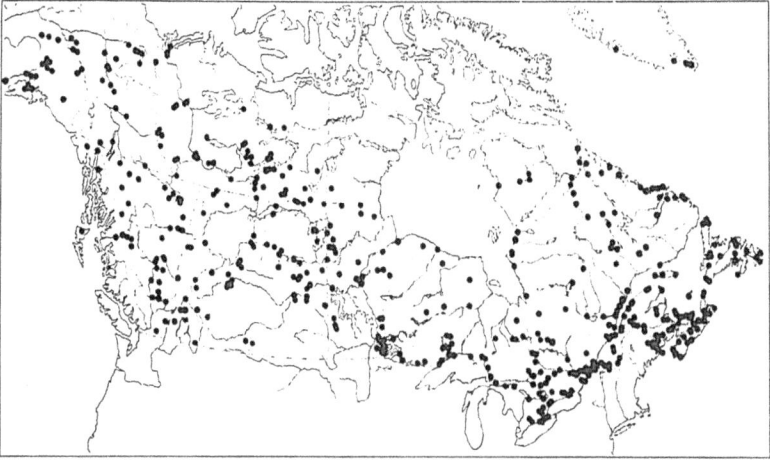

Map 41 *Equisetum sylvaticum*

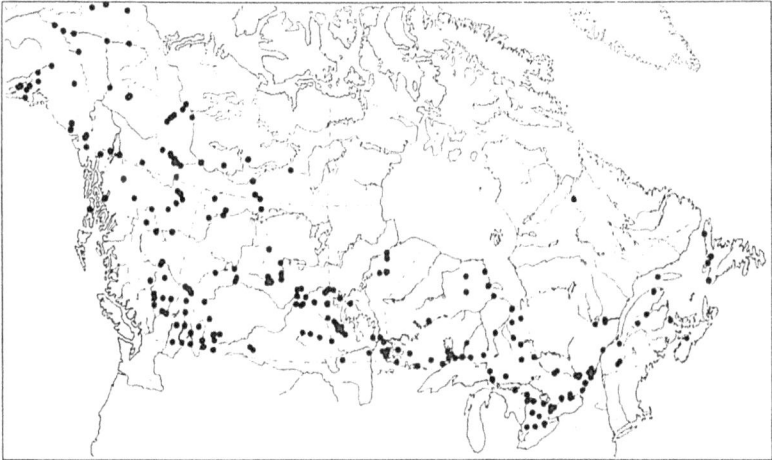

Map 42 *Equisetum pratense*

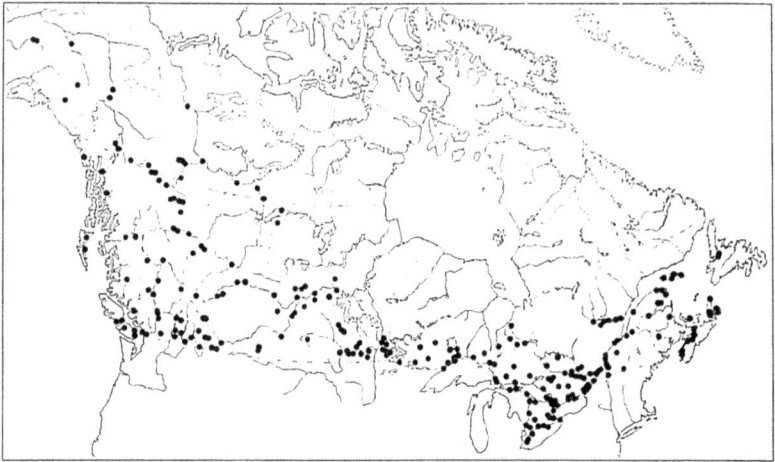

Map 43   *Equisetum hyemale* ssp. *affine*

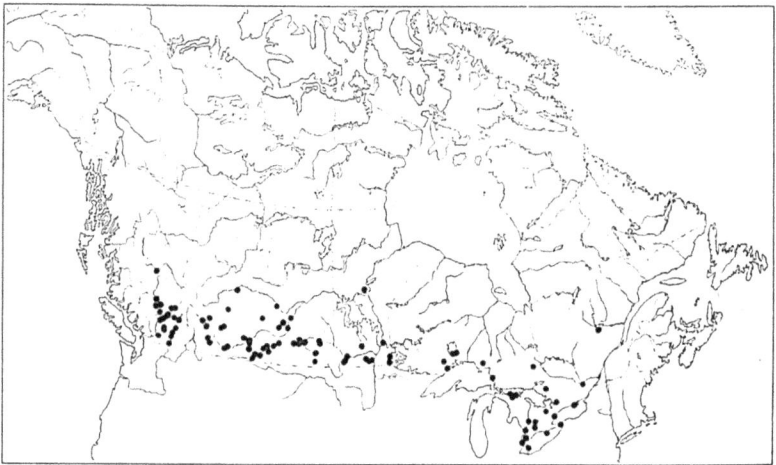

Map 44   *Equisetum laevigatum*

Maps

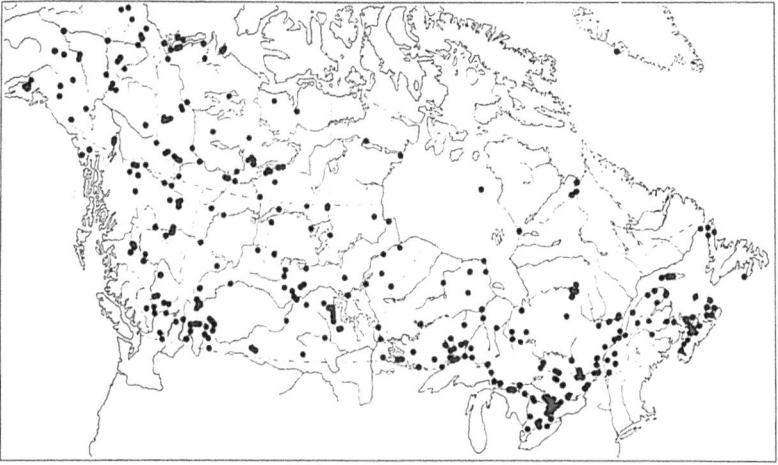

Map 45   *Equisetum scirpoides*

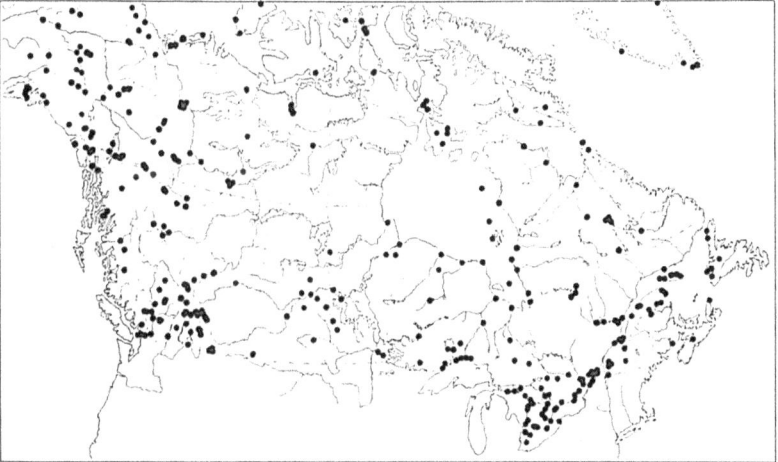

Map 46   *Equisetum variegatum* ssp. *variegatum*

Map 47   *Equisetum variegatum* ssp. *alaskanum*

Map 48   *Ophioglossum vulgatum* var. *pseudopodum*

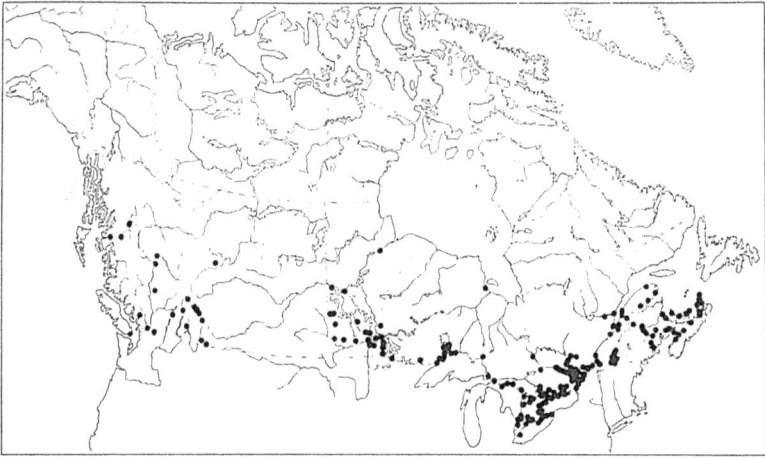

Map 49   *Botrychium virginianum* var. *virginianum*

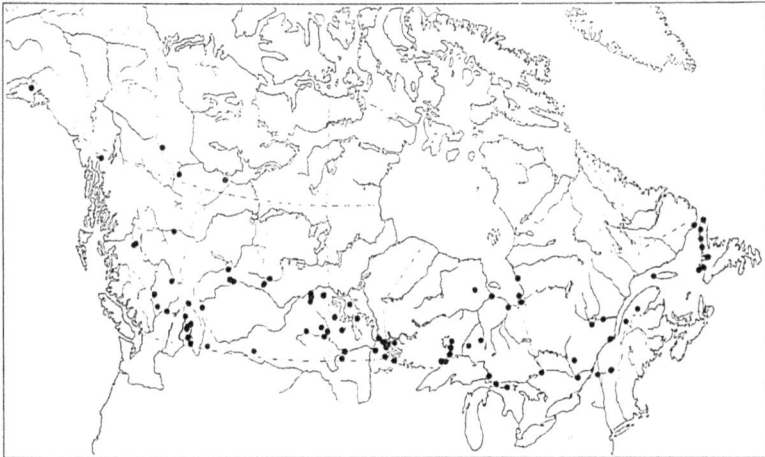

Map 50   *Botrychium virginianum* var. *europaeum*

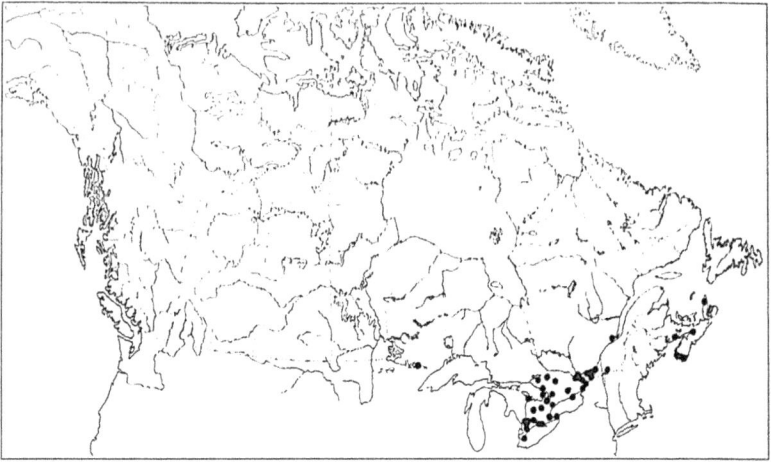

Map 51 *Botrychium dissectum*

Map 52 *Botrychium obliquum*

Maps

Map 53 *Botrychium oneidense*

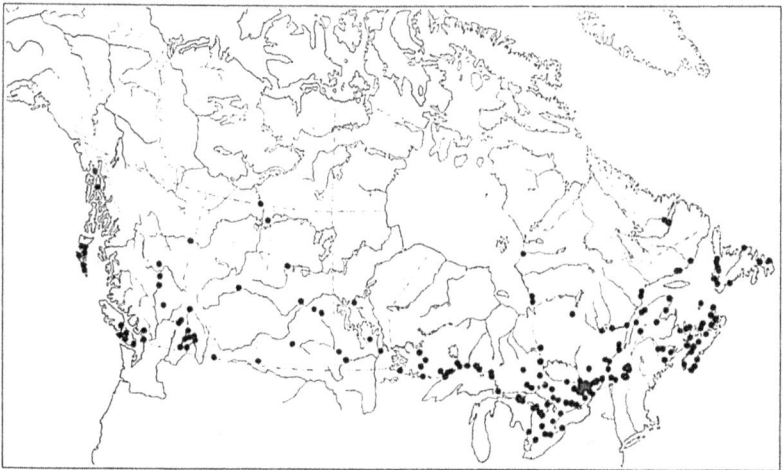

Map 54 *Botrychium multifidum* (s. l.)

Map 55 *Botrychium rugulosum*

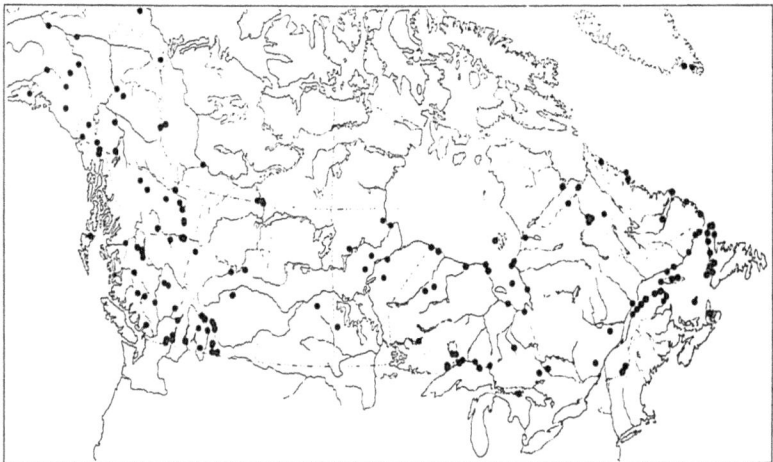

Map 56 *Botrychium lunaria*

Map 57 *Botrychium minganense*

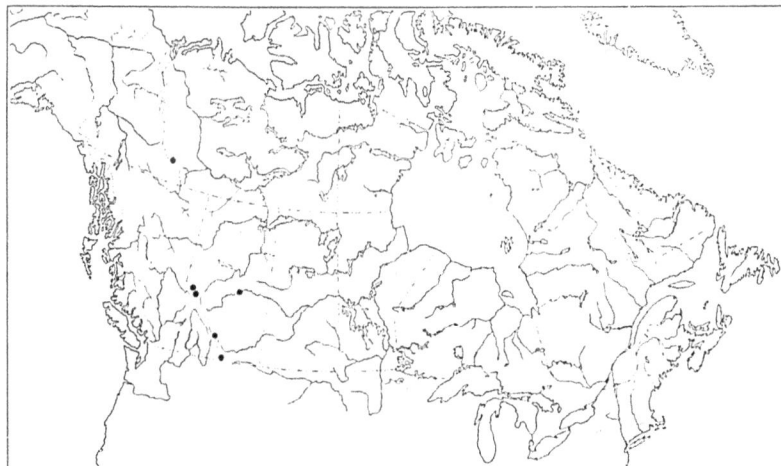

Map 58 *Botrychium dusenii*

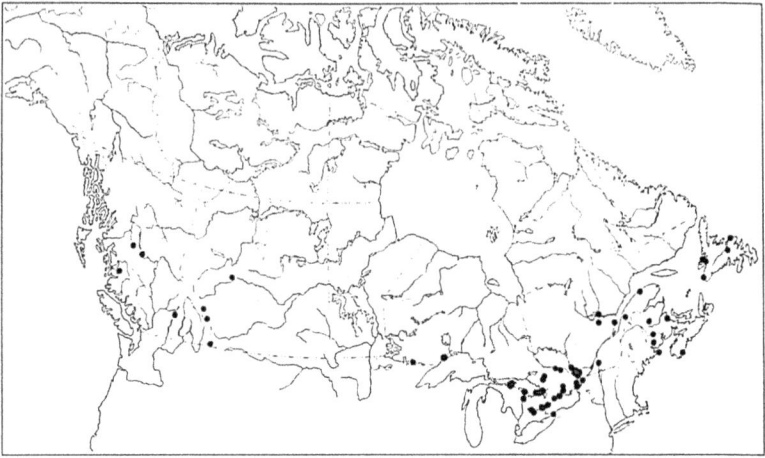

Map 59 *Botrychium simplex* (s. l.)

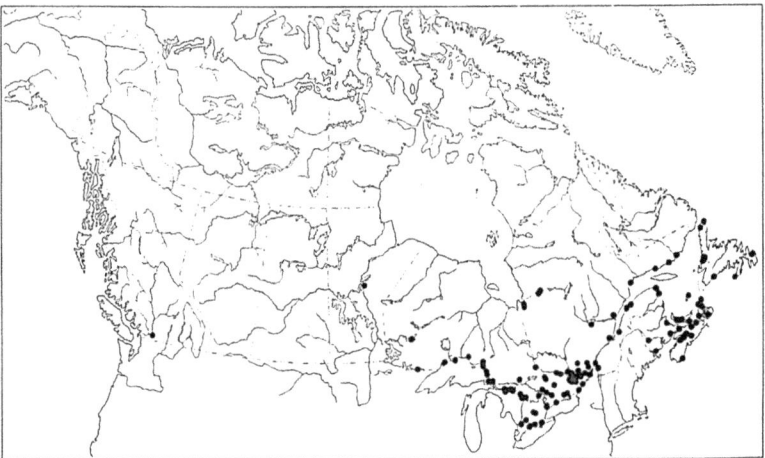

Map 60 *Botrychium matricariifolium*

Maps

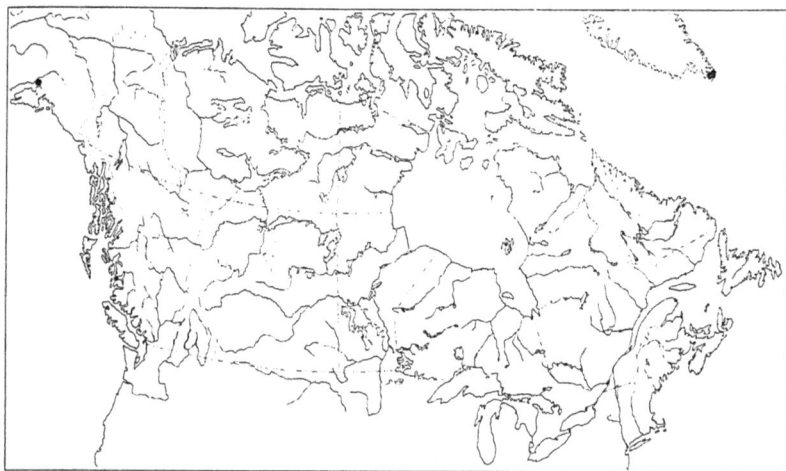

Map 61   *Botrychium boreale* ssp. *boreale*

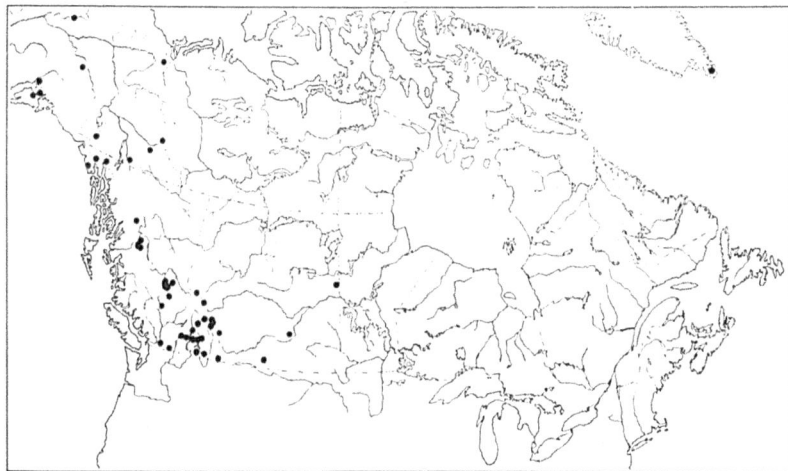

Map 62   *Botrychium boreale* ssp. *obtusilobum*

Map 63   *Botrychium lanceolatum* var. *lanceolatum*

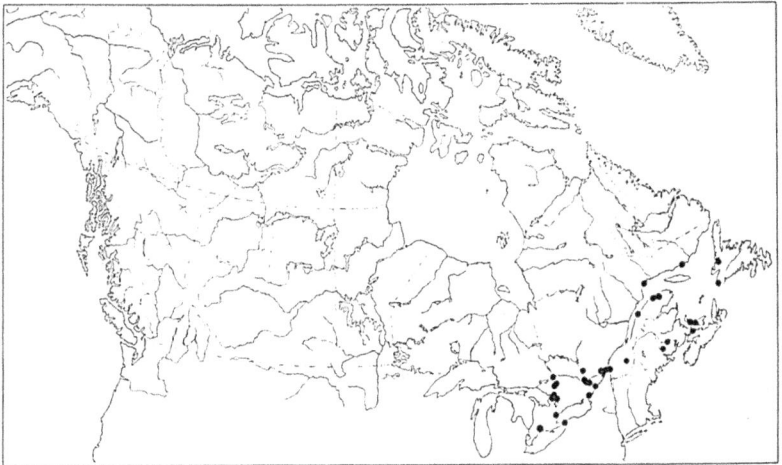

Map 64   *Botrychium lanceolatum* var. *angustisegmentum*

Maps

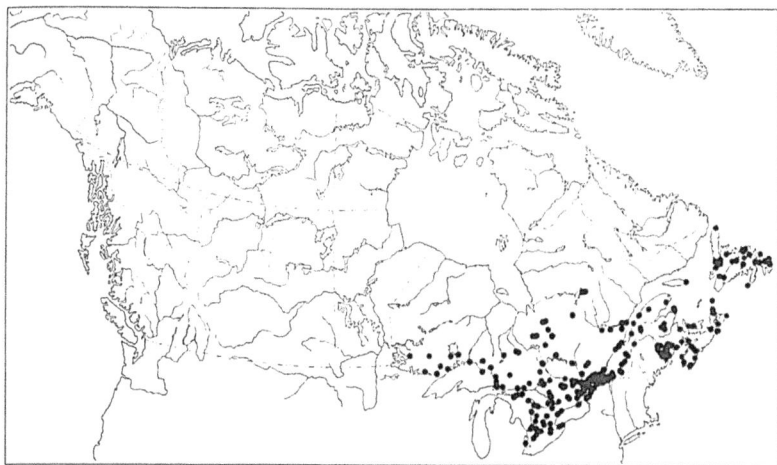

Map 65   *Osmunda regalis* var. *spectabilis*

Map 66   *Osmunda claytoniana*

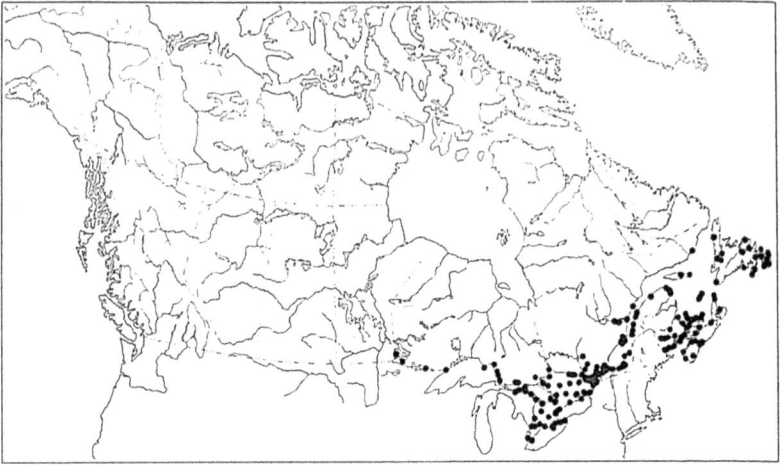

Map 67   *Osmunda cinnamomea*

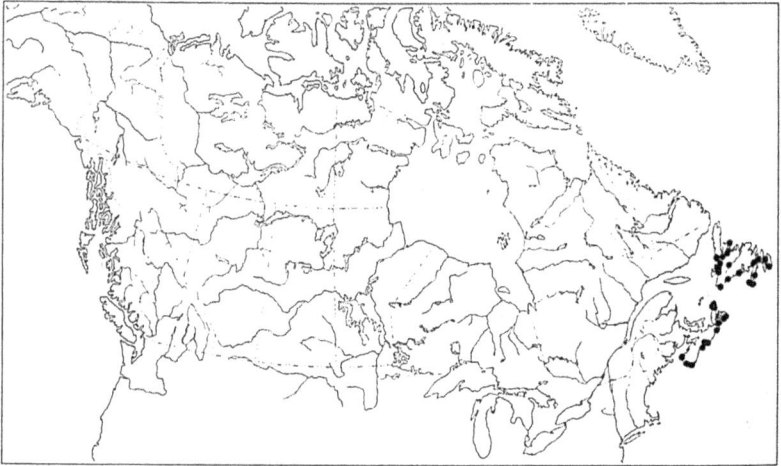

Map 68   *Schizaea pusilla*

Map 69   *Mecodium wrightii*

Map 70   *Dennstaedtia punctilobula*

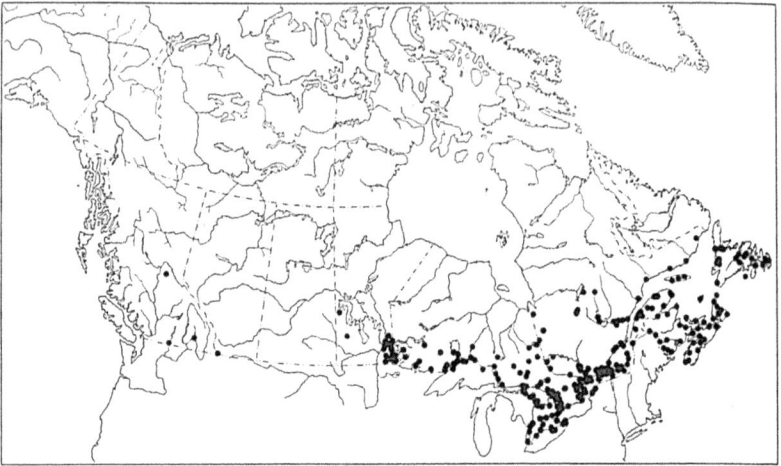

Map 71 *Pteridium aquilinum* var. *latiusculum*

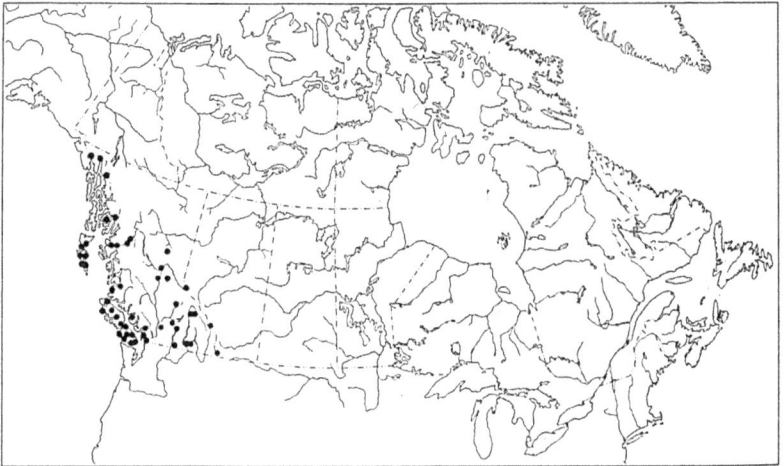

Map 72 *Pteridium aquilinum* var. *pubescens*

Map 73   *Cheilanthes feei*

Map 74   *Cheilanthes gracillima*

Map 75 *Aspidotis densa*

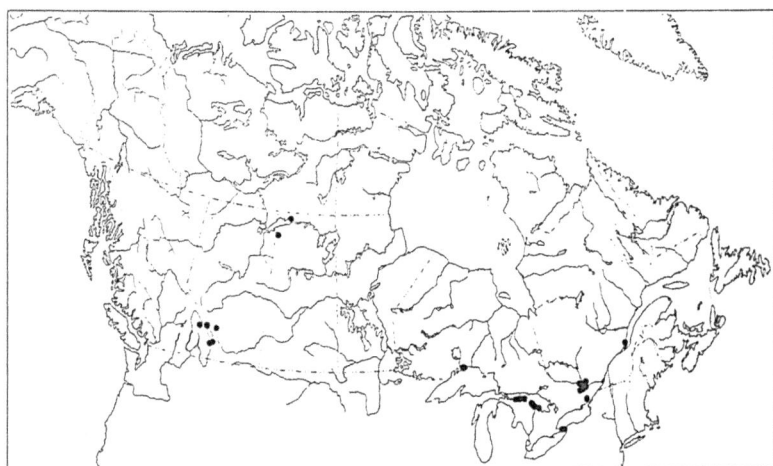

Map 76 *Pellaea atropurpurea*

Maps

Map 77 *Pellaea glabella* var. *glabella*

Map 78 *Pellaea glabella* var. *nana*

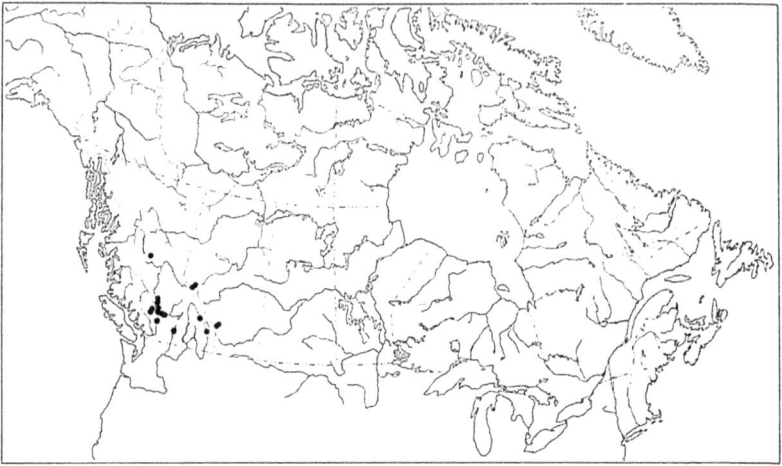

Map 79   *Pellaea glabella* var. *simplex*

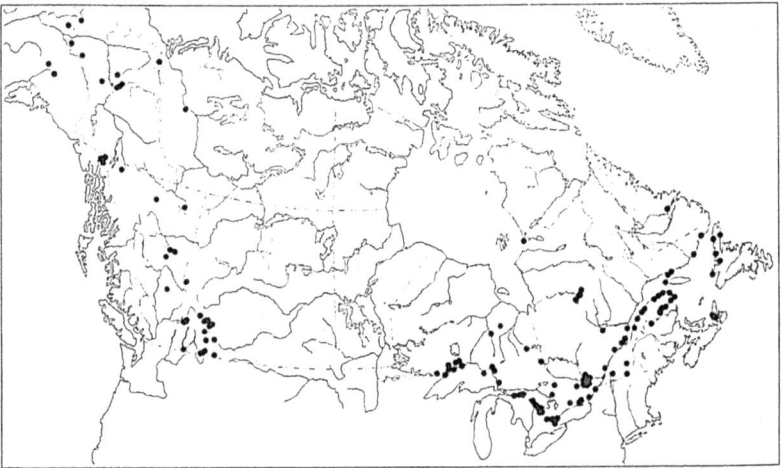

Map 80   *Cryptogramma stelleri*

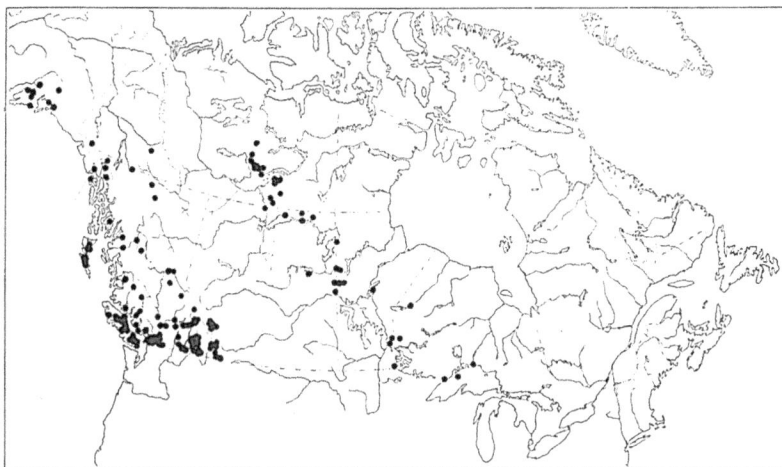

Map 81   *Cryptogramma crispa* var. *acrostichoides*

Map 82   *Cryptogramma crispa* var. *sitchensis*

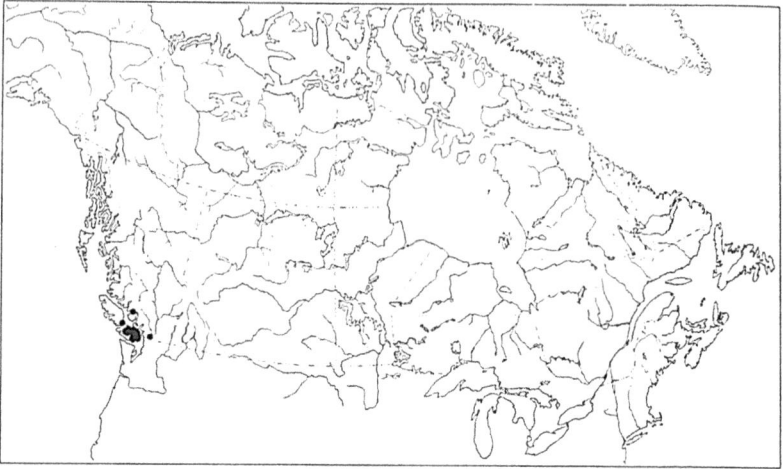

Map 83   *Pityrogramma triangularis*

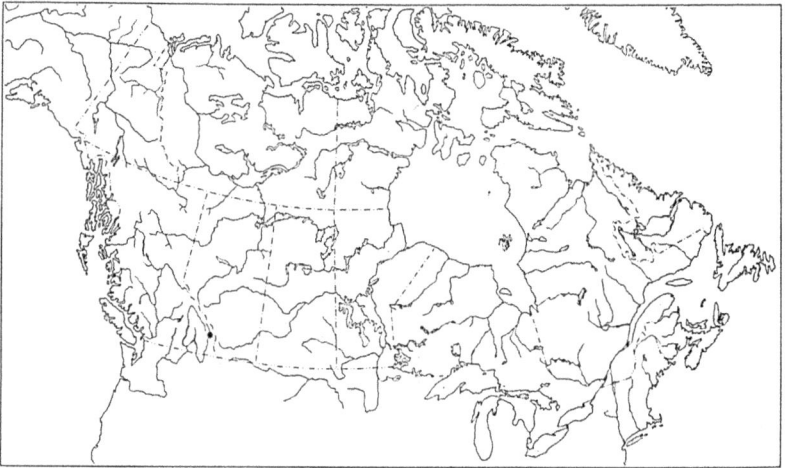

Map 84   *Adiantum capillus-veneris*

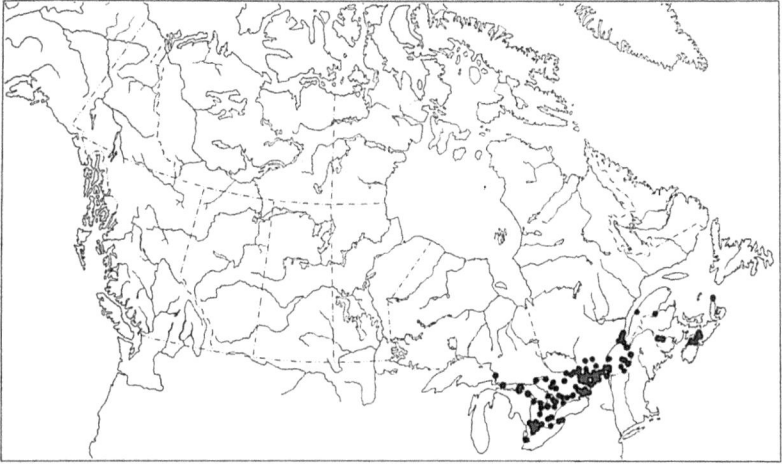

Map 85   *Adiantum pedatum* ssp. *pedatum*

Map 86   *Adiantum pedatum* ssp. *aleuticum*

Map 87 *Adiantum pedatum* var. *subpumilum*

Map 88 *Adiantum pedatum* ssp. *calderi*

Maps

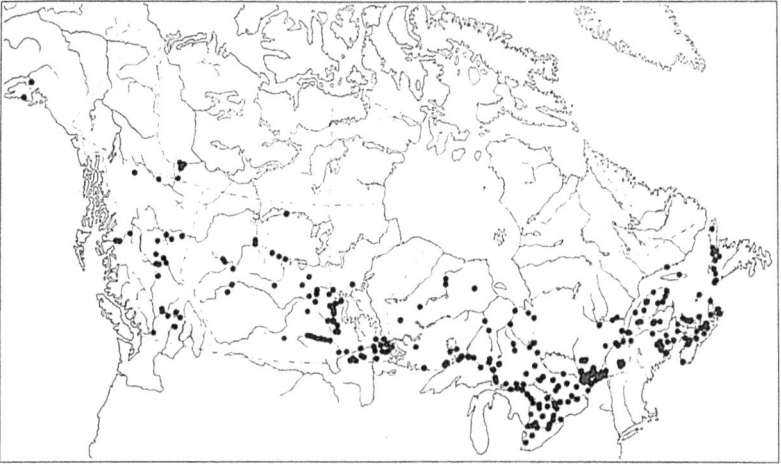

Map 89   *Matteuccia struthiopteris* var. *pensylvanica*

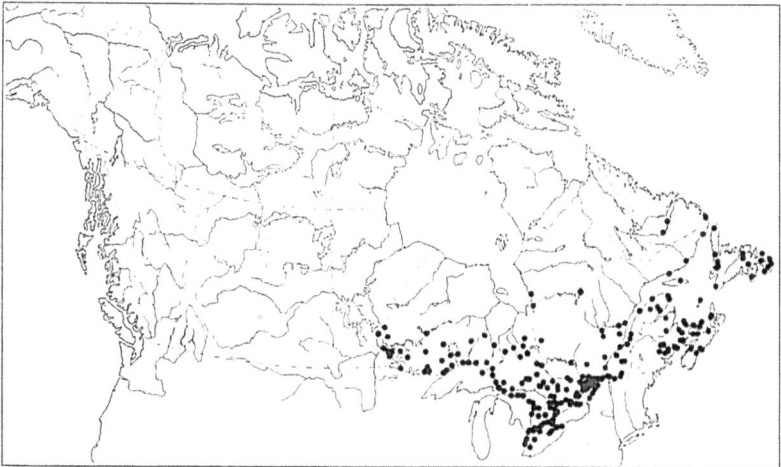

Map 90   *Onoclea sensibilis*

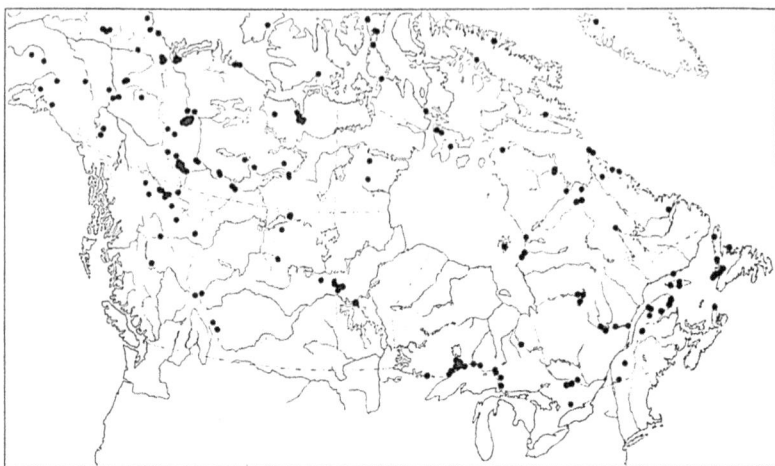

Map 91 *Woodsia glabella*

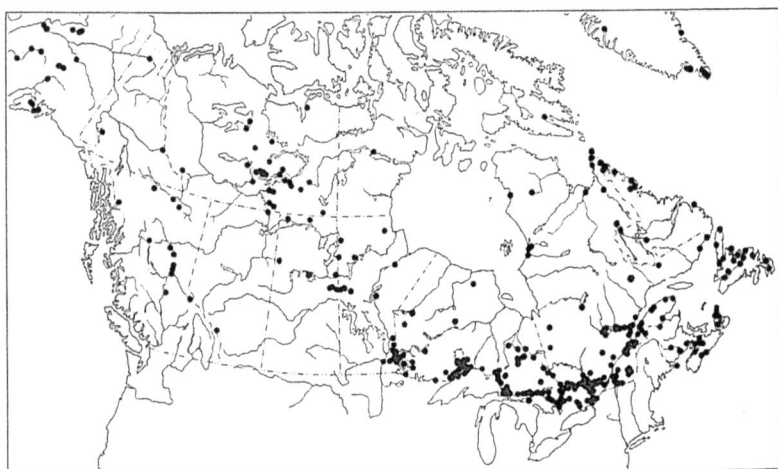

Map 92 *Woodsia ilvensis*

Maps

Map 93 *Woodsia alpina*

Map 94 *Woodsia oregana*

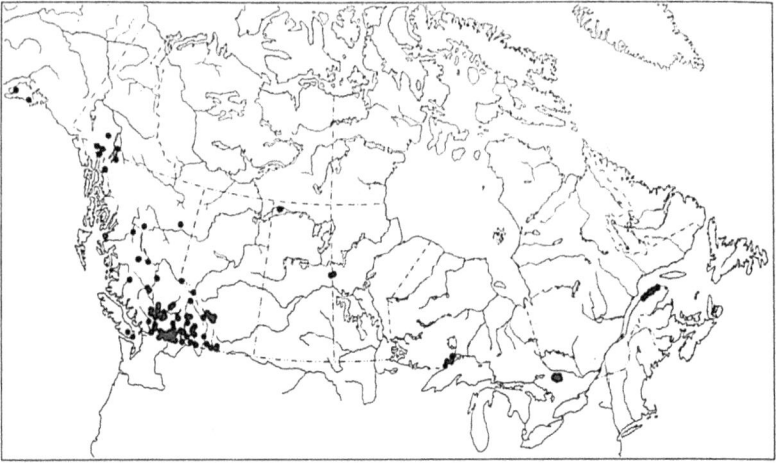

Map 95 *Woodsia scopulina*

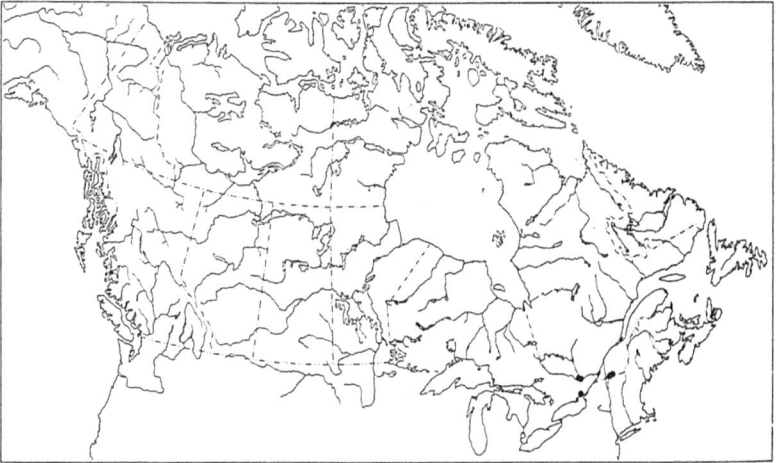

Map 96 *Woodsia obtusa*

Map 97 *Polystichum acrostichoides*

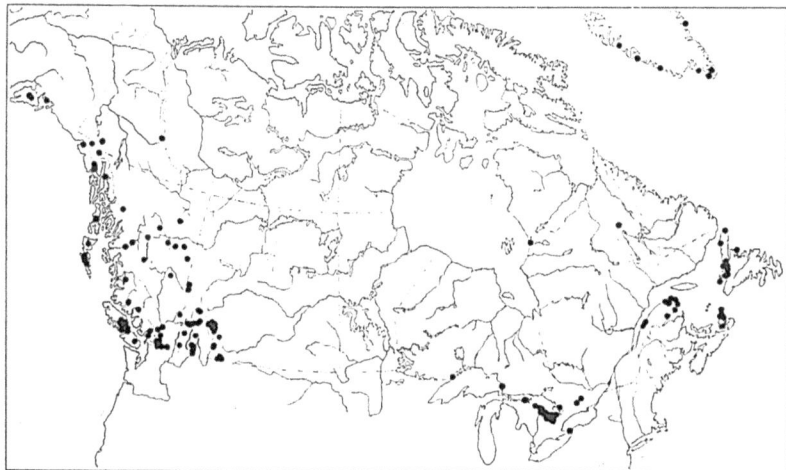

Map 98 *Dryopteris lonchitis*

Map 99   *Polystichum lemmonii*

Map 100   *Polystichum imbricans*

Maps

Map 101   *Polystichum kruckebergii*

Map 102   *Polystichum scopulinum*

Map 103　*Polystichum munitum*

Map 104　*Polystichum andersonii*

Map 105   *Polystichum californicum*

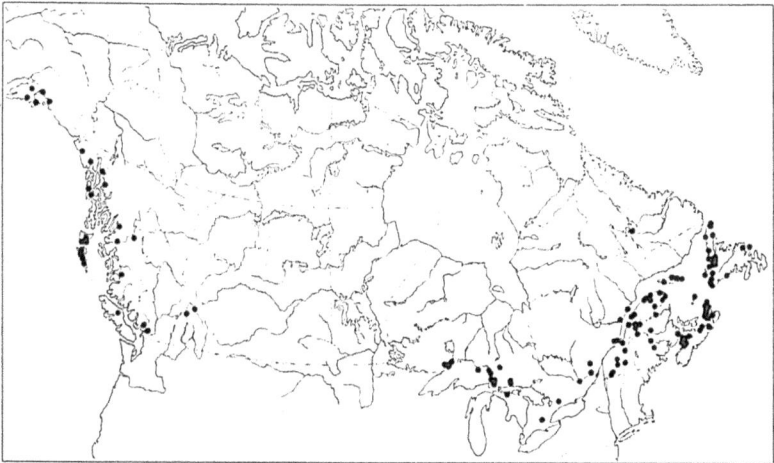

Map 106   *Polystichum braunii*

Map 107 *Polystichum setigerum*

Map 108 *Dryopteris arguta*

366   *Maps*

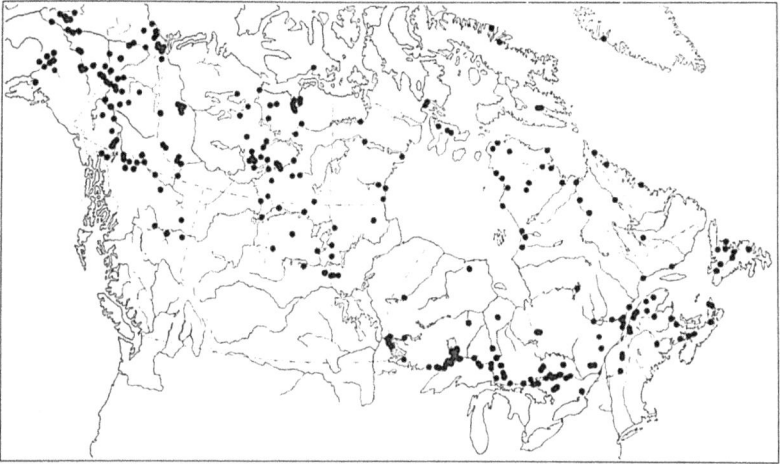

Map 109 *Dryopteris fragrans*

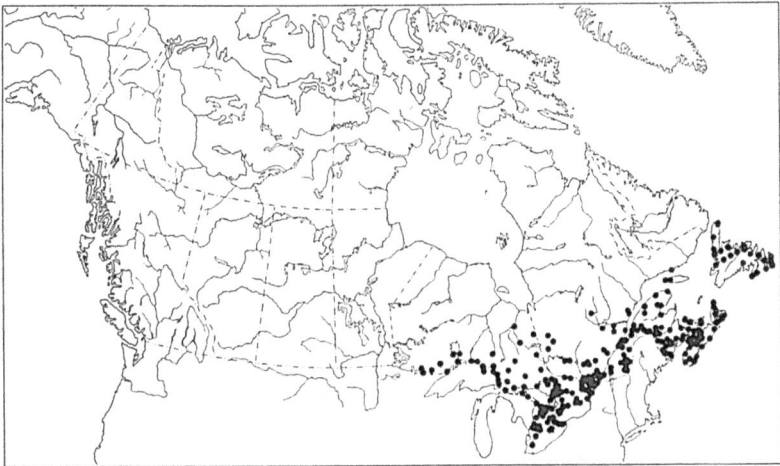

Map 110 *Dryopteris intermedia*

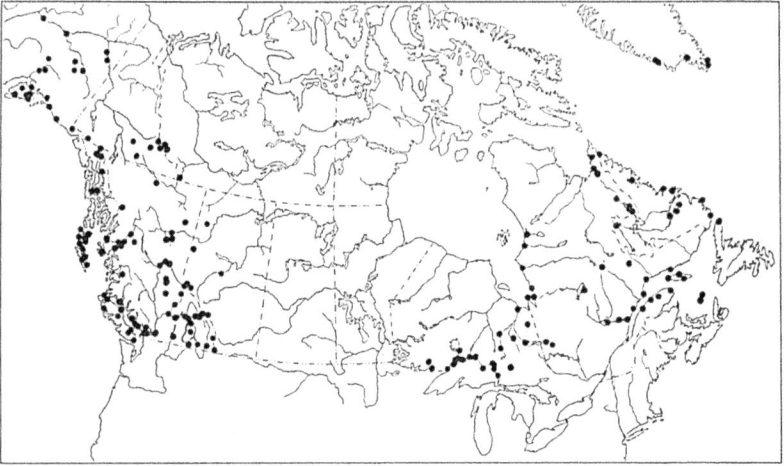

Map 111   *Dryopteris expansa*

Map 112   *Dryopteris campyloptera*

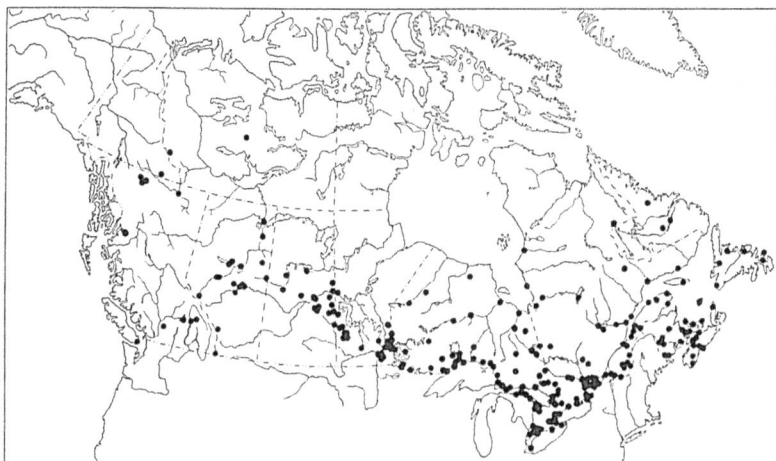

Map 113 *Dryopteris carthusiana*

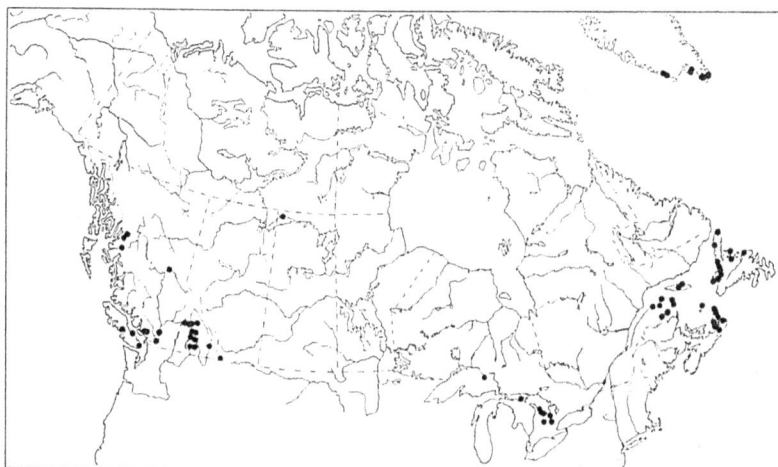

Map 114 *Dryopteris filix-mas*

Map 115   *Dryopteris marginalis*

Map 116   *Dryopteris goldiana*

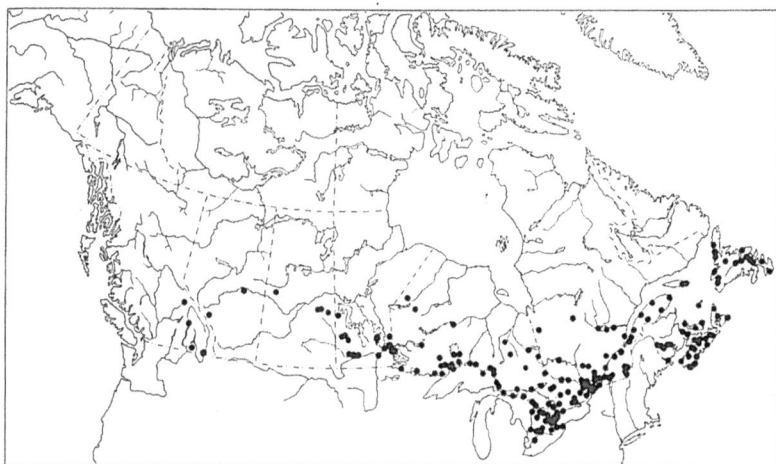

Map 117 *Dryopteris cristata*

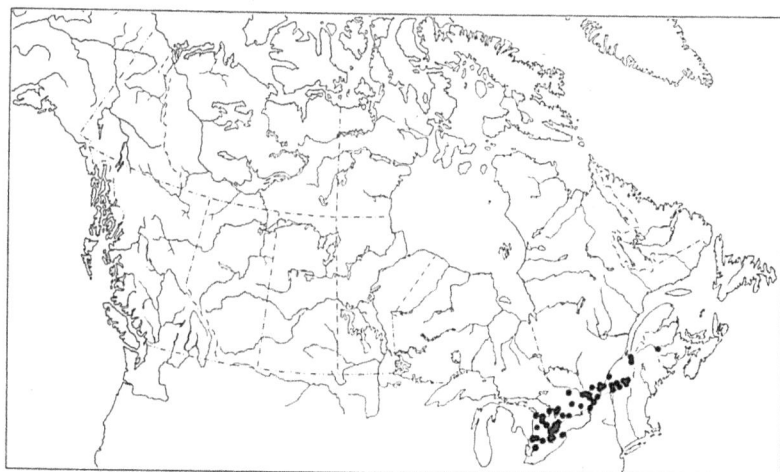

Map 118 *Dryopteris clintoniana*

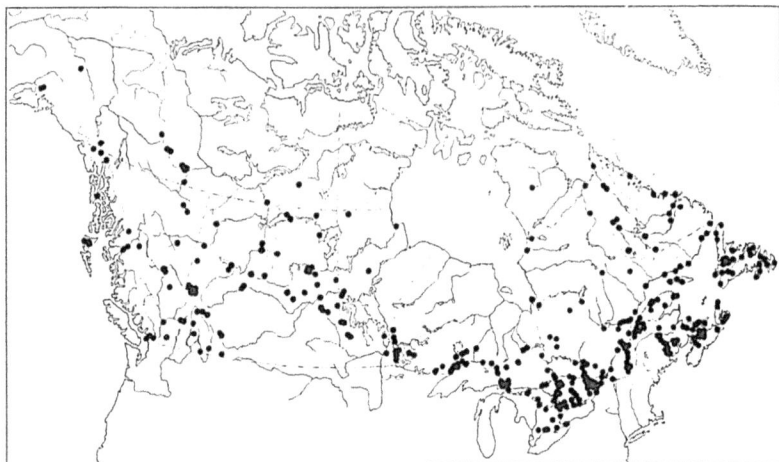

Map 119   *Gymnocarpium dryopteris* ssp. *dryopteris*

Map 120   *Gymnocarpium dryopteris* ssp. *disjunctum*

Maps

Map 121   *Gymnocarpium jessoense* ssp. *parvulum*

Map 122   *Gymnocarpium robertianum*

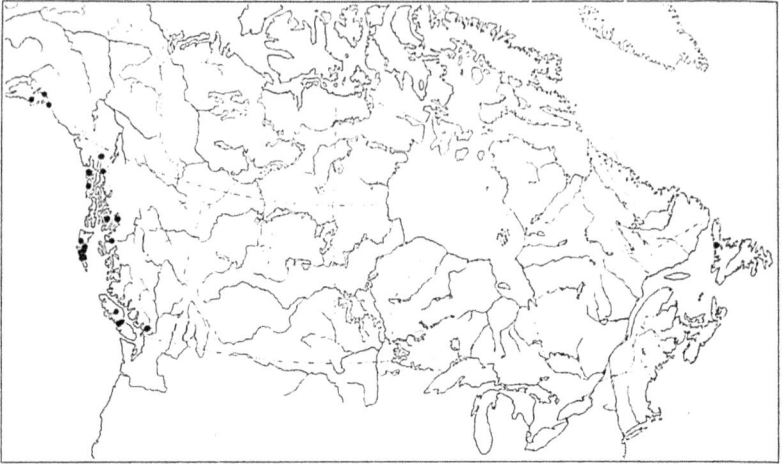

Map 123 *Thelypteris limbosperma*

Map 124 *Thelypteris nevadensis*

Maps

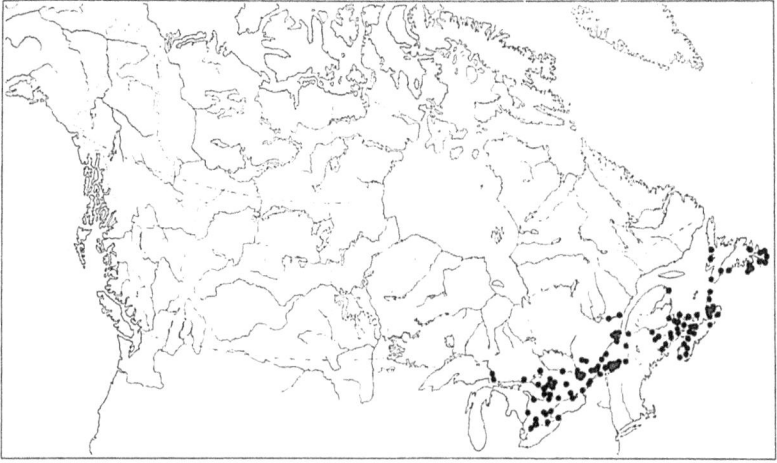

Map 125   *Thelypteris noveboracensis*

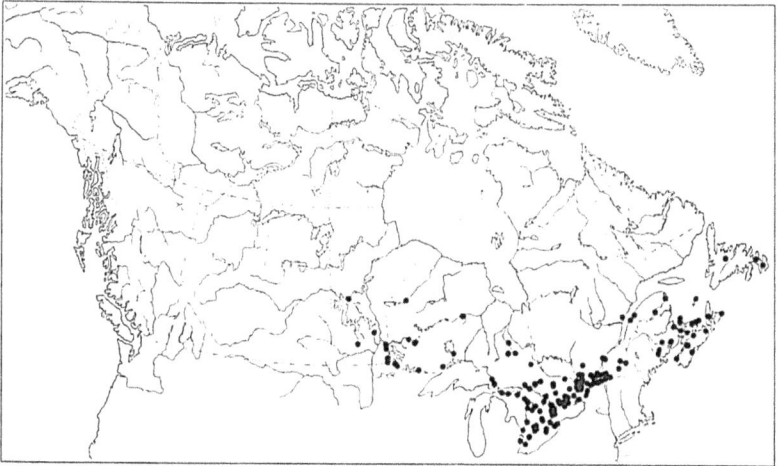

Map 126   *Thelypteris palustris* var. *pubescens*

Map 127 *Thelypteris simulata*

Map 128 *Phegopteris hexagonoptera*

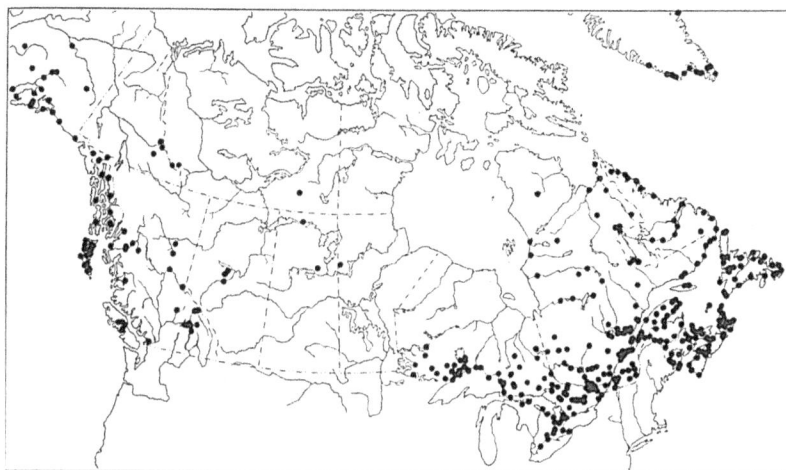

Map 129 *Phegopteris connectilis*

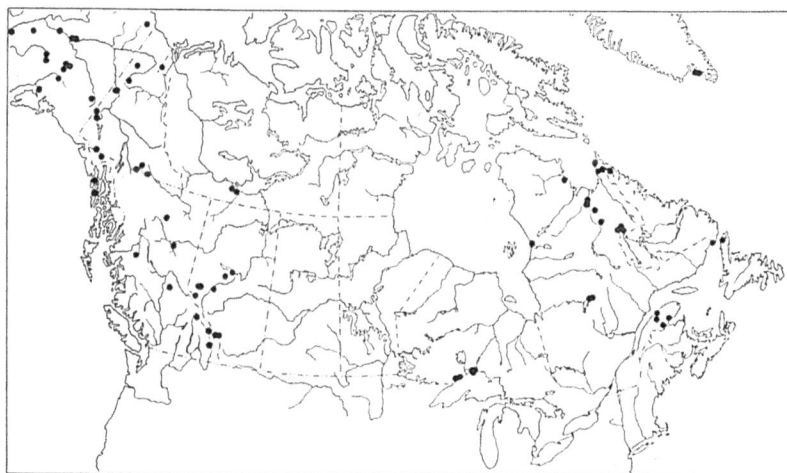

Map 130 *Cystopteris montana*

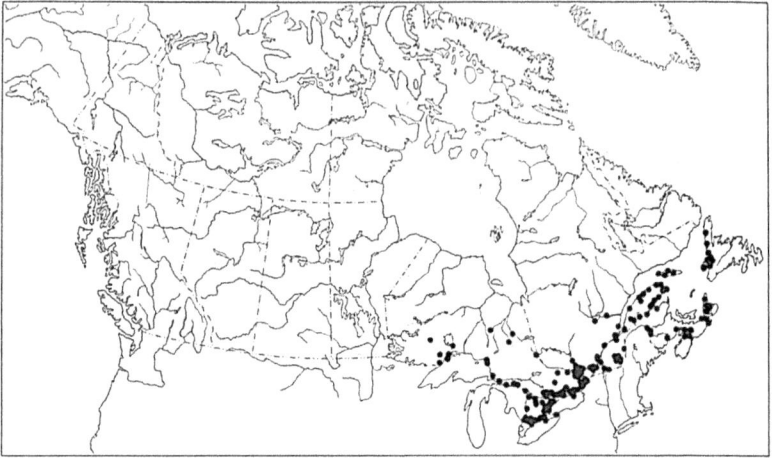

Map 131 *Cystopteris bulbifera*

Map 132 *Cystopteris protrusa*

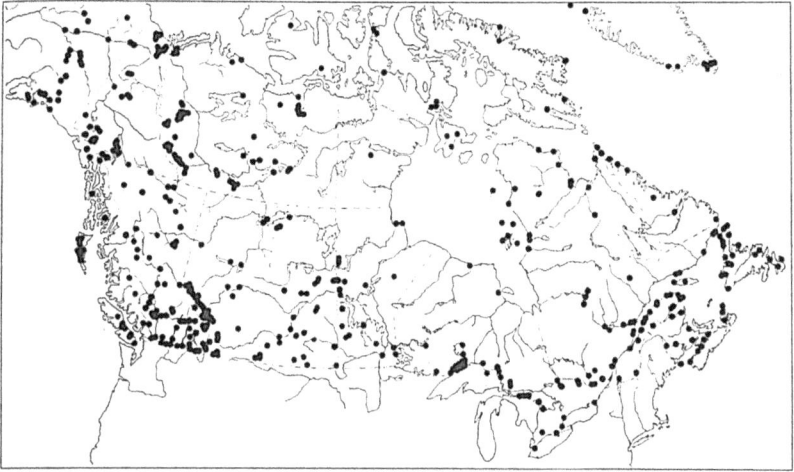

Map 133 *Cystopteris fragilis* var. *fragilis*

Map 134 *Cystopteris fragilis* var. *mackayii*

Maps

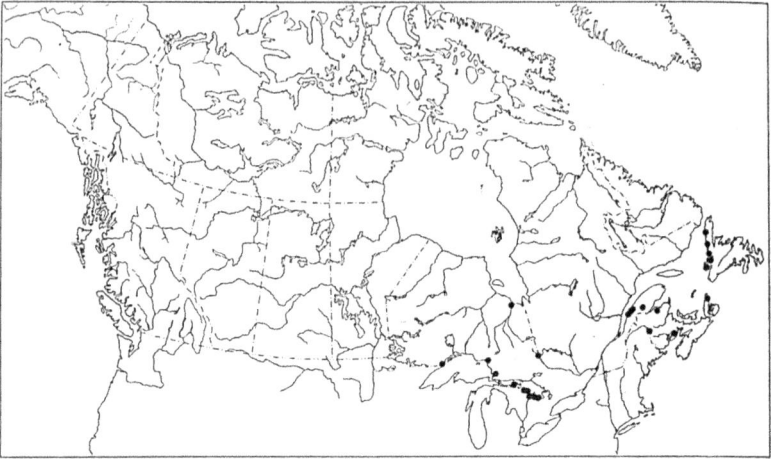

Map 135   *Cystopteris laurentiana*

Map 136   *Athyrium alpestre* ssp. *americanum*

Map 137   *Athyrium filix-femina var. cyclosorum*

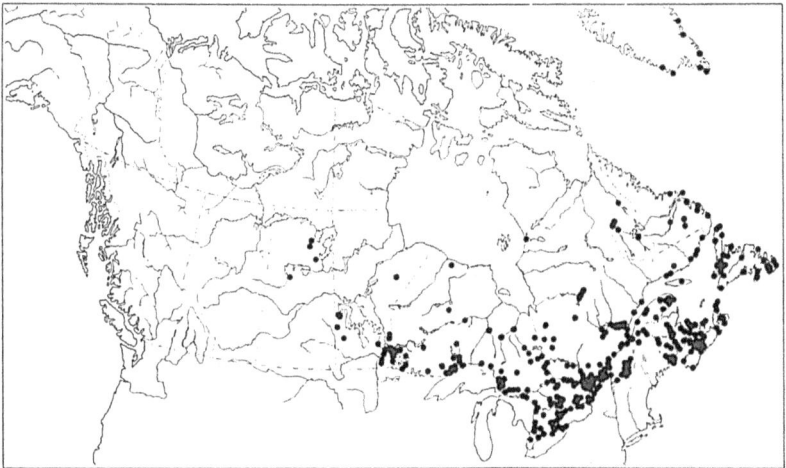

Map 138   *Athyrium filix-femina var. michauxii*

Map 139 *Athyrium pycnocarpon*

Map 140 *Athyrium thelypterioides*

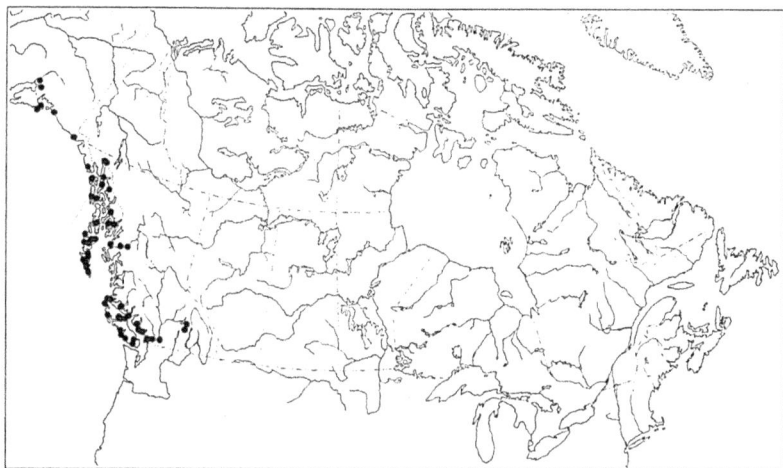

Map 141 *Blechnum spicant*

Map 142 *Woodwardia areolata*

Map 143   *Woodwardia fimbriata*

Map 144   *Woodwardia virginica*

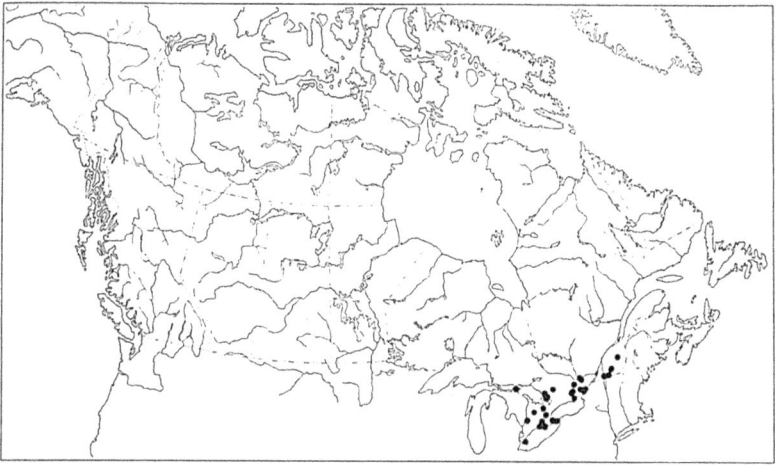

Map 145 *Asplenium platyneuron*

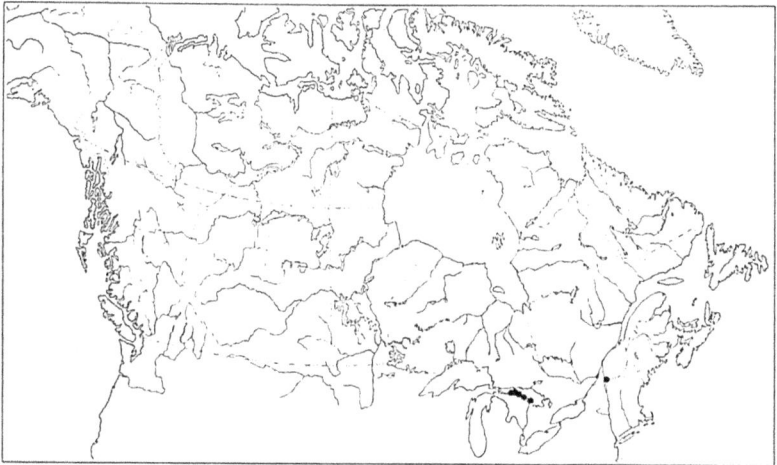

Map 146 *Asplenium ruta-muraria*

Map 147 *Asplenium trichomanes*

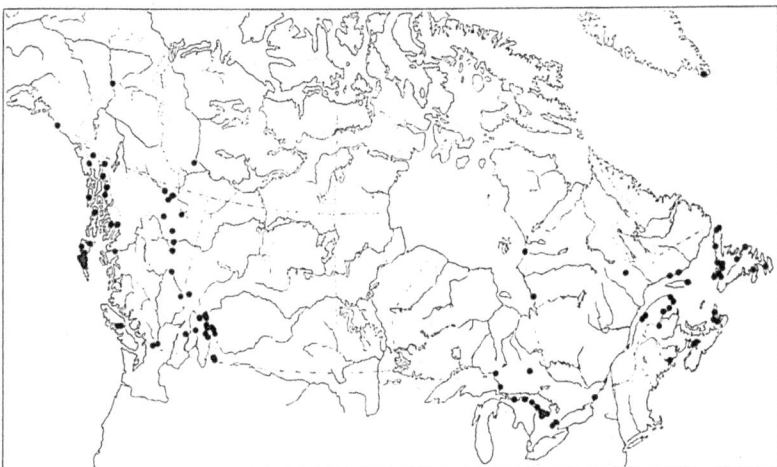

Map 148 *Asplenium viride*

Map 149 *Camptosorus rhizophyllus*

Map 150 *Phyllitis scolopendrium* var. *americana*

Map 151 *Polypodium scouleri*

Map 152 *Polypodium glycyrrhiza*

Map 153 *Polypodium amorphum*

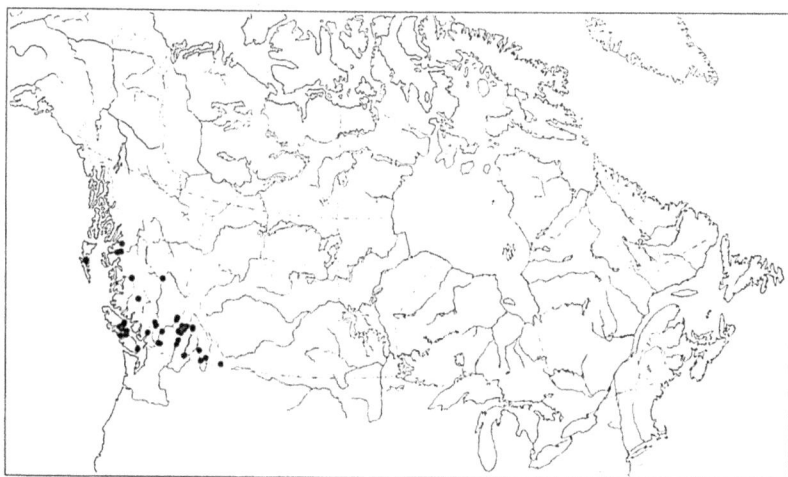

Map 154 *Polypodium hesperium*

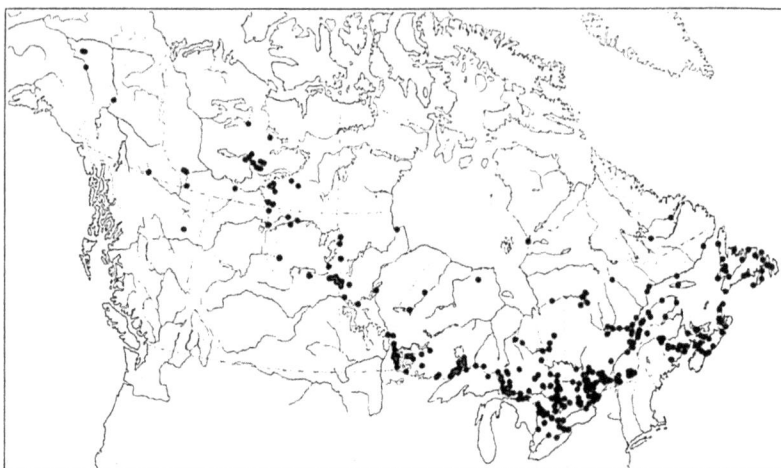

Map 155 *Polypodium virginianum* (s. l.)

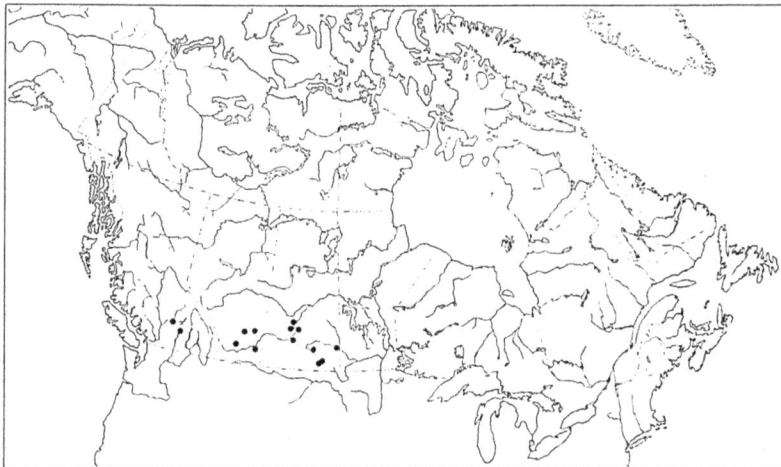

Map 156 *Marsilea vestita*

Map 157 *Marsilea quadrifolia*

Map 158 *Azolla caroliniana*

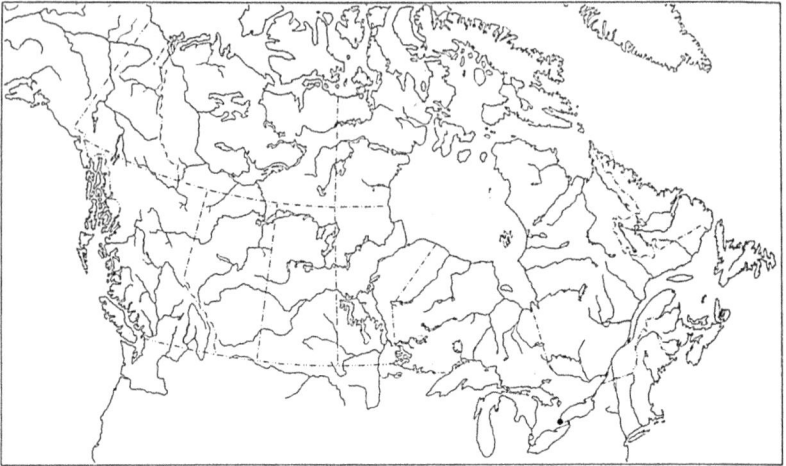

Map 159   *Azolla mexicana*

GLOSSARY

abaxial On the side of an organ away from the axis; dorsal.

acrid Sharp and harsh or unpleasantly pungent in taste or odor.

acuminate Tapering to a slender point.

adaxial Toward the axis; ventral.

adnate Grown together or attached; applied only to unlike organs, as stipules adnate to the petiole.

allopolyploidy The doubling or higher multiplication of chromosome sets from various species or genera either spontaneously or experimentally induced (allotetraploid, allohexaploid).

anastomosing Connecting by crossveins and forming a network.

annual Of one year's duration.

antheridium In cryptogams the organ corresponding to an anther.

apiculate Ending abruptly in a small, usually sharp tip.

apogamous Developed without fertilization; parthenogenetic.

appressed Lying close to or parallel to an organ, as hairs appressed to a leaf or leaves appressed to a stem.

arborescent Of large size and more or less tree-like, but without the clear distinction of a single trunk.

archegonium The organ in the higher cryptogams that corresponds to a pistil in the flowering plants.

areole A small space marked out upon or beneath a surface.

articulate Jointed; having nodes, joints, or places where separation may naturally take place.

ascending Growing obliquely upward (of stems); directed obliquely forward with respect to the organ to which they are attached (of parts of a plant).

attenuate Gradually tapering to a very slender point.

auricle A small, ear-shaped projecting lobe or appendage at the base of an organ.

auriculate Having an auricle.

autopolyploidy The presence of more than two (diploidy) of the monoploid chromosome sets characteristic of the species.

axil The angle formed between any two organs.

bipinnate Doubly or twice pinnate.

bivalvate Having two valves.

blade The expanded part of a frond.

bract A more-or-less modified leaf subtending a flower, or belonging to an inflorescence, or sometimes appearing cauline.

bristle A stiff hair, or any slender body, that may be likened to a hog's bristle.

bristly Provided with bristles.

caespitose Growing in dense tufts; usually applied only to small plants.

caudate Having a slender tail-like terminal appendage.

chaff A small thin scale or bract that becomes dry and membranous.

chaffy Having or resembling chaff.

ciliate Having marginal hairs.

clinal Series of changes in form; a gradient of biotypes along an environmental transition.

concolorous Uniform in color.

confluent Flowing or running together.

continuous Marked by uninterrupted extension in space, time, or sequence.

cordate Heart-shaped; sometimes applied to whole organs, but more often to the base only.

coriaceous Leathery in texture.

corm The enlarged fleshy base of a stem, bulb-like but solid; a solid bulb.

costa A rib; a midrib or mid nerve.

crenate Dentate with teeth much rounded.

crenulate Finely crenate.

cuneate Wedge-shaped; narrowly triangular with the acute angle pointed downward.

cuspidate Tipped with a cusp or a sharp and firm point.

cytotype Any variety (race) of a species whose chromosome complement differs quantitatively or qualitatively from the standard complement of that species.

deciduous Falling after completion of the normal function; not evergreen.

decompound More than once compound or divided.

decumbent Prostrate at base, either erect or ascending elsewhere.

decurrent Extending downward from the point of intersection.

deflexed Bent or turned abruptly downward.

deltoid Broadly triangular.

dentate Toothed along the margin, the apex of each tooth sharp and directed outward.

denticulate Minutely dentate.

dichotomous Forking more or less regularly into branches of about equal size.

dimorphic Occurring in two forms.

diploid With $2n$ chromosomes per cell.

distinct Separate; not united; evident.

divergent Inclining away from each other.

dorsal Located on or pertaining to the back or outer surface of an organ.

ecilate Without cilia.

echinate Provided with prickles.

elater Appendage of spores of horsetails, formed from the outermost wall layer, coiling and uncoiling as air is dry or moist, possibly assisting in spore dispersal.

ellipse A regular oval.

elliptical Oval in outline; having narrowed to rounded ends and being widest at or about the middle; of, relating to, or shaped like an ellipse.

emersed Standing out of or rising above a surface (as of a fluid).

endophyte A plant living within another plant.

entire With a continuous, unbroken margin.

epiphyte A plant growing attached to another plant, but not parasitic.

erose Irregularly cut or toothed along the margin.

extirpate To destroy completely; eradicate.

falcate Scythe-shaped; curved and flat, tapering gradually.

fastigiate Erect and close together.

fertile Capable of normal reproductive functions.

fibrillose With fine fibers.

filiform Thread-like; long, slender, and terete.

fimbriate Fringed.

flabellate, flabelliform Fan-shaped or broadly wedge-shaped.

flaccid Flabby; lacking in stiffness.

flexuous Curved alternately in opposite directions.

frond The expanded leaf-like portion of a fern, including stipe, rachis, and pinnae.

gametophyte In the life cycle, the generation in which sexual organs are produced.

gemma A bud or body analogous to a bud by which some plants propagate themselves.

genome The basic chromosome set of an organism.

glabrous Lacking pubescence; smooth.

glandular Containing or bearing glands.

glaucous Gray, grayish green, or bluish green, with a thin coat of fine removable particles that are often waxy in texture; covered or whitened with a bloom.

globose Spherical or nearly so.

herbaceous Without a persistent woody stem above ground; dying back to the ground at the end of the growing season; leaf-like in color and texture.

hyaline Transparent or translucent.

hybrid Produced by dissimilar parents; a cross-breed of two species.

hybridization The production of a hybrid.

imbricate Overlapping, either vertically or spirally, where the lower piece covers the base of the next higher.

indusium The covering of the sorus.

internode The portion of a stem or other structure between two nodes.

interspecific Between two different species.

isoenzyme A phase of an enzyme (a protein that even in low concentration speeds up, enables, or controls chemical reactions in living organisms without being used up in the reactions).

lacerate Having an irregularly jagged margin; irregularly cut as if torn.

lamina A sheet or plate; the flat, expanded portion of a structure such as a leaf or petal.

lanceolate Shaped like a lance head, much longer than wide and widest below the middle.

lateral Situated on or arising from the side of an organ.

ligulate Having a ligule; having the nature of a ligule.

ligule In *Isoetes*, a small triangular or elongate delicate tissue extending slightly above the sporangium.

linear Narrow and elongate, with parallel sides.

loculate Having or divided into loculi.

locule A cavity or one of the cavities within an ovary, a fruit, or an anther.

lunate Of the shape of a half-moon or crescent.

macrospore The larger kind of spore in *Selaginellaceae* and *Isoetes*, and in other genera.

marcescent Withering and persistent.

medial Being or occurring in the middle.

megasporangium, macrosporangium Sporangium within which megaspores are formed. In flowering plants known as the ovary.

megaspore or macrospore The larger of the two kinds of spores produced by heterosporous ferns; the first cell of a female gametophyte generation of these plants and of seed plants.

meiosis The reduction divisions that result in the production of four cells from a single one, the number of chromosomes per cell being reduced from $2n$ to n.

meiotic Characterized by meiosis.

membranous Thin and pliable, as an ordinary leaf, in contrast to chartaceous, coriaceous, or succulent.

microsporangium Sporangium within which microspores are formed. In flowering plants, the pollen sac.

microspore The smaller kind of spore in *Selaginellaceae* and *Isoetes,* and in other genera.

mucro A short and small abrupt tip.

mucronate Tipped with a mucro.

mycorrhiza An association of a fungus with the root of a higher plant.

node The place upon a stem that normally bears a leaf or whorl of leaves; the solid constriction in the culm of a grass; a knot-like or knob-like enlargement.

oblanceolate Lanceolate with the broadest part above the middle.

oblique Unequal-sided; slanting.

oblong Two to three times longer than broad and with nearly parallel sides.

obtuse Blunt or rounded at the end.

orbicular Essentially circular.

ovate Egg-shaped; having an outline like that of an egg, with the broader end basal.

palmate Having three or more lobes, nerves, leaflets, or branches arising from one point; digitate.

panicle A loose, irregularly compound inflorescence with pedicellate flowers.

paniculate Arranged in a panicle.

papillose Bearing minute nipple-shaped projections.

paraphyses Filaments of sterile cells among sporangia.

pedunculate Born upon a peduncle.

peltate Shield-shaped and attached to the support by the lower surface.

pendulous Hanging or drooping.

perennial A plant that continues its growth from year to year.

petiolate Having a petiole.

petiole The basal stalk-like portion of an ordinary leaf, in contrast with the expanded blade; the support of a leaf.

pinna One of the primary divisions of a pinnate or pinnately compound frond.

pinnate Compound; having branches, lobes, or leaflets arranged on two sides of a common rachis.

pinnatifid Pinnately cleft.

pinnule A secondary pinna.

polyphyletic Of individuals derived in the course of evolution from two interbreeding populations or phyletic stocks.

precocious Appearing or developing very early.

proliferating, proliferous Producing buds and plantlets from leaves or as other offshoots.

prostrate Lying flat upon the ground.

prothallus A cellular, usually flat, thallus-like growth, resulting from the germination of a spore, upon which sexual organs and eventually new plants are developed.

puberulent Minutely or sparsely pubescent with scarcely elongate hairs.

pubescent Bearing hairs on the surface.

raceme A simple inflorescence of pediceled flowers upon a common more or less elongated axis.

racemose In racemes; or resembling a raceme.

rachis The upper part of the petiole, bearing the pinnae and continuous with the stipe.

recurved Curved downward or backward.

reflexed Abruptly bent downward or backward.

reniform Kidney-shaped; wider than long, rounded in general outline, and with a wide basal sinus.

reticulate In the form of a network.

revolute Rolled backward, so that the upper surface of the organ is exposed and the lower side more or less concealed.

rhizome An underground usually horizontal stem; a rootstock.

rhombic Having the outline of an equilateral parallelogram.

rhomboid A solid with a rhombic outline.

rhomboidal Having the shape of a rhomboid.

salient Something that projects outward or upward from its surroundings.

scabrous Rough to the touch, owing to the structure of the epidermes or the pressure of short, stiff hairs.

scarious Thin, dry, and membranous; not green.

scurfy Covered with scale-like or bran-like particles.

serrate Toothed along the margin, the apex of each tooth sharp (compare crenate) and directed forward (compare dentate).

serrulate Finely serrate.

sessile Without a stalk of any kind.

seta A bristle.

setiform Like a bristle.

siliceous Composed of or abounding in silica.

sinus The cleft or recess between two lobes.

sorus (pl. sori) A heap or cluster of sporangia bearing the spores.

spatulate Shaped like a spatula; maintaining its width or somewhat broadened toward the rounded summit; spoon-shaped.

spicule A minute, slender, pointed, usually hard body.

spinule A small spine.

spinulose Bearing small spines over the surface.

sporangium   The globular organ in which the spores are produced.

spore   An asexual reproductive cell that germinates into a prothallus, which in turn gives rise to sexual reproduction.

sporocarp   The fruit cases of certain cryptogams that contain sporangia or spores.

sporophyll   A specialized organ for the production of spores in sporangia.

sporophyte   In the life cycle, the generation in which spores are produced.

sporulation   The formation of spores.

stellate, stelliform   Star-shaped.

stipe   The lower part of the petiole, which does not bear pinnae.

stoma (pl. stomata)   A minute orifice or mouth-like opening between two guard cells in the epidermis, particularly on the lower surface of the leaves, through which gaseous interchange between the atmosphere and the intercellular spaces of the parenchyma is effected.

stramineous   Straw-colored.

strigose   Having appressed, sharp, straight, and stiff hairs pointing in the same direction.

strobile   An inflorescence resembling a spruce or fir cone, partly made up of imbricated bracts or scales.

sub- (prefix)   Slightly; more or less; somewhat.

subcylindrical   Slightly or somewhat cylindrical.

submersed   Growing, or adapted to growing, under water.

subulate   Awl-shaped.

succulent   Juicy; fleshy.

sympatric   With areas of distribution that coincide or overlap.

tangential   Of, relating to, or of the nature of a tangent.

ternate   Arranged in threes.

tetraploid   With $4n$ chromosomes per cell.

tomentose   Woolly, with an indument of crooked, matted hairs.

trapezoid   A quadrilateral having only two sides parallel.

trigonous   Three-angled.

tripinnate   Three times pinnate.

triploid   With $3n$ chromosomes per cell.

truncate   Ending abruptly, as if cut off.

tuberculate   Bearing small processes or tubercules.

tubercule   A small tuber or tuber-like (not necessarily subterranean) body, often formed as the result of a symbiotic relation of organisms.

ultramafic   Rock types in which the elemental composition is largely silicates of iron and magnesium.

vallecula Applied to the grooves in the intervals between the ridges, as in the stems of *Equisetum*. Vallecular: pertaining to such grooves.

velum The membranous indusium in *Isoetes*.

ventral Belonging to the anterior or inner face of an organ, as opposed to dorsal; adaxial.

verticillate Arranged in a whorl.

vicariad One of two or more related organisms that occur in similar environments but in distinct and often widely separated areas.

villous Covered densely with fine long hairs but not matted.

whorl Leaves or other plant parts arranged in a circle around the stem.

xeric Characterized by, relating to, or requiring only a small amount of moisture.

REFERENCES

Alston, A.H.G. 1955. The heterophyllous Selaginellae of continental North America. Bull. Br. Mus. (Nat. Hist.) Bot. 1(8):221–274.

Alt, K.S.; Grant, V. 1960. Cytotaxonomic observations on the gold-back fern. Brittonia 12:153–70.

Argus, G.W.; White, D.J. 1977. The rare vascular plants of Ontario. Syllogeus 14:1–63.

Argus, G.W.; White, D.J. 1978. The rare vascular plants of Alberta. Syllogeus 17:1–46.

Argus, G.W.; White, D.J. 1983. Atlas of the rare vascular plants of Ontario, Part 2. National Museum of Natural Sciences, Ottawa, Ont.

Barclay-Estrup, P.; Hess, G.V. 1974. Adder's tongue fern, *Ophioglossum vulgatum* L. in northwestern Ontario. Can. Field-Nat. 88:217–219.

Beitel, J.M. 1979a. The clubmosses *Lycopodium sitchense* and *L. sabsinifolium* in the Upper Great Lakes region. Mich. Bot. 18:3–13.

Beitel, J.M. 1979b. Clubmosses (*Lycopodium*) in North America. Fiddlehead Forum 6:1–8.

Beitel, J.M.; Wagner, F.S. 1982. The chromosomes of *Lycopodium lucidulum*. Am. Fern J. 72:33–35.

Billington, C. 1952. Ferns of Michigan. Cranbrook Inst. Sci. Bull. 32. 240 pp.

Blasdell, R.F. 1963. A monographic study of the fern genus *Cystopteris*. Mem. Torrey Bot. Club 21:1–102.

Bloom, W. 1955. Comparative viability of sporocarps of *Marsilea quadrifolia* in relation to age. Trans. Ill. State Acad. Sci. 47:72–76.

Bloom, W. 1961. Heat resistance of sporocarps of *Marsilea quadrifolia*. Am. Fern J. 51:95–97.

Boivin, B. 1952. Two variations of *Pteridium aquilinum*. Am. Fern J. 42:131–133.

Boivin, B. 1960a. Centurie de plantes canadiennes III. Nat. Can. (Que.) 87:25–49.

Boivin, B. 1960b. A new *Equisetum*. Am. Fern J. 50:107–109.

Boivin, B. 1962. Études ptéridologiques. II: *Gymnocarpium* Newman. Bull. Soc. Bot. Fr. 109:127–128.

Boivin, B. 1966a. Études ptéridologiques. III: Variations du *Woodsia oregana*. Bull. Soc. Bot. Fr. 113:407–409.

Boivin, B. 1966*b*. Notes sur les *Lycopodium* du Canada. Nat. Can. (Que.) 93:355–359.

Boivin, B. 1968. Énumération des plantes du Canada. Provancheria 6.

Bottarelli, F. 1968. On poisoning: *Pteris aquilina, Equisetum arvense* and *E. palustre*. Veterinaria (Milano) 17:308–322.

Bouchard, A.; Barabé, D.; Hay, S. 1977. An isolated colony of *Oreopteris limbosperma* (All.) Holub in Gros Morne National Park, Newfoundland, Canada. Nat. Can. (Que.) 104:239–244.

Bouchard, A.; Barabé, D.; Dumais, M.; Hay, S. 1983. The rare vascular plants of Quebec. Syllogeus 48:1–75.

Bouchard, A.; Hay, S.G. 1974. Addition à la flore de Terre-Neuve: *Lycopodium alpinum* L. Nat. Can. (Que.) 101:803.

Bouchard, A.; Hay, S.G. 1976. *Thelypteris limbosperma* in eastern North America. Rhodora 78:552–553.

Braun, M. 1938. Index to North American ferns. Published by the author, Orleans, Mass. 217 pp.

Britton, D.M. 1953. Chromosome studies in ferns. Am. J. Bot. 40:575–583.

Britton, D.M. 1964. Chromosome numbers of ferns in Ontario. Can. J. Bot. 42:1349–1354.

Britton, D.M. 1965. Hybrid wood ferns in Ontario. Mich. Bot. 4:3–9.

Britton, D.M. 1967. Diploid *Dryopteris dilatata* from Quebec. Rhodora 69:1–4.

Britton, D.M. 1968. The spores of four species of spinulose wood ferns (*Dryopteris*) in eastern North America. Rhodora 70:340–347.

Britton, D.M. 1972*a*. Spore ornamentation in the *Dryopteris spinulosa* complex. Can. J. Bot. 50:1617–1621.

Britton, D.M. 1972*b*. The spores of *Dryopteris clintoniana* and its relatives. Can. J. Bot. 50:2027–2029.

Britton, D.M. 1972*c*. Spinulose wood ferns, *Dryopteris*, in western North America. Can. Field-Nat. 86:241–247.

Britton, D.M. 1974. The significance of chromosome numbers in ferns. Ann. Mo. Bot. Gard. 61:310–317.

Britton, D.M. 1976. The distribution of *Dryopteris spinulosa* and its relatives in eastern Canada. Am. Fern J. 66:69–74.

Britton, D.M. 1977. The fern *Woodsia obtusa* (Spreng.) Torrey in Ontario. Can. Field-Nat. 91:84–85.

Britton, D.M.; Jermy, A.C. 1974. The spores of *Dryopteris filix-mas* and related taxa in North America. Can. J. Bot. 52:1923–1926.

References

Britton, D.M.; Legault, A.; Rigby, S.J. 1967. *Pellaea atropurpurea* (L.) Link and *Pellaea glabella* Mett in Quebec. Nat. Can. (Que.) 94:761–763.

Britton, D.M.; Rigby, S.J. 1968. In search of the purple cliff-brake. Ont. Nat. 5:5–7, 12.

Britton, D.M.; Soper, J.H. 1966. The cytology and distribution of *Dryopteris* species in Ontario. Can. J. Bot. 44:63–78.

Britton, D.M.; Widén, C.-J. 1974. Chemotaxonomic studies on *Dryopteris* from Quebec and eastern North America. Can. J. Bot. 52:627–638.

Britton, D.M.; Widén, C.-J.; Brunton, D.F.; Keddy, P.A. 1975. A new hybrid woodfern, *Dryopteris* × *algonquinensis* D.M. Britton, from Algonquin Park, Ontario. Can. Field-Nat. 89:163–171.

Brown, D.F.M. 1964. A monographic study of the fern genus *Woodsia*. Nova Hedwigia 16:1–154.

Bruce, J.G. 1975. Systematics and morphology of subgenus *Lepidotis* of the genus *Lycopodium* (Lycopodiaceae). Ph.D. thesis, University of Michigan, Ann Arbor, Mich.

Bruce, J.G. 1976. Gametophytes and subgeneric concepts in *Lycopodium*. Am. J. Bot. 63:919–924.

Brunton, D.F. 1972. More slender cliff-brake in the Ottawa District. Trail & Landscape 6(3): 92–93.

Brunton, D.F. 1978. The holly fern in Alberta. Blue Jay 36:82–83.

Brunton, D.F. 1979. Taxonomy, distribution, and ecology of the cliff-brake ferns (*Pellaea*: Polypodiaceae) in Alberta. Can. Field-Nat. 93:288–295.

Brunton, D.F.; Lafontaine, J.D. 1974. The distribution of *Pellaea* in Quebec and eastern Ontario. Nat. Can. (Que.) 101:937–939.

Buck, W.R. 1977. A new species of *Selaginella* in the *S. apoda* complex. Can. J. Bot. 55:366–371.

Burgess, T.J.W. 1886. Recent additions to Canadian Filicineae, with new stations for some of the species previously recorded. Trans. Roy. Soc. Can., Sect. IV:9–18.

Burman, W.A. 1909. Flora of Manitoba. Pages 156–182 *in* British Association for the Advancement of Science: A handbook to Winnipeg and the province of Manitoba. Winnipeg, Man.

Butters, F.K. 1917. The genus *Athyrium* and the North American ferns allied to *Athyrium filix-femina*. Rhodora 19:170–207.

Calder, J.A.; Taylor, R.L. 1968. Flora of the Queen Charlotte Islands, Part 1. Can. Dep. Agr. Res. Branch Monogr. 4(1). 659 pp.

Campbell, C.A.; Britton, D.M. 1977. Pteridophytes of the Regional Municipality of Waterloo, Ontario. Can. Field-Nat. 91:262–268.

References

Campbell, R. 1898–1899. Canadian ferns. Can. Hort. Mag. 2:81–87, 131–137, 157–165, 183–192, 209–218, 235–244, 261–270.

Carlson, T.M.; Wagner, W.H. 1982. The North American distributions of the genus *Dryopteris*. Contrib. Univ. Mich. Herb. 15:141–162.

Catling, P.M. 1975. Alpine woodsia (*Woodsia alpina* (Bolton) S.F. Gray) in southeastern Ontario. Can. Field-Nat. 89:177–178.

Ching, R.C. 1963. A reclassification of the family Thelypteridaceae from the mainland of Asia. Acta Phytotaxon. Sin. 8:289–335.

Clausen, R.T. 1938. A monograph of the Ophioglossaceae. Mem. Torrey Bot. Club 19(2):1–177.

Cody, W.J. 1963. *Woodwardia* in Canada. Am. Fern J. 53:17–27.

Cody, W.J. 1978. Ferns of the Ottawa District (rev. ed.). Agric. Can. Publ. 974. 111 pp.

Cody, W.J. 1979. Vascular plants of restricted range in the continental Northwest Territories. Syllogeus 23:1–57.

Cody, W.J. 1980. Fougères du district d'Ottawa. Agric. Can. Publ. 974. 112 pp.

Cody, W.J. 1982. *Adiantum pedatum* ssp. *calderi*, a new subspecies in northeastern North America. Rhodora 85:93–96.

Cody, W.J.; Crompton, C.W. 1975. The biology of Canadian weeds. 15: *Pteridium aquilinum* (L.) Kuhn. Can. J. Plant Sci. 55:1059–1072.

Cody, W.J.; Hall, I.V.; Crompton, C.W. 1977. The biology of Canadian weeds. 26: *Dennstaedtia punctilobula* (Michx.) Moore. Can. J. Plant Sci. 57:1159–1168.

Cody, W.J.; Lafontaine, J.D. 1975. The fern genus *Woodsia* in Manitoba. Can. Field-Nat. 89:66–69.

Cody, W.J.; Mulligan, G.A. 1982. Chromosome numbers of some Canadian ferns and fern allies. Nat. Can. (Que.) 109:273–275.

Cody, W.J.; Wagner, V. 1981. The biology of Canadian Weeds. 49: *Equisetum arvense* L. Can. J. Plant Sci. 61:123–133.

Copeland, E.B. 1947. Genera Filicum. Chronica Botanica Co., Waltham, Mass. 247 pp.

Cordes, L.D.; Krajina, V.J. 1968. *Mecodium wrightii* on Vancouver Island. Am. Fern J. 58:181.

Cranfill, R. 1980. Ferns and fern allies of Kentucky. Kentucky Nature Preserves Commission, Scientific and Technical Series, No. 1. 284 pp.

Cruise, J. 1972. Spring harvest. Ont. Nat. 11:18–23.

Czerepanov, S.K. 1981. Plantae Vasculares URSS. Nauka, Leningrad. 509 pp.

DeBenedictis, V.M.M. 1969. Apomixis in ferns with special reference to sterile hybrids. Ph.D. dissertation, University of Michigan, Ann Arbor, Mich. 203 pp.

Douglas, G.W.; Argus, G.W.; Dickson, H.L.; Brunton, D.F. 1981. The rare vascular plants of the Yukon. Syllogeus 28:1–61.

Duckett, J.G. 1979. An experimental study of the reproductive biology and hybridization in the European and North American species of *Equisetum*. Bot. J. Linn. Soc. 79:205–229.

Eaton, D.C. 1879. The ferns of North America, Vol. 1. Cassino, Salem, Mass. 352 pp.

Eifert, V.S.; Metcalfe, B. 1963. Native ferns of eastern North America. Federation of Ontario Naturalists, Toronto, Ont. 64 pp.

Emmott, J.I. 1964. A cytogenetic investigation in a *Phyllitis-Asplenium* complex. New Phytol. 63:306–318.

Erskine, D.S. 1961. The plants of Prince Edward Island. Agric. Can. Publ. 1088. 270 pp.

Evans, A.M. 1970. A review of systematic studies of the pteridophytes of the southern Appalachians. Pages 117–146 *in* Holt, P.C. ed. The distributional history of the biota of the southern Appalachians, Part II. Flora Research Division, Monogr. 2. Virginia Polytechnic Institute and State University, Blacksburg, Va.

Evans, A.M.; Wagner, W.H. 1964. *Dryopteris goldiana* × *intermedia*: A natural woodfern cross of noteworthy morphology. Rhodora 66:255–266.

Fernald, M.L. 1928. The American representatives of *Asplenium ruta-muraria*. Rhodora 30:37–43.

Fernald, M.L. 1945. Botanical specialties of the Seward Forest and adjacent areas of southeastern Virginia. Rhodora 47:93–142.

Fernald, M.L. 1950. Gray's manual of botany (8th ed.). American Book Co., New York. 1632 pp.

Fraser-Jenkins, C.R. 1980. Nomenclatural notes on *Dryopteris*: 4. Taxon 29:607–612.

Gibby, M. 1977. The origin of *Dryopteris campyloptera*. Can. J. Bot. 55:1419–1428.

Gibby, M.; Walker, S. 1977. Further cytogenetic studies and reappraisal of the diploid ancestry in the *Dryopteris carthusiana* complex. Fern Gaz. 11:315–324.

Gleason, H.A.; Cronquist, A. 1963. Manual of the vascular plants of northeastern United States and adjacent Canada. Van Nostrand Co., Princeton, N.J. 810 pp.

Greenwood, E.W. 1967. Mass occurrences of the fern *Ophioglossum vulgatum* in the Ottawa District, Ontario. Can. Field-Nat. 81:186–188.

Gussow, H.T. 1912. Field horsetail (*Equisetum arvense* L.) Pages 210–211 *in* Agric. Can. Exp. Farm Rep. Ottawa, Ont.

Hagenah, D.J. 1961. Spore studies in the genus *Cystopteris*. I: The distribution of *Cystopteris* with non-spiny spores in North America. Rhodora 63:181–193.

Hagenah, D.J. 1963. Pteridophytes of the Huron Mountains. Mich. Bot. 2:73–93.

Harms, V.L. 1978. *Athyrium filix-femina* new to Saskatchewan. Am. Fern J. 68:119–120.

Harms, V.L. 1983. The lady fern, *Athyrium filix-femina*, in Saskatchewan. Am. Fern J. 73:117–21.

Hauke, R.L. 1961. The smooth scouring rush and its complexities. Am. Fern J. 50:185–193.

Hauke, R.L. 1963. A taxonomic monograph of the genus *Equisetum* subgenus *Hippochaete*. Beih. Nova Hedwigia 8:1–123.

Hauke, R.L. 1966. A systematic study of *Equisetum arvense* Beih. Nova Hedwigia 13:81–109.

Hauke, R.L. 1978. A taxonomic monograph of *Equisetum* subgenus *Equisetum*. Beih. Nova Hedwigia 30:385–455.

Hersey, R.E.; Britton, D.M. 1981. A cytological study of three species and a hybrid taxon of *Lycopodium* (Section *complanata*) in Ontario. Can. J. Genet. Cytol. 23:497–504.

Hickey, R.J. 1977. The *Lycopodium obscurum* complex in North America. Am. Fern J. 67:45–48.

Hickey, R.J.; Beitel, J.M. 1979. A name change for *Lycopodium flabelliforme*. Rhodora 81:137–140.

Hinds, H.R. 1983. The rare vascular plants of New Brunswick. Syllogeus 50:1–38.

Hitchcock, C.L.; Cronquist, A.; Ownbey, M.; Thompson, J.W. 1969. Vascular plants of the Pacific Northwest, Part 1. University of Washington Press, Seattle, Wash. 914 pp.

Holmgren, P.K.; Keuken, W.; Schofield, E.K. 1981. Index Herbariorum, Part I (7th ed.). Regnum Veg. 106:1–452.

Holtum, R.E. 1971. Studies in the family Thelypteridaceae. III: A new system of genera in the Old World. Blumea 19:2–52.

Holub, J. 1964. *Lycopodiella*, novy rod radu Lycopodiales. Preslia (Prague) 36:16–22.

Holub, J. 1969. *Oreopteris*, a new genus of the family Thelypteridaceae. Folia Geobot. Phytotaxon. 4:33–53.

Holub, J. 1975. *Diphasiastrum*, a new genus in Lycopodiaceae. Preslia (Prague) 47:97–110.

Hultén, E. 1967. Comments on the flora of Alaska and Yukon. Ark. Bot. 7(1):1–147.

Hultén, E. 1968. Flora of Alaska and neighboring territories. Stanford University Press, Stanford, Calif. 1008 pp.

Iwatsuki, I. 1961. The occurrence of *Mecodium wrightii* in Canada. Am. Fern J. 51:141–144.

Jermy, A.C.; Arnold, H.R.; Farrell, L.; Perring, F.H. 1978. Atlas of ferns of the British Isles. Botanical Society of the British Isles and the British Pteridological Society, London. 1–101.

Jermy, A.C.; Crabbe, J.A.; Thomas, B.A. 1973. The phylogeny and classification of the ferns. Bot. J. Linn. Soc. 67 (Suppl. 1). 284 pp.

Jermy, A.C.; Page, C.N. 1980. Additional field characters separating the subspecies of *Asplenium trichomanes* in Britain. Fern Gaz. 12:112–113.

Johnson, A.W.; Packer, J.G. 1968. Chromosome numbers in the flora of Ogotoruk Creek, N.W. Alaska. Bot. Not. 121:403–456.

Knobloch, I.W. 1967. Chromosome numbers in *Cheilanthes*, *Notholaena*, *Llavea* and *Polypodium*. Am. J. Bot. 54:461–464.

Knobloch, I.W. 1976. Pteridophyte hybrids. Mich. State Univ. Mus. Biol. Ser. 5(4):277–352.

Kott, L. 1980a. *Polystichum braunii* in Waterloo County. Ont. Field Biol. 34:47–48.

Kott, L. 1980b. The taxonomy and biology of the genus *Isoetes* L. in northeastern North America. Ph.D. thesis. University of Guelph, Guelph, Ont. 234 pp.

Kott, L.S. 1981. *Isoetes acadiensis*, a new species from eastern North America. Can. J. Bot. 59:2592–2594.

Kott, L.S.; Bobbette, R.W.S. 1980. *Isoetes eatonii*, a quillwort new for Canada. Can. Field-Nat. 94:163–166.

Kott, L.S.; Britton, D.M. 1980. Chromosome numbers for *Isoetes* in northeastern North America. Can. J. Bot. 58:980–984.

Kott, L.S.; Britton, D.M. 1982a. Comparison of chromatographic spot patterns of some North American *Isoetes* species. Am. Fern J. 72:15–18.

Kott, L.S.; Britton, D.M. 1982b. A comparative study of sporophyte morphology of the three cytotypes of *Polypodium virginianum* in Ontario. Can. J. Bot. 60:1360-1370.

Kruckeberg, A.L. 1982. Noteworthy collections, British Columbia. Madrono 29:271.

Kruckeberg, A.R. 1964. Ferns associated with ultramafic rocks in the Pacific Northwest. Am. Fern J. 54:113-126.

Lafontaine, J.D. 1973. Range extension of the blunt-lobed woodsia, *Woodsia obtusa* (Spreng.) Torr. (Polypodiaceae) in Canada. Can. Field-Nat. 87:56.

Lafontaine, J.D.; Brunton, D.F. 1972. The purple cliff-brake, *Pellaea atropurpurea* (L.) Link, in western Quebec. Can. Field-Nat. 86:297-298.

Lang, F.A. 1969. A new name for a species of *Polypodium* from northwestern North America. Madrono 20:53-60.

Lang, F.A. 1971. The *Polypodium vulgare* complex in the Pacific Northwest. Madrono 21:235-254.

Lawson, G. 1864. Synopsis of Canadian ferns and filicoid plants. Edinb. New Philos. J. n.s. 19:273-290.

Lawson, G. 1889. The school fern-flora of Canada. Pages 221-251 *in* A. Gray, Botany for young people and common schools: How plants grow, a simple introduction to structural botany with a popular flora. Mackinlay, Halifax, N.S.

Legault, A.; Blais, V. 1968. Le *Cheilanthes siliquosa* Maxon dans le nord-est Americain. Nat. Can. (Que.) 95:307-316.

Lellinger, D.B. 1968. A note on *Aspidotis*. Am. Fern J. 58:140-141.

Lellinger, D.B. 1981. Notes on North American ferns. Am. Fern J. 71:90-94.

Löve, A.; Löve, D. 1961. Some chromosome numbers of Icelandic ferns and fern-allies. Am. Fern J. 51:127-128.

Löve, A.; Löve, D. 1966a. Cytotaxonomy of the alpine vascular plants of Mount Washington. Univ. Colo. Stud. Ser. Biol. 24:1-74.

Löve, A.; Löve, D. 1966b. The variation of *Blechnum spicant*. Bot. Tidsskr. 62:186-196.

Löve, A.; Löve, D. 1968. Cytotaxonomy of *Blechnum spicant*. Collect. Bot. (Barc.) 7:665-676.

Löve, A.; Löve, D. 1976. Pages 483-500 *in* A. Löve, IOPB chromosome number reports LIII. Taxon 25:483-500.

Löve, A.; Löve, D. 1977. New combinations in ferns. Taxon 26:324-326.

Löve, A.; Löve, D.; Kapoor, B.M. 1971. Cytotaxonomy of a century of Rocky Mountain orophytes. Arc. Alp. Res. 3:139-165.

Löve, A.; Löve, D.; Pichi Sermolli, R.E.G. 1977. Cytotaxonomical atlas of the Pteridophyta. Cramer, Vaduz, Liechtenstein. 398 pp.

Lovis, J.D. 1964. The taxonomy of *Asplenium trichomanes* in Europe. Br. Fern Gaz. 9:147–160.

Lovis, J.D. 1977. Evolutionary patterns and processes in ferns. Adv. Bot. Res. 4:229–415.

Lovis, J.D.; Melzer, H.; Reichstein, T. 1965. *Asplenium* × *adulteriniforme* hybr. nov. = diploides *Asplenium trichomanes* L. × A. *viride* Hudson. Bauhinia 2(3):231–237.

Lucansky, T.W. 1981. Chain ferns of Florida. Am. Fern J. 71:101–108.

Macoun, J. 1890. Catalogue of Canadian plants. Part V: Acrogens. Brown, Montreal, Que.

Macoun, J.; Burgess, T.J.W. 1884. Canadian Filicineae. Trans. R. Soc. Can., Sect. IV:163–226.

Maher, R.V.; Argus, G.W; Harms, V.L.; Hudson, J.H. 1979. The rare vascular plants of Saskatchewan. Syllogeus 20:1–55.

Maher, R.V., White, D.J.; Argus, G.W.; Keddy, P.A. 1978. The rare vascular plants of Nova Scotia. Syllogeus 18:1–37.

Manton, I. 1950. Problems of cytology and evolution in the Pteridophyta. Cambridge University Press, Cambridge, England. 316 pp.

Manton, I.; Sledge, W.A. 1954. Observations on the cytology and taxonomy of the pteridophyte flora of Ceylon. Philos. Trans. R. Soc. Lond. Biol. Sci. 238:127–185.

Manton, I.; Vida, G. 1968. Cytology of the fern flora of Tristan da Cunha. Proc. R. Soc. Lond. Biol. Sci. 170:361–379.

Marie-Victorin, Frère. 1923. Les Filicinées du Québec. Contrib. Inst. Bot. Univ. Montreal 2:1–98.

Marie-Victorin, Frère. 1925. Les Lycopodinées du Québec et leurs formes mineures. Contrib. Inst. Bot. Univ. Montreal 3:1–117.

Marquis, R.J.; Voss, E.G. 1981. Distributions of some western North American plants disjunct in the Great Lakes region. Mich. Bot. 20:53–82.

McCord, D.R. 1864. Notes on the habitats and varieties of some Canadian ferns. Can. Nat. Geol. 1:354–362.

Mehra, P.N.; Loyal, D.S. 1959. Cytological studies in *Marsilea* with particular reference to *Marsilea minuta* L. Res. Bull. Panjab Univ. Sci. 10:357–374.

Mickel, J.T. 1979. How to know the ferns and fern allies. Brown, Dubuque, Iowa. 229 pp.

Montgomery, J.D. 1982. *Dryopteris* in North America. Part II: The hybrids. Fiddlehead Forum 9:23–30.

Montgomery, J.D.; Paulton, E.M. 1981. *Dryopteris* in North America. Fiddlehead Forum 8:25–31.

Moran, R.C. 1981. × *Asplenosorus shawneensis*, a new natural hybrid between *Asplenium trichomanes* and *Camptosorus rhizophyllus*. Am. Fern J. 71:85–89.

Moran, R.C. 1982. The *Asplenium trichomanes* complex in the United States and adjacent Canada. Am. Fern J. 72:5–11.

Moran, R.C. 1983. *Cystopteris tenuis* (Michx.) Desv.: A poorly understood species. Castanea 48(3):218–223.

Morton, C.V. 1950. Notes on the ferns of the eastern United States. Am. Fern J. 40:213-225, 241–252.

Morton, C.V. 1970. Recent fern literature. Am. Fern J. 60:126–127.

Moss, E.H. 1959. Flora of Alberta. University of Toronto Press, Toronto, Ont. 546 pp.

Muenscher, W.C. 1955. Weeds (2nd ed.). Macmillan, New York. 577 pp.

Mulligan, G.A.; Cinq-Mars, L.; Cody, W.J. 1972. Natural interspecific hybridization between sexual and apogamous species of the beech fern genus *Phegopteris* Fée. Can. J. Bot. 50:1295–1300.

Mulligan, G.A.; Cody, W.J. 1968. Pages 285–288 *in* A. Löve, IOPB chromosome number reports XVII. Taxon 17.

Mulligan, G.A.; Cody, W.J. 1969. The highest chromosome number known to occur in a North American plant. Can. Field-Nat. 83:277–278.

Mulligan, G.A.; Cody, W.J. 1979. Chromosome numbers in Canadian *Phegopteris*. Can. J. Bot. 57:1815–1819.

Ogden, E.B. 1948. The ferns of Maine. University of Maine Studies, Second Series, 62. 128 pp.

Parnis, E.M. Ferns of the Montreal region. McGill University Herbarium, Sainte-Anne de Bellevue, Que. 11 pp.

Pfeiffer, N. 1922. Monograph on the Isoetaceae. Trans. Acad. Sci. (St. Louis) 9(2):79–233.

Pohl, R. 1955. Toxicity of ferns and *Equisetum*. Am. Fern J. 45:95–97.

Porsild, A.E. 1945. The so-called *Woodsia alpina* in North America. Rhodora 47:145–148.

Presl, K.B. 1851. Epimeliae Botanicae. Abh. K. Boehm. Gesell. Wiss. 5(6):361–624.

Pryer, K.M. 1981. Systematic studies in the genus *Gymnocarpium* Newm. in North America. M.Sc. thesis, University of Guelph, Guelph, Ont. 166 pp.

Pursh, F. 1814. Flora Americae septentrionalis. White, Cochrane, London. 2 vols. 751 pp.

Rapp, W.F. 1954. The toxicity of *Equisetum*. Am. Fern J. 44:148–154.

Reed, C.F. 1954. Index Marsileata and Salviniata. Bol. Soc. Broteriana 28:5–61.

Reeves, T. 1977. The genus *Botrychium* (Ophioglossaceae) in Arizona. Am. Fern J. 67:33–39.

Reichstein, T. 1981. Hybrids in European Aspleniaceae (Pteridophyta). Bot. Helvitica 91:89–139.

Reznicek, T.A. 1972. *Woodsia scopulina* in Algonquin Park, Nipissing District, Ontario. Can. Field-Nat. 86:368–369.

Richter, H.E. 1961. *Equisetum palustre* poisoning in horses. Wien. tieraerztl. Monatsschr. 48:761–762.

Rigby, S.J. 1973. Chromosome pairing in obligately apogamous ferns: *Pellaea atropurpurea* and *Pellaea glabella* var. *glabella*. Rhodora 75:122–131.

Rigby, S.J., Britton, D.M. 1970. The distribution of *Pellaea* in Canada. Can. Field-Nat. 84:137–144.

Roland, A.E. 1941. The ferns of Nova Scotia. Proc. N.S. Inst. Sci. 20(3):64–120.

Roland, A.E.; Smith, E.C. 1969. The flora of Nova Scotia. Nova Scotia Museum, Halifax, N.S.

Rousseau, C. 1974. Géographie floristique du Québec/Labrador. Les Presses de l'Université Laval, Québec, Que. 799 pp.

Sarvela, J. 1978. A synopsis of the fern genus *Gymnocarpium*. Ann. Bot. Fenn. 15:101–106.

Sarvela, J. 1980. *Gymnocarpium* hybrids from Canada and Alaska. Ann. Bot. Fenn. 17:292–295.

Sarvela, J.; Britton, D.M.; Pryer, K. 1981. Studies on the *Gymnocarpium robertianum* complex in North America. Rhodora 83:421–431.

Scoggan, H.J. 1957. Flora of Manitoba. Natl. Mus. Can. Bull. 140. 619 pp.

Scoggan, H.J. 1978. The flora of Canada. Natl. Mus. Nat. Sc. Publ. Bot. 7(2):93–545.

Shivas, M.G. 1961. Contributions to the cytology and taxonomy of species of *Polypodium* in Europe and America. I: Cytology. J. Linn. Soc. Lond. Bot. 58:13–25.

References   411

Smith, A.R. 1971. Chromosome numbers of some New World species of *Thelypteris*. Brittonia 23:354–360.

Smith, A.R. 1975. The California species of *Aspidotis*. Madrono 23:15–24.

Soper, J.H. 1949. The vascular plants of southern Ontario. Department of Botany, University of Toronto, Toronto, Ont. 95 pp.

Soper, J.H. 1954. The hart's-tongue fern in Ontario. Am. Fern J. 44:129–147.

Soper, J.H. 1955. *Asplenium cryptolepis* on Manitoulin Island. Am. Fern J. 45:97–104.

Soper, J.H. 1963. Ferns of Manitoulin Island, Ontario. Am. Fern J. 53:28-40, 71-81, 109–123.

Soper, J.H. 1964. The slender cliff brake. Ont. Nat. 2:20–23.

Soper, J.H.; Maycock, P.F. 1963. A community of arctic-alpine plants on the east shore of Lake Superior. Can. J. Bot. 41:183–198.

Soper, J.H.; Rao, S. 1958. *Isoetes* in eastern Canada. Am. Fern J. 48:97–102.

Strasburger, E. 1907. Apogamie bei *Marsilea*. Flora 97:123–191.

Svenson, H.K. 1944. The New World species of *Azolla*. Am. Fern J. 34:69–84.

Tatuno, S.; Takei, M. 1969. Karyological studies in Hymenophyllaceae. I: Chromosomes of the genus *Hymenophyllum* and *Mecodium* in Japan. Bot. Mag. Tokyo 82:121–129.

Taylor, R.L.; Brockman, R.P. 1966. Chromosome numbers of some western Canadian plants. Can. J. Bot. 44:1093–1103.

Taylor, R.L.; Mulligan, G.A. 1968. Flora of the Queen Charlotte Islands, Part 2. Can. Dep. Agr. Res. Branch Monogr. 4(2) 148 pp.

Taylor, T.M.C. 1947. New species and combinations in *Woodsia*, section *Perrinia*. Am. Fern J. 37:84–88.

Taylor, T.M.C. 1963. The ferns and fern-allies of British Columbia. B.C. Prov. Mus. Nat. Hist. Handb. 12. 172 pp.

Taylor, T.M.C. 1967. *Mecodium wrightii* in British Columbia and Alaska. Am. Fern J. 57:1–6.

Taylor, T.M.C. 1970. Pacific Northwest ferns and their allies. University of Toronto Press, Toronto, Ont. 247 pp.

Taylor, T.M.C.; Lang, F.A. 1963. Chromosome counts in some British Columbia ferns. Am. Fern J. 53:123–126.

Taylor, W.C.; Mohlenbrock, R.H.; Murphy, J. 1975. The spores and taxonomy of *Isoetes butleri* and *I. melanopoda*. Am. Fern J. 65:33–38.

References

Tryon, A.F. 1957. A revision of the fern genus *Pellaea* section *Pellaea*. Ann. Mo. Bot. Gard. 44:125–193.

Tryon, A.F. 1968. Comparisons of sexual and apogamous races in the fern genus *Pellaea*. Rhodora 70:1–24.

Tryon, A.F. 1972. Spores, chromosomes and relations of the fern *Pellaea atropurpurea*. Rhodora 74:220–241.

Tryon, A.F.; Britton, D.M. 1958. Cytotaxonomic studies of the fern genus *Pellaea*. Evolution 12:137–145.

Tryon, A.F.; Tryon, R. 1973. *Thelypteris* in northeastern North America. Am. Fern J. 63:65–76.

Tryon, A.F.; Tryon, R. 1974. Geographic patterns in temperate American ferns and some relationships in *Thelypteris*. Am. Fern J. 64:99–104.

Tryon, A.F.; Tryon, R.; Badre, F. 1980. Classification, spores and nomenclature of the marsh fern. Rhodora 82:461–474.

Tryon, R.M. 1941. Revision of the genus *Pteridium*. Rhodora 43:1–31, 37–67.

Tryon, R.M. 1942. A new *Dryopteris* hybrid. Am. Fern J. 32:81–85.

Tryon, R.M. 1948. Some woodsias from the north shore of Lake Superior. Am. Fern J. 38:159–170.

Tryon, R.M. 1955. *Selaginella rupestris* and its allies. Ann. Mo. Bot. Gard. 42:1–99.

Tryon, R.M. 1960. A review of the genus *Dennstaedtia* in America. Contrib. Gray Herb. Harv. Univ. 187:23–52.

Tryon, R.M.; Britton, D.M. 1966. A study of variation in the cytotypes of *Dryopteris spinulosa*. Rhodora 68:59–92.

Tryon, R.M.; Fassett, N.C.; Dunlop, D.W.; Diemer, M.E. 1953. The ferns and fern allies of Wisconsin (2nd ed.). University of Wisconsin, Madison, Wisc. 158 pp.

Tschermak-Woess, E.; Dolezal-Janisch, R. 1959. Uber die karyologische anatomie einiger Pteridophyten sowie auffallende unterschiede in kernvolumen bei *Cyrtomium falcatum*. Oesterr. Bot. Z. 106:315–324.

Vincent, G. 1981. *Phegopteris hexagonoptera*, espèce rare et menacée. Bull. Soc. Anim. Jard. Inst. Bot. 6:2–24.

von Euw, J.; Lounasmaa, M.; Reichstein, T.; Widén, C.-J. 1980. Chemotaxonomy in *Dryopteris* and related fern genera. Review and evaluation of analytical methods. Stud. Geobot. 1:275–311.

Wagner, D.H. 1979. Systematics of *Polystichum* in western North America north of Mexico. Pteridologia 1:1–64.

Wagner, W.H. 1954. Reticulate evolution in the Appalachian aspleniums. Evolution 8:103–118.

References 413

Wagner, W.H. 1955. Cytotaxonomic observations on North American ferns. Rhodora 57:219–240.

Wagner, W.H. 1959. American grapeferns resembling *Botrychium ternatum.* Am. Fern J. 49:97–103.

Wagner, W.H. 1960a. Periodicity and pigmentation in *Botrychium* subg. *sceptridium* in the northeastern United States. Bull. Torrey Bot. Club 87:303–325.

Wagner, W.H. 1960b. Evergreen grapeferns and the meanings of infraspecific categories as used in North American pteridophytes. Am. Fern J. 50:32–45.

Wagner, W.H. 1961. Roots and taxonomic difference between *Botrychium oneidense* and *B. dissectum.* Rhodora 63:164–175.

Wagner, W.H. 1962. Plant compactness and leaf production in *Botrychium multifidum* "ssp. *typicum*" and "forma *dentatum.*" Am. Fern J. 52:1–18.

Wagner, W.H. 1963. A biosystematic study of United States ferns. Am. Fern J. 53:1–16.

Wagner, W.H. 1966a. Two new species of ferns from the United States. Am. Fern J. 56:3–17.

Wagner, W.H. 1966b. New data on North American oak ferns, *Gymnocarpium.* Rhodora 68:121–138.

Wagner, W.H. 1970. Evolution of *Dryopteris* in relation to the Appalachians. Pages 147–192 in P.C. Holt, ed. Distributional history of the biota of the southern Appalachians. Part II: Flora. Research Division, Monogr. 2. Virginia Polytechnic Institute and State University, Blacksburg, Va.

Wagner, W.H. 1971. The southeastern adder's-tongue, *Ophioglossum vulgatum* var. *pycnostichum,* found for the first time in Michigan. Mich. Bot. 10:67–74.

Wagner, W.H. 1973a. *Asplenium montanum* × *platyneuron:* A new primary member of the Appalachian spleenwort complex for Crowder's Mountain, N.C. J. Elisha Mitchell Sci. Soc. 89:218–223.

Wagner, W.H. 1973b. Reticulation of holly ferns (*Polystichum*) in the western United States and adjacent Canada. Am. Fern J. 63:99–115.

Wagner, W.H.; Boydston, K.E. 1978. A dwarf coastal variety of maidenhair fern, *Adiantum pedatum.* Can. J. Bot. 56:1726–1729.

Wagner, W.H.; Chen, K.L. 1964. Pages 99–110 in A. Löve and O.T. Solbrig, IOPB chromosome number reports I. Taxon 13.

Wagner, W.H.; Chen, K.L. 1965. Abortion of spores and sporangia as a tool in the detection of *Dryopteris* hybrids. Am. Fern J. 55:9–29.

Wagner, W.H.; Farrar, D.R.; Chen, K.L. 1965. A new sexual form of *Pellaea glabella* var. *glabella* from Missouri. Am. Fern J. 55:171–178.

Wagner, W.H.; Farrar, D.R.; McAlpine, B.W. 1970. Pteridology of the Highlands Biological Station area, southern Appalachians. J. Elisha Mitchell Sci. Soc. 86:1–27.

Wagner, W.H.; Hagenah, D.J. 1956. A diploid variety in the *Cystopteris fragilis* complex. Rhodora 58:79–87.

Wagner, W.H.; Johnson, D.M. 1981. Natural history of the ebony spleenwort, *Asplenium platyneuron* (Aspleniaceae), in the Great Lakes area. Can. Field-Nat. 95:156–166.

Wagner, W.H.; Lord, L.P. 1956. The morphological and cytological distinctness of *Botrychium minganense* and *B. lunaria* in Michigan. Bull. Torrey Bot. Club 83:261–280.

Wagner, W.H.; Rawlings, D.E. 1962. A sampling of *Botrychium* subg. *Sceptridium* in the vicinity of Leonardtown, St. Mary's Co., Md. Castanea 27:132–142.

Wagner, W.H.; Wagner, F.S. 1966. Pteridophytes of the Mountain Lake area, Giles Co. Virginia: Biosystematic studies 1964-1965. Castanea 31:121–140.

Wagner, W.H.; Wagner, F.S. 1981. New species of moonworts, *Botrychium* subg. *Botrychium* (Ophioglossaceae), from North America. Am. Fern J. 71:20–30.

Wagner, W.H.; Wagner, F.S. 1982a. The taxonomy of *Dryopteris* × *poyseri* Wherry. Mich. Bot. 21:75–88.

Wagner, W.H.; Wagner, F.S. 1982b. *Botrychium rugulosum* (Ophioglossaceae): A newly recognized species of evergreen grapefern in the Great Lakes area of North America. Contrib. Univ. Mich. Herb. 15:315–324.

Wagner, W.H.; Wagner, F.S. 1983a. Genus communities as a systematic tool in the study of New World *Botrychium* (Ophioglossaceae). Taxon 32:51–63.

Wagner, W.H.; Wagner, F.S. 1983b. The moonworts of the Rocky Mountains; *Botrychium hesperium* and a new species formerly confused with it. Am. Fern J. 73:53–62.

Wagner, W.H.; Wagner, F.S. 1983c. Western Canada. *Botrychium* Newsl. No. 4. University of Michigan Herbarium, Ann Arbor, Mich. 10 pp.

Wagner, W.H.; Wagner, F.S.; Miller, C.N., Jr.; Wagner, D.H. 1978. New observations on the royal fern hybrid, *Osmunda* × *ruggii*. Rhodora 80:92–106.

Wagner, F.S. 1983. The *Botrychium lanceolatum* group in western North America (Abstr.). Am. J. Bot. 70(5):95.

Walker, S. 1961. Cytogenetic studies in the *Dryopteris spinulosa* complex, II. Am. J. Bot. 48:607-614.

Walker, T.G. 1979. The cytogenetics of ferns. Pages 229-415 *in* A.F. Dyer. The experimental biology of ferns. Academic Press, London.

Waterway, M.J.; Lei, T.T. 1982. *Polystichum lonchitis* in central Quebec-Labrador. Am. Fern J. 72:85-87.

Weatherby, C.A. 1936. A list of varieties and forms of the ferns of eastern North America. Am. Fern J. 26:97-98.

Wherry, E.T., Jr. 1961. The fern guide. Doubleday, Garden City, N.Y. 318 pp.

White, D.J.; Johnson, K.L. 1980. The rare vascular plants of Manitoba. Syllogeus 27:1-52.

Widen, C.-J.; Britton, D.M. 1971. A chromatographic and cytological study of *Dryopteris dilatata* in North America and eastern Asia. Can. J. Bot. 49:247-258.

Wilce, J.H. 1965. Section *Complanata* of the genus *Lycopodium*. Nova Hedwigia 19:1-233.

Wilce, J.H. 1972. Lycopod spores, I: General spore patterns and the generic segregates of *Lycopodium*. Am. Fern J. 62:65-79.

Zhukova, P.G.; Petrovski, V.V. 1972. Kromosomnye chisla nekotorykh tsvetkovykh rastenii ostrova Vrangelia, II. Bot. Zh. (Leningr.) 57:554-567.

Addenda to References

Work on this publication was completed in 1983. Since then, some important publications on ferns and fern allies have been published. Among them are the following:

Alverson, E.R. 1988. Biosystematics of North American parsley ferns, *Cryptogramma* (Adiantaceae). Am. J. Bot. 75:136-137 (Abstr.).

Argus, G.W.; Pryer, K.M.; White, D.J.; Keddy, C.J., eds. 1982-1987. Atlas of the rare vascular plants of Ontario. Four parts. National Museum of Natural Sciences, Ottawa, Ont.

Barrington, D.S. 1986. The morphology and cytology of *Polystichum* × *potteri* hybr. nov. (= *P. acrostichoides* × *P. braunii*). Rhodora 88:297-313.

Beitel, J. 1986. The *Huperzia selago* (Lycopodiaceae) complex in the Pacific Northwest. Am. J. Bot. 73(5):733-734.

Blondeau, M.; Cayouette, J. (1987). Extensions d'aire dans la flore vasculaire du Nouveau-Québec. Nat. Can. (Que.) 114:117-126.

Bouchard, A.; Barabé, D.; Bergeron, Y.; Dumais, M.; Hay, S. 1985. La phytogéographie des plantes vasculaires rares du Québec. Nat. Can. (Que.) 112:283–300.

Bouchard, A.; Barabé, D.; Dumais, M.; Hay, S. 1983. The rare vascular plants of Quebec. Syllogeus 48:1–75.

Britton, D.M. 1984a. Biosystematic studies on pteridophytes in Canada: Progress and problems. Pages 543–560 in Grant, W.F., ed. Plant Biosystematics. Academic Press, Don Mills, Ont. 674 pp.

Britton, D.M. 1984b. Checklist of Ontario pteridophytes. Part I: Fern allies. Plant Press 2(4):95–99.

Britton, D.M. 1985. Checklist of Ontario pteridophytes. Part II: Ferns. Plant Press 3(1):14–23.

Britton, D.M.; Anderson, A.B. 1986. The ferns of Manitoulin Island: Notes and a new record. Plant Press 4:60–61.

Britton, D.M.; Catling, P.M.; Norris, J.; Varga, S. (in press). Isoetes engelmannii an addition to the flora of Canada. Can. Field-Nat.

Britton, D.M.; Stewart, W.G.; Cody, W.J. 1985. Cystopteris protrusa, creeping fragile fern, an addition to the flora of Canada. Can. Field-Nat. 99(3):380–382.

Brunton, D.F. 1986a. Status of the southern maidenhair Fern, Adiantum capillus-veneris (Adiantaceae), in Canada. Can. Field-Nat. 100:404–408.

Brunton, D.F. 1986b. Status of the mosquito fern, Azolla mexicana (Salviniaceae), in Canada. Can. Field-Nat. 100(3):409–413.

Bryan, F.A.; Soltis, D.E. 1987. Electrophoretic evidence for allopolyploidy in the fern Polypodium virginianum. Syst. Bot. 12(4):553–561.

Catling, P.M. 1985. Notes on the occurrence, ecology and identification of the Massachusetts fern, Thelypteris simulata, in Ontario. Can. Field-Nat. 99(3):300–307.

Catling, P.M.; Erskine, D.S.; MacLaren, R.B. 1985. The plants of Prince Edward Island [D.S. Erskine] with new records, nomenclatural changes, and corrections and deletions. Agric. Can. Publ. 1798. 272 pp.

Cayouette, J. 1984. Additions et extensions d'aire dans la flore vasculaire du Nouveau-Québec. Nat. Can. (Que.) 111:263–274.

Ceska, A. 1986. An annotated list of rare and uncommon plants of the Victoria area. Victoria Nat. 43(5):1–14.

Cody, W.J. 1983. Adiantum pedatum ssp. calderi, a new subspecies in northeastern North America. Rhodora 85:93–96.

Cody, W.J.; Britton, D.M. 1984. *Polystichum lemmonii*, a rock shield-fern new to British Columbia. Can. Field-Nat. 98(3):375.

Cody, W.J.; Britton, D.M. 1985. Male fern, *Dryopteris filix-mas*, a phytogeographically important discovery in northern Saskatchewan. Can. Field-Nat. 99(1):101–102.

Cody, W.J.; Schueler, F.W. 1988. A second record of the mosquito fern, *Azolla caroliniana*, in Ontario. Can. Field-Nat. 102:545–546.

Cranfill, R.; Britton, D.M. 1983. Typification within the *Polypodium virginianum* complex (Polypodiaceae). Taxon 32(4):557–560.

Crist, K.C.; Farrar, D.R. 1983. Genetic load and long-distance dispersal in *Asplenium platyneuron*. Can. J. Bot. 61(6):1809–1814.

Derrick, L.N.; Jermy, A.C.; Paul, A.M. 1987. Checklist of European Pteridophytes. Sommerfeltia 6:1–94.

Deshaye, J.; Morisset, P. 1985. La flore vasculaire du lac à l'Eau Claire, Nouveau-Québec. Provancheria 18:1–52.

Fraser-Jenkins, C.R. 1986. A classification of the genus *Dryopteris* (Pteridophyta: Dryopteridaceae). Bull. Br. Mus. (Nat. Hist.) Bot. Ser. 14(3):183–218.

Gastony, G.J. 1986. Electrophoretic evidence for the origin of fern species by unreduced spores. Am. J. Bot. 73(11):1563–1569.

Gawler, S.C. 1983. Note on *Adiantum pedatum* L. ssp. *calderi* Cody. Rhodora 85:389–390.

Goltz, J.P.; Britton, D.M.; Whiting, R.E. 1984. *Phegopteris hexagonoptera* (Michx.) Fée (southern or broad beech fern) discovered in the district of Muskoka. Plant Press 2(2):39–40.

Haufler, C.H. 1985. Pteridophyte evolutionary biology: The electrophoretic approach. Proc. R. Soc. Edinb. Sect. B (Biol.) 86B:315–323.

Haufler, C.H; Soltis, D.E. 1986. Genetic evidence suggests that homosporous ferns with high chromosome numbers are diploid. Proc. Nat. Acad. Sci. USA 83:4389–4393.

Haufler, C.H.; Windham, M.D. 1988. Exploring the origin of Polypodium vulgare, again. Am. J. Bot. 75:139. (Abstr.).

Haufler, C.H.; Windham, M.D.; Britton, D.M.; Robinson, S.J. 1985. Triploidy and its evolutionary significance in Cystopteris protrusa. Can. J. Bot. 63(10):1855–1863.

Hickey, J.R. 1986. *Isoetes* megaspore surface morphology: Nomenclature, variation and systematic importance. Am. Fern J. 76:1–16.

Hinds, H.R. 1983. The rare vascular plants of New Brunswick. Syllogeus 50:1–38.

Hinds, H.R. 1986. Flora of New Brunswick. Primrose Press, Fredericton, N.B. 460 pp.

Holtum, R.E. 1981. The genus *Oreopteris* (Thelypteridaceae). Kew Bull. 36(2):223–226.

Knobloch, I.W.; Gibby, M.; Fraser-Jenkins, C. 1984. Recent advances in our knowledge of pteridophyte hybrids. Taxon 33(2):256–270.

Kott, L.S. 1982. A comparative study of spore germination of some *Isoetes* species of northeastern North America. Can. J. Bot. 60(9):1679–1687.

Kott, L.S.; Britton, D.M. 1982. A comparative study of sporophyte morphology of the three cytotypes of *Polypodium virginianum* in Ontario. Can. J. Bot. 60(8):1360–1370.

Kott, L.S.; Britton, D.M. 1983. Spore morphology and taxonomy of *Isoetes* in northeastern North America. Can. J. Bot. 61(12):3140–3163.

Lavoie, G. 1984. Contribution à la connaissance de la flore vasculaire de la Moyenne-et-Basse-Côte-Nord, Québec/Labrador. Provancheria 17:1–149.

Lellinger, D.B. 1985. A field manual of the ferns and fern allies of the United States and Canada. Smithsonian Institution Press, Washington, D.C. 389 pp.

Markham, K.R.; Moore, N.A.; Given, D.R. 1983. Phytochemical reappraisal of taxonomic subdivisions of Lycopodium (Pteridophyta-Lycopodiaceae) based on flavonoid glycoside distribution. N.Z. J. Bot. 21:113–120.

McNeill, J.; Pryer, K. 1985. The status and typification of *Phegopteris* and *Gymnocarpium*. Taxon 34(1):136–156.

Mills, A. 1985. Revised status of the holly fern, *Polystichum lonchitis*, in Ontario. Can. Field-Nat. 99(2):252–254.

Mokry, F.; Rasbach, H.; Reichstein, T. 1986. *Asplenium adulterinum* Milde ssp. *presolanense* subsp. nova (Aspleniaceae, Pteridophyta). Bot. Helv. 96(1):7–18.

Moran, R.C. 1983*a*. *Cystopteris tenuis* (Michx.) Desv.: A poorly understood species. Castanea 48:218–223.

Moran, R.C. 1983*b*. *Cystopteris* × *wagneri*: A new naturally occurring hybrid between *C.* × *tennesseensis* and *C. tenuis*. Castanea 48:224–229.

Morton, J.K.; Venn, J.M. 1984. The flora of Manitoulin Island and the adjacent islands of Lake Huron, Georgian Bay and the North Channel. Department of Biology, University of Waterloo, Waterloo, Ont. 181 pp.

Moss, E.H. 1983. Flora of Alberta. 2nd ed. Revised by Packer, J.G. University of Toronto Press, Toronto, Ont. 687 pp.

References   419

Munro, D. 1988. A disjunct station of *Asplenium ruta-muraria* with *Pellaea atropurpurea* and *Pellaea glabella* in eastern Ontario. Am. Fern J. 78(4):136–138.

Packer, J.G.; Bradley, C.E. 1984. A checklist of the rare vascular plants in Alberta. Prov. Mus. Alberta, Nat. Hist. Occ. Pap. No. 5. 112 pp.

Paris, C. 1986. A biosystematic investigation of the *Adiantum pedatum* complex in eastern North America. Am. J. Bot. 73(5):738–739.

Paris, C.A; Windham, M.D. 1988. A biosystematic investigation of the *Adiantum pedatum* complex in eastern North America. Syst. Bot. 13(2):240–255.

Perkins, S.K.; Peters, G.A.; Lumpkin, T.A.; Calvert, H.E. 1985. Scanning electron microscopy of perine architecture as a taxonomic tool in the genus *Azolla* Lamarck. Scanning Electron Microsc. 4:1719–1734.

Prange, R.K.; Von Aderkas, P. 1985. The biological flora of Canada 6. *Matteuccia struthiopteris* (L.) Todaro, Ostrich Fern. Can. Field-Nat. 99(4):517–532.

Pryer, K.M.; Britton, D.M. 1983. Spore studies in the genus *Gymnocarpium*. Can. J. Bot. 61(2):377–388.

Pryer, K.M.; Britton, D.M.; McNeill, J. 1983. A numerical analysis of chromatographic profiles in North American taxa of the fern genus *Gymnocarpium*. Can. J. Bot. 61(10):2592–2602.

Pryer, K.M.; Windham, M.D. 1988. A re-examination of *Gymnocarpium dryopteris* (L.) Newman in North America. Am. J. Bot. 75:142. (Abstr.).

Soltis, D.E. 1986. Genetic evidence for diploidy in *Equisetum*. Am. J. Bot. 73(5):740.

Soltis, D.E; Soltis, P.S. 1987a. Breeding system of the fern *Dryopteris expansa*: Evidence for mixed mating. Am. J. Bot. 74(4):504–509.

Soltis, D.E.; Soltis, P.S. 1987b. Polyploidy and breeding systems in homosporous pteridophyta: A reevaluation. Am. Nat. 130:219–232.

Soltis, P.S.; Soltis, D.E. 1987. Population structure and estimates of gene flow in the homosporous fern *Polystichum munitum*. Evolution 41(3):620–629.

Straley, G.B.; Taylor, R.L.; Douglas, G.W. 1985. The rare vascular plants of British Columbia. Syllogeus 59:1–165.

Taylor, W.C.; Luebke, N.T. 1986a. Germinating spores and growing sporelings of aquatic *Isoetes*. Am. Fern J. 76:21–24.

References

Taylor, W.C.; Luebke, N.T. 1986*b*. Interspecific hybridization and the taxonomy of North American *Isoetes*. Am. J. Bot. 73:740–741. (Abstr.).

Taylor, W.C.; Luebke, N.T.; Smith, M.B. 1985. Speciation and hybridization in North American *Isoetes*. Proc. R. Soc. Edinb. Sect. B (Biol.) 86B:259–263.

Tryon, R. 1986. The biogeography of species, with special reference to ferns. Bot. Rev. 52(2):117–156.

Vincent, G. 1981. *Phegopteris hexagonoptera*, espèce rare et menacée. Bull. Soc. Anim. Jard. Inst. Bot. 6(2):2–24.

von Aderkas, P. 1983. Studies of gametophytes of *Matteucia struthiopteris* (ostrich fern) in nature and in culture. Can. J. Bot. 61(12):3267–3270.

von Aderkas, P. 1984. Economic history of ostrich fern, *Matteuccia struthiopteris*, the edible fiddlehead. Econ. Bot. 38(1):14–23.

Wagner, F.S. 1987. Evidence for the origin of the hybrid cliff fern, *Woodsia Xabbeae* (Aspleniaceae: Athyrioideae). Syst. Bot. 12(1):116–124.

Wagner, F.S. 1988*a*. Moonworts recently discovered in the Great Lakes Area. Fiddlehead Forum 15(1):2–3.

Wagner, F.S. 1988*b*. Natural hybrids between woodferns, *Dryopteris*, and holly ferns, Polystichum. Am. J. Bot. 75:143–144. (Abstr.).

Wagner, W.H.; Rouleau, E. 1984. A western holly fern, *Polystichum* × *scopulinum* in Newfoundland. Am. Fern J. 74:33–36.

Wagner, W.H.; Wagner, F.S. 1983*a*. Genus communities as a systematic tool in the study of New World *Botrychium* (Ophioglossaceae). Taxon 32(1):51–63.

Wagner, W.H.; Wagner, F.S. 1983*b*. Two moonworts of the Rocky Mountains: *Botrychium hesperium* and a new species formerly confused with it. Am. Fern J. 73:53–62.

Wagner, W.H.; Wagner, F.S. 1984. Western Canada. *Botrychium* Newsl. No. 4. University of Michigan Herbarium, Ann Arbor, Mich. 10 pp.

Wagner, W.H.; Wagner, F.S. 1986. Three new species of moonworts (*Botrychium* subg. *Botrychium*) endemic in western North America. Am. Fern J. 76:33–47.

Wagner, W.H.; Wagner, F.S. 1988. Detecting *Botrychium* hybrids in the Lake Superior Region. Mich. Bot. 27(3): 75–80.

Wagner, W.H.; Wagner, F.S.; Haufler, C.; Emerson, J.K. 1984. A new nothospecies of moonwort (Ophioglossaceae, *Botrychium*). Can. J. Bot. 62:629–634.

References

Waterway, M.J. 1986. A reevaluation of *Lycopodium porophilum* and its relationship to *L. lucidulum* (Lycopodiaceae). Syst. Bot. 11(2):263–276.

Webber, J.M. 1984. *Marsilea quadrifolia* (four leaf waterfern) in Ontario. Plant Press 2(3):68–70.

Werth, C.R.; Guttman, S.I.; Eshbaugh, W.H. 1985. Recurring origins of allopolyploid species in *Asplenium*. Science 228:731–733.

Widén, C.-J.; Britton, D.M. 1985. Phloroglucinol derivatives of *Dryopteris tokyoensis* and the missing genome in *D. cristata* and *D. carthusiana* (Dryopteridiaceae). Ann. Bot. Fenn. 22:213–218.

Wollenweber, E.; Dietz, V.H.; Shilling, G.; Favre-Bonvin, J.; Smith, D.M. 1985. Flavonoids from chemotypes of the goldback fern, *Pityrogramma triangularis*. Phytochemistry (Oxf.) 24:965–971.

INDEX

Accepted names that are part of the Canadian flora are set in bold italics; synonyms are set in light italics.

www.ingramcontent.com/pod-product-compliance
Lightning Source LLC
Chambersburg PA
CBHW051708020426
42333CB00014B/898